BOGATIN'S
PRACTICAL GUIDE to
BEST MEASUREMENT PRACTICES
for DIGITAL OSCILLOSCOPES

ERIC BOGATIN

E-ISBN: 978-1-63081-964-4
Cover design by Iain Hill

© 2025
Artech House
685 Canton Street
Norwood, MA 02062

ARTECH HOUSE

BOSTON | LONDON
artechhouse.com

Table of Contents

Chapter 1
Measurements Are in Your Future

Electrical measurements are the anchor to reality. We use them as a window into the world of electrical signals, which are invisible to our senses. In this respect, the world of electrical signals, voltages, and currents is an abstract world, one in which we have to use our imagination and our engineer's mind's eye to see.

We use instruments to measure and help us visualize that which we cannot directly sense. In the vast majority of cases, the only electrical quality we can actually measure is a *voltage difference* between two points, which may vary in time.

We use *transducers* or *sensors* to convert physical qualities such as temperature, pressure, magnetic field, vibration, or even acceleration into a voltage signal. We use other circuit elements such as resistors or circuits such as voltage dividers to convert a current or resistance into a voltage.

Measuring a voltage signal with an acceptable signal-to-noise ratio (SNR), with minimal measurement artifacts displayed in real time, is an essential capability needed in every lab for every measurement. Success at good quality measurements is just as much about the specific instrument you use in your measurement as it is about following the *best measurement and analysis practices*.

This book introduces you to the best measurement and analysis practices to record the "true," "actual," or "real" signal as it is at the *device under test* (DUT), reduce the *uncertainty* associated with the measurement process, and reduce the *noise* and the *measurement artifacts*. To achieve these quality measurements, I introduce you to the *oscilloscope*, often affectionately called a *scope*, the most important instrument for measuring and displaying voltage signals. You will learn the most important best measurement practices and how to get the most value out of the scope you have.

While there is a wide spectrum of scope instruments available on the market from dozens of vendors, they all have similar basic features. Many of them have similar advanced features that will help you get to the answer you are looking for faster. This is sometimes referred to as "faster time to insight."

Their main differences are in their specific capabilities, such as bandwidth, sample rate, number of channels, and voltage resolution. The user interfaces are often similar between scopes from the same vendor in the same family but different between scopes from different vendors. However, being familiar with the user interface on one specific scope will allow you to look for, find, and master all the controls on the specific oscilloscope you are using. Mastering one scope's user interface will give you a sense of what to expect in a different scope's user interface.

1.1   DSO: Sampling Scopes or Real-Time Scopes

Almost all *oscilloscopes*, or *scopes* for short, manufactured since 2001, are *digital storage scopes* (DSO) that use an analog-to-digital converter (ADC) to convert the analog voltage signal at their input into a digital signal and then process this digital signal with an ASIC, FPGA, or software on a PC. This is why many scopes today are referred to as DSOs. **Figure 1.1** shows an example of one family of DSO scopes available from Teledyne LeCroy.

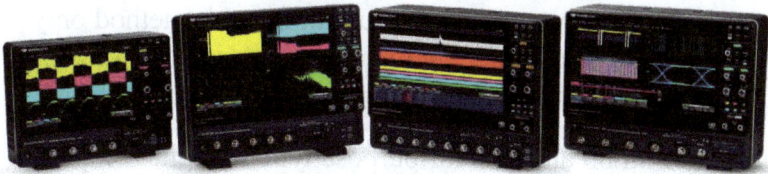

Figure 1.1 An example of a family of real time scopes with 12-bit vertical resolution and consecutive sampling rates as high as 20 GSps, courtesy of Teledyne LeCroy.

Unfortunately, the three-letter acronym (TLA), DSO, is also used to refer to a digital *sampling* oscilloscope. A sampling scope samples the signal with a very short acquisition time window that might be as short as 10 psec.

During this short time window, the instantaneous voltage at the input channel of the scope charges up a small capacitor, which holds the voltage in a sample-and-hold circuit while it is read by a slower ADC. This extra time to read each recorded voltage means the ADC can be an ADC with a slow sample rate but high resolution, at 14 or even 16-bit vertical resolution.

A sampling scope, as its name implies, only samples the signal waveform every so often at a specific time delay from a trigger event, on the order of every 100 nsec.

This means it does not record a continuous stream of consecutive measurements of a fast-changing signal waveform but only sees a few selected points in the waveform during each acquisition buffer. As the acquisition window's time delay from the trigger event advances in small increments, different parts of the repetitive waveform are sampled on each cycle.

This process of sampling the measurement is sometimes a feature found in real time scopes, where it is called equivalent time sampling (ETS).

If the signal is repetitive, a sampling scope can recreate the complete waveform with a very short time resolution and very high vertical voltage resolution. Since the sampling window can be very short, the time resolution of a sampling scope can be on the order of a few psec at 16-bit vertical resolution. But this method only works on a repetitive signal, not on a transient signal or one that may have some random changes to it.

The more common and general-purpose type of digital scope is a *real time* scope. This type of scope will measure consecutive time points at a higher consecutive sample rate, but each point is measured over a longer duration and generally has a lower vertical bit resolution than the acquisition window of a sampling scope.

This means that while a real time scope may not have the horizontal or vertical resolution of a sampling scope, it only requires one acquisition sweep to measure the entire signal waveform. A real time scope can measure any arbitrary, non-repetitive, transient waveform in addition to repetitive signals.

For repetitive signals, multiple consecutive acquisition buffers can be synchronously added and averaged together to reduce the random or asynchronous noise and increase the vertical resolution, easily matching the resolution of sampling scopes. This increases the effective number of bits (ENOB).

This technique of averaging consecutive buffers, synchronously triggered, is sometimes referred to as a digital lock-in technique. This is named for the lock-in amplifier method of improving the signal-to-noise ratio of signals synchronous with an external reference. This method is described in detail in a later chapter.

With advances in the sampling rate and time resolution of real time scopes, their time resolution rivals that of sampling scopes, though not as high a vertical bit-resolution and usually at a higher price than a sampling scope.

Going forward in this book, the term DSO will always refer to a real time digital storage oscilloscope.

1.2   The Digital Multimeter

The simplest voltmeter that should be in everyone's lab is the digital multimeter (DMM), such as the one shown in **Figure 1.2**.

Figure 1.2 Examples of two popular DMMs with a price range of $7 (from Sparkfun) on the left and $150 (from Fluke) on the right. Both are measuring the same battery voltage, reading the same voltage to within 10 mV out of 9 V or agreeing to within 0.11%.

Fundamentally, a DMM measures a voltage with an update roughly every second. This is a general workhorse instrument used in any lab.

Generally, a DMM *only* measures a voltage. Everything else displayed is converted from voltage measurements into other terms, like current, resistance, or RMS values. You use it to measure a variety of electrical quantities. It is most commonly used to measure DC or constant voltage, current, and resistance and to check continuity, but it is limited in its capabilities.

If you care about how a signal's voltage varies over time, either over a very short period of time, like over a few nsec, or its pattern over a longer period of time, like many seconds, a DMM just won't do it. You need to graduate to the next-level instrument, an oscilloscope. One of the important purposes of a scope is to display the patterns in the signal's voltage over time, $V(t)$. This can span from less than a nanosecond to many seconds.

A DMM and a scope both measure voltages. However, the DMM only measures and displays the one averaged value of the voltage of the waveform over about a second and displays the numerical value on the screen. It tells you nothing of the underlying pattern the voltage makes.

An example of this important difference is shown in **Figure 1.3**, where the same signal from a microcontroller's pin is being measured with a DMM and a scope. The DMM displays the DC or average value, while the scope displays the instantaneous value of the voltage at discrete time intervals and displays the pattern.

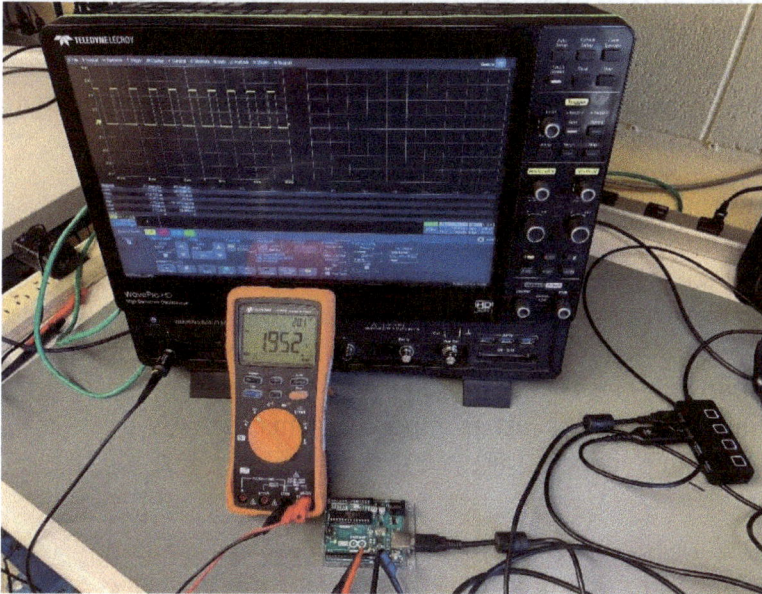

Figure 1.3 The voltage from a microcontroller digital I/O pin was simultaneously measured with a DMM and a scope. The DMM records a constant, average voltage. The scope reveals the actual pattern of 0 V to 5 V with a 40% duty cycle.

If you only care about the voltage of the signal averaged over roughly 1 second, then you might be able to get away with a DMM. This instrument will measure the signal's voltage averaged over a second and update it about every second.

If this is good enough for your application, then a $10 DMM would meet your needs. For this reason, a DMM is an essential instrument to have in any lab.

However, when you measure a voltage source with a DMM you have no idea what the behavior of this voltage waveform is over time periods shorter than 1 second. There could be 100 kHz oscillations at the output of your amplifier, and you will have no idea. There could be 10 kHz noise from the switch mode power supply (SMPS), which is 50% of the DC value, and you would have no idea.

This is why it is so important to measure all unknown signals with a scope first to verify that the signal is constant. Only then will you have confidence in how you interpret the voltage reading with a DMM. Always apply your engineering mind's eye to consider if the voltage you are measuring is a constant voltage or a modulated voltage. If you have a modulated voltage, use a scope first.

This is even the case when measuring a DC power source. The most important purpose of a power source is to provide a DC voltage at some rated available current draw. You may think it is only necessary to measure the DC voltage of the power source, and a DMM would be the right instrument. While this is an important metric, knowing the voltage noise or stability of the voltage on the power rail is sometimes just as important. This can only be measured with a scope.

As a general principle, you should always measure power rails with a scope. The DMM will give you an average voltage value, which is important, but will tell you nothing about the noise on the power rail.

Often, this noise is as important as the DC value. **Figure 1.4** shows an example of the measured voltage rail noise on a switch mode power supply, SMPS, and the resulting jitter on a clock powered with this rail. The jitter in the clock shows the impact from the power rail noise. This is an example of why it is important always to measure power rails with a scope so that you can characterize its noise.

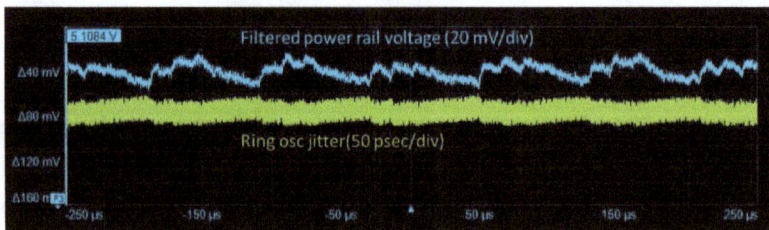

Figure 1.4 Voltage noise on a 5 V power rail and the jitter on a clock from this power rail noise. The impact on the jitter from the power rail noise is seen in the bottom green trace.

When only the DC or average component of the voltage from the DUT is measured with a DMM, there is a lot of potentially valuable information left on the table. A scope will reveal both the average value and the transient value of a voltage signal.

Given this limitation of the DMM in only measuring slowly varying voltages, a DMM is the right choice for you if you only want to measure:

- The voltage of a DC voltage source, like a battery

- The resistance of a resistor

- The continuity of a connection

- The DC current draw in a circuit

But for *all* other electrical measurements, you should consider using a scope first.

1.3   Why an Oscilloscope

An oscilloscope fundamentally measures the voltage at its inputs as a function of time. We generally refer to this input voltage as a *signal*. If you care about how a signal changes over time, an oscilloscope is the tool for you.

Measuring signals with a scope will open up a new and valuable window into the behavior of your signals and should be your go-to instrument to measure all signals.

An example of the behavior of a signal measured from a microphone is shown in **Figure 1.5**.

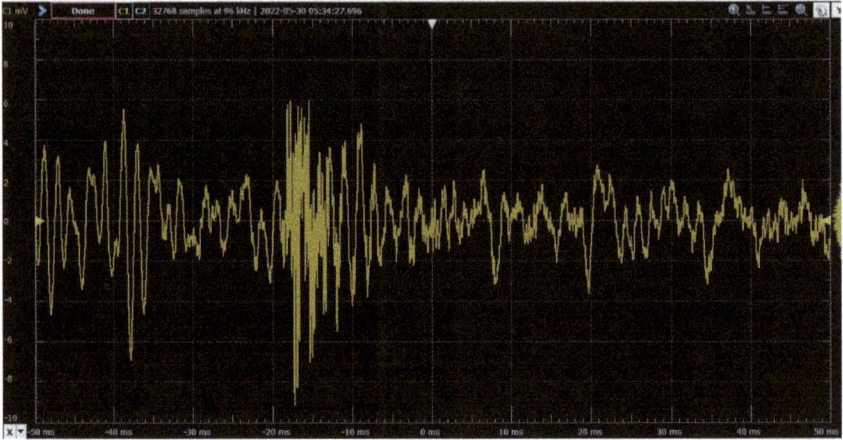

Figure 1.5 Measured V(t) from a microphone from random sounds in my office and displayed on a scope interface using my computer's sound card as the scope. There are 9,600 voltage measurement points displayed on this screen.

However, a scope is good for not just changing signals. It can provide just as good a measurement of a DC voltage as a DMM, when used correctly.

The numerical value of DC voltage measurements is described with two terms: the precision and the accuracy of the measurement.

It is always tempting to think that the more digits displayed for a measurement, the more accurate the measurement. This is *not* the case. The number of digits displayed is a measure of the voltage resolution. It is a rough measure of the precision or reproducibility of the measurement.

Accuracy is about how close a measurement will be to a National Institute of Standards and Technology (NIST) traceable measurement. For low-cost DMMs, such as the 830L, sold under many brand names, the rated accuracy is typically within 0.5% of a NIST traceable measurement. Some benchtop DMMs, such as the Keysight 34460A, with up to 6 digits of resolution, have a rated NIST traceable accuracy of less than 0.01%.

Sometimes, precision is important when looking for small changes in the measurement. A low-cost DMM typically has 4 digits or resolution. On the 1 V scale, this means measuring to +/- 1 mV. Using averaging, a scope can display just as many digits in recording a DC voltage as a DMM. A benchtop DMM can typically display a voltage with 5 or 6 digits.

As a simple example, an external 1 V DC signal was used as a source. It was measured simultaneously with a low-cost DMM, a high-performance benchtop DMM, and a Digilent AD2 scope with averaging. This measurement setup is shown in **Figure 1.6**.

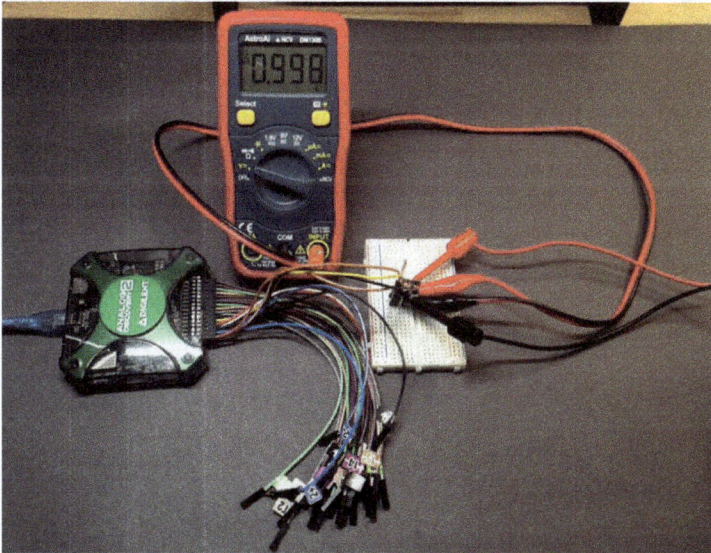

Figure 1.6 Measuring the same DC voltage with a scope, a low-end DMM, and a benchtop DMM. The Keithley 196 DMM is not shown.

The measured DC values were:

Scope: 0.9973 V

AstroAI DM130b: 0.998 V

Keithley 196 DMM: 0.9982 V

Using the Keithley 196 DMM as the reference, the low-cost DMM was within the last digit, or +/- 1 mV, or 0.1%. The scope was off by 0.7 mV out of 1 V or 0.07%. This exceeds the rated absolute accuracy of 2%.

When a high precision or high accuracy value of the DC signal is important, you can either use a DMM in parallel with the scope or set up your scope to measure the DC component with a higher effective number of bits (ENOB), usually by averaging an entire screen buffer and even multiple acquisition buffers. This means that for voltage measurements, a scope, when properly used, can provide as precise a voltage measurement as a DMM.

1.4   Analog Scopes

The first scopes, developed long before the digital revolution, were based on an electron beam that would be scanned across a phosphorescent screen. Two pairs of plates through which the electron beam traveled deflected the beam vertically and horizontally. The electron beam generator and the deflection plates were known as an electron gun. **Figure 1.7** shows an example of an electron gun and its vertical and horizontal plates.

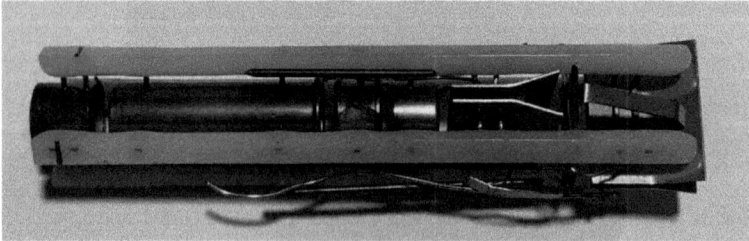

Figure 1.7 An electron gun used to create and drive an electron beam in an analog scope. The beam enters from the left and is collimated, then passes through the vertical and horizontal deflector plates on the right.

When the beam of electrons hit the screen, the screen would glow brightly. A voltage ramp applied to the horizontal plates of the electron gun would deflect the electron beam across the screen from left to right as fast as the linear ramp would sweep. This sets the position of the electron beam on the horizontal axis as a function of time.

While it was scanned horizontally, the input signal to the scope would be amplified and used to drive a voltage between the two vertical plates in the gun. The amplified input voltage would move the beam vertically while it was scanned horizontally. The resulting image on the screen would be the voltage as a function of time.

Of course, the electron beam had to be inside a glass vacuum tube to make its undistorted way from the gun to the screen. This system was commonly referred to as a *cathode ray tube* (CRT). These early analog scopes were referred to as CRT scopes. An example of a CRT scope is shown in **Figure 1.8**.

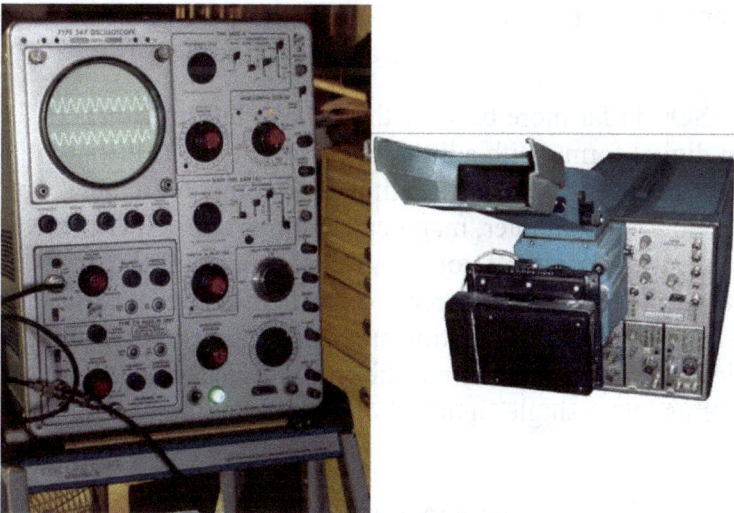

Figure 1.8. An example of a Tektronix 547 analog scope and the camera attachment used to capture a waveform on the screen of an analog scope.

These early scopes did one thing: display the input voltage waveform over time on the front scale. They often had a limit to how fast the displayed signal could change, related to the *bandwidth* of the measurement. The measurements were turned into hard copies using a camera to take a picture of the front screen.

Modern scopes work on an entirely different principle, do not require a camera, and are far more versatile. Today, the only value of an analog scope is to look cool in a science fiction movie. They have become the prop for all mad scientists.

1.5   DSOs Are More Powerful Than Analog Scopes

The original CRT-based scopes basically had one function: they displayed the voltage over some time period on the front screen. What you did with that information was up to you. You could take a picture and extract measurements like peak-to-peak value or average value from the picture using a ruler and scales.

Today's DSOs do far more because the voltage information is already in digital form. With a little processing in the firmware of an FPGA or in the software of the embedded computer in the scope or with an external computer, many calculations can be performed automatically on this data. If you can think of the algorithm to apply to the collection of measured V(t) data, most DSOs can perform the algorithm's calculation and display the processed information or convert the entire buffer of thousands of measurements into a single figure of merit measurement, often in real time.

All DSOs have the following important controls and features:

- Vertical scaling and offset
- Horizontal scaling and offset
- Triggering
- Cursors
- FFT real time spectral analysis
- Ability to save a waveform and compare to others on the same screen

- Ability to export a screen capture picture or csv files of the displayed data buffer

All DSOs, even the lowest-end scopes, have the feature of performing a fast Fourier transform (FFT) to convert the measured voltage over time, V(t), into the frequency domain to display the real time spectral components of the signal. **Figure 1.9** shows the screen of even a free scope that uses the sound card in your PC as the ADC with these different capabilities.

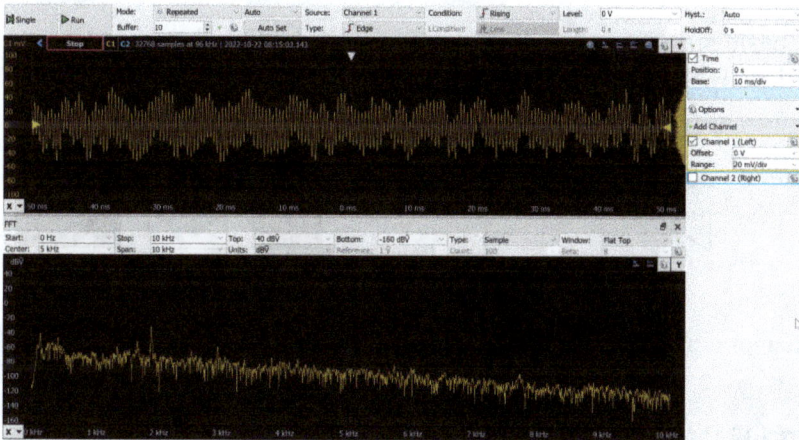

Figure 1.9 Example of the measurements of the sound of a whistle using the Digilent Waveform software, using my PC sound card as the actual scope instrument. The top screen is the voltage waveform recorded by the sound card's ADC, while the bottom screen is the calculated FFT spectrum of this signal, showing the peak at about 2 kHz.

1.6   Internal Structure of ALL DSOs

Today, all scopes are based on converting the analog signal from the *device under test* (DUT) into a digital signal using an *analog-to-digital converter (ADC)* and displaying the processed signal on a screen, which is typically a *liquid crystal display* (LCD). There are no more vacuum tubes in these modern scopes.

These new instruments, based on an ADC circuit that converts an analog input signal into digital bits, are referred to as digital storage oscilloscopes precisely because they store the digital version of the analog signal.

These are special cases of the general instrument referred to as a digitizing waveform recorder (DWR). The IEEE S1057 specification, ratified and released by the IEEE in 2017, covers the details of how DWR devices are described and how their features are measured.

This specification can be found here: https://ieeexplore.ieee.org/document/8291741.

The basic structure of virtually every modern digital storage oscilloscope is shown in **Figure 1.10**.

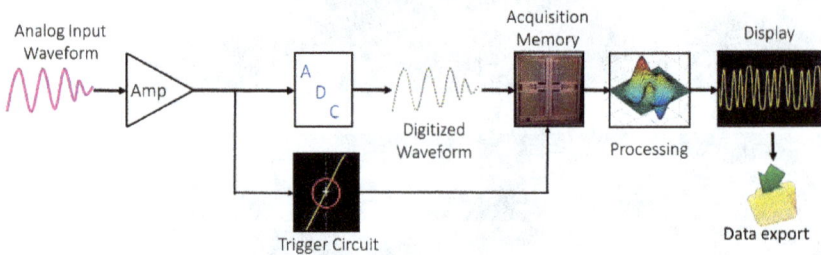

Figure 1.10 The internal structure of every modern DSO. The major hardware components are the preamp or analog front end, the analog-to-digital converter, the trigger circuit, and the acquisition memory. The other functions are often accomplished in software or firmware. Courtesy of Teledyne LeCroy.

After the analog input signal is transformed into a digital signal and placed in memory, the processing can all be done in software.

Since the data is often streamed into a memory buffer as it is taken, the actual measured values can all be displayed, even the ones measured before the trigger event. This means a DSO scope can "look back in time" to see the signal that led up to the trigger event.

1.7   Turning Measurements into Information and into Action

Any instrument will return raw measurements. In a scope, this is the voltage as a function of time. Often, in a scope measurement, 1,000 to 100,000 individual voltage measurements can be recorded and displayed on the front screen. This is a lot of measurements.

Other than looking for a specific pattern, rarely are we able to do anything with the raw measurements directly. Instead, we want to convert the raw measurements into information. We do this by matching the pattern we see on the front screen to an ideal signal waveform with a well-defined set of *parameters or figures of merit*, which describe or characterize the important features of the waveform. The figures of merit are the valuable information we use to describe the important features of the measurement.

For example, if the signal we care about is a DC signal, such as shown in **Figure 1.11**, the ideal pattern we match the measurement to is a constant voltage, with some DC value, a peak-to-peak voltage excursion, an RMS value, and possibly the amplitude and frequency of a repetitive signal such as 60 Hz pick up or some switching frequency.

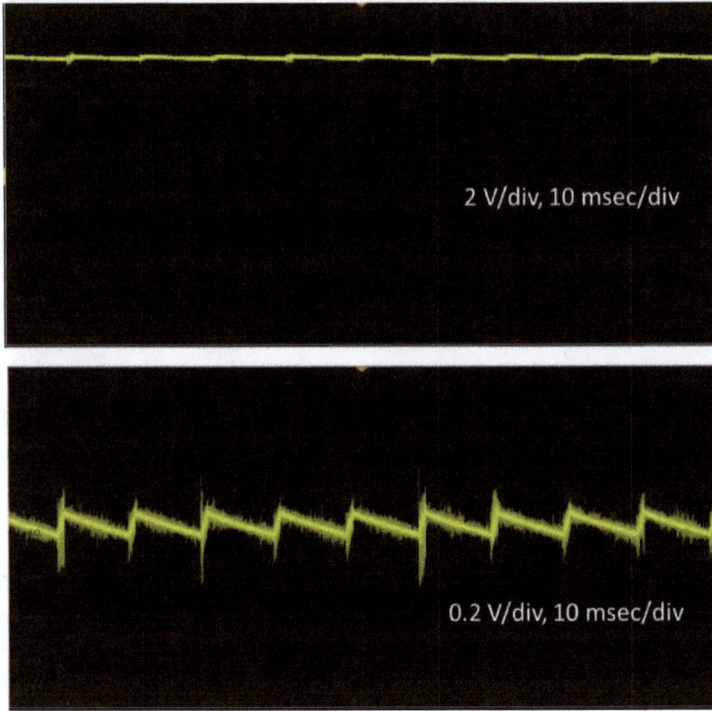

Figure 1.11 Example of the measured voltage from an AC to DC 5 V power source. The top trace is on a coarse scale; the bottom trace is zoomed in.

These five numbers, the average value, the peak-to-peak, the RMS, and the amplitude and frequency of any ripple pattern, are the figures of merit of the waveform. To read these figures of merit directly off the front screen, we can readjust the scales to zoom in and make it easier to read the values directly. These values extracted by eye from the front screen in this example are:

Average = 5.40 V +/- 0.05 V

Peak-to-peak = 0.4 V +/- 0.1 V

RMS = 0.15 V +/- 0.05 V

Ripple frequency = 100 Hz +/- 20 Hz

Ripple amplitude = 0.1 V +/- 0.05 V

If all we had were these five figures of merit, we could recreate a fair semblance of the measured waveform. The five figures of merit could substitute for the 10,000 raw measurements.

Once we convert the thousands of raw measurements into a handful of figures of merit as the information contained in the waveform, we can use this information to make a decision. There is almost always a consequence as a result of a measurement. There is some decision to be made or some action to take. After all, if there is no consequence from the results of the measurement, why do the measurement?

In this example, we could decide that this power supply, in the special case of no current load, has a low enough peak-to-peak noise of 0.4 V/5.4 V = 7% that it is acceptable to use in our application.

Or we may conclude that this DC voltage value of 5.4 V is too far outside the voltage range for our application, which might require 5.0 V with an acceptable tolerance of +/- 0.02 V.

The process we will go through for all our measurements is:

1. Use the best measurement practices to collect the raw measurements.

2. Apply the most important consistency test, Rule #9: Is the measurement consistent with what we expected?

3. Perform whatever other consistency tests we can think of to gain confidence in the quality of the measurements.

4. Analyze the uncertainty in the measurement and reduce the uncertainty as much as practical to get closer to the actual value present.

5. Turn the raw measurements into information as a few figures of merit.

6. Evaluate the information to make a decision and answer the "So what?" question.

1.8   Signals, Noise, and Artifacts

We call the voltage waveform we care about the *signal*. We refer to any other voltage in the measurement as *noise*. Sometimes, this noise is part of the measured voltage from the DUT and is as important as the signal. Sometimes, though, it is unwanted distortion, interference, artifact, or noise added to the signal we care about.

When we look at the voltage on a power rail, for example, we expect to see a DC voltage, like 5 V. Any variation in this DC level would be noise. This noise can be part of the voltage rail, as switching noise from a switch mode power supply (SMPS). In this case, we might want to characterize this noise from the DUT. It would tell us about the quality of the SMPS. The DUT noise is important information about the DUT.

But sometimes, the noise arises from the measurement system, such as when it comes from interference radio frequency (RF) pick up in the cables between the DUT and the scope. It could also be distortion from the analog signal chain in the probe-scope-measurement system or amplifier and digitizing noise from the scope. In this case, the measurement system introduces the noise.

Every scientist and engineer believes in an *absolute reality* when it comes to the physical world. There exists an actual, true signal from the DUT that exists independently of how it is measured.

Everyone using the same techniques should measure exactly the same signal to within some level of *measurement uncertainty*. The uncertainty is fundamentally limited to the *calibration* of the instrument to a NIST traceable standard, and then due to contributions from:

✓ Interference from external sources

✓ Random or asynchronous noise from the measurement system or the DUT

✓ Distortion due to the processing of the signal by the measurement system

✓ Measurement artifacts due to the interactions of the DUT's equivalent circuit model and the measurement system's equivalent circuit model

The uncertainty of a measurement from the measurement system defines how much the measurement of the DUT could be different from the actual voltage present, due to limitations in accuracy, measurement artifacts, or other sources of noise. Knowing the uncertainty of a measurement is just as important as knowing the measurement itself.

Identifying the sources of the uncertainty and reducing them based on their root cause is part of understanding your measurement system. The process of identifying the sources of noise and reducing their impact is part of best measurement practices.

A distortion of the signal from the DUT introduced by the measurement system and measurement process is called a *measurement artifact*. The guiding principle in all measurements is to be aware of the potential artifacts and know how to avoid them. The best way of doing this is to analyze the equivalent electrical circuit model of the DUT and the measurement system and evaluate the circuit's impact on the signal from the DUT. This is called *situational awareness*.

Situational awareness is about being aware of how the finite limitations of the instrument and the interactions between the electrical properties of the DUT and the electrical properties of the measurement system will distort the actual signals present: what might they be and how to reduce them.

One of the most important measurement artifacts arises from reflection noise from fast transient signals from the DUT, the DUT signal's source impedance, the cable connecting to the scope, and the scope's input impedance. This important artifact is covered in detail in a later chapter.

You should be aware of the equivalent circuit model of the measuring instrument and how this equivalent circuit model will interact with the circuit model of the DUT and its signal source. Some of the features to include in the equivalent circuit model of your instrument are:

- ✓ The input resistance of the instrument
- ✓ The input capacitance of the instrument
- ✓ The channel-to-channel crosstalk
- ✓ The bandwidth of the scope
- ✓ The bandwidth of the probes and cables
- ✓ The input impedance of the probes and cables and how it loads the DUT
- ✓ The series resistance of the probes and cables
- ✓ Reflections in the cable connections between the DUT output impedance and the scope's input impedance
- ✓ The vertical voltage resolution
- ✓ The horizontal time resolution and sample rate of the ADC
- ✓ The aliasing of the DUT signal and sampling rate of the ADC
- ✓ The voltage noise floor of the scope amplifier

Generally, an instrument such as a DMM or a scope, will always display a measurement when you push the run button. But, if this measurement has an artifact, it is worse than no measurement. It is a measurement that may be misleading. If you are not aware of the potential for artifacts, you may use this measurement believing it to be "correct" and interpret wrong information, from which you make wrong decisions.

You can never know if a measurement is the absolute signal on the DUT or has an artifact component. All you can do is check the consistency of the measurement with other measurements you can perform. Are all the results from all the measurements you can think of consistent with your interpretation of the measurement from the DUT?

This is generally accomplished by thinking of the DUT and the measurement instrument as a system. We create equivalent circuit models for the DUT and the instrument, and we evaluate, based on these models, how the signal from the DUT will be distorted by the combination of the DUT circuit and the instrument circuit.

The better you understand the equivalent circuit model and properties of your measurement system and your DUT, the better you can be aware of, control, and reduce the measurement artifacts.

1.9   Best Measurement Practices

While there is no foolproof, guaranteed way to avoid a measurement artifact, there are a few best measurement practices to follow to reduce the chance of introducing an artifact in your measurement.

Sometimes, it is difficult to know for sure in every measurement that you are, in fact, measuring the actual, real, objective signal from the DUT, not contaminated by a measurement artifact.

This is what best measurement practices are all about: using a process that reduces possible measurement artifacts and includes methods from which you gain confidence that you are measuring the objective reality of the DUT.

The more measurements that are consistent with each other, the more confident you will be that the measurements may be close to reality.

There are many ways of screwing up a measurement or experiment and only a few ways of doing it right. Whenever you do a specific measurement or experiment, or a complete project, you want to do what you can to reduce the risk of screwing it up and increase the chance of obtaining accurate results from which you will draw correct conclusions and take appropriate actions.

Here are six general habits to follow to reduce the risk of introducing an artifact and increase the quality of your results in an experiment, measurement, or project, which will be demonstrated over and over again in the measurement examples in this book:

1. Read the datasheet for all the components.

2. Read the manual for all the instruments. Sometimes these are boring. Sometimes, there is a lot of content you won't initially use. That's OK. Skim through it so that you are aware of what is there, and when you need the information, you know where to find it.

3. If you have a new instrument, like a DMM or a scope, initially measure something for which you know the answer or know what to expect to get familiar with the measurement process. The only way to know if you are using the instrument correctly is if you measure what you expect to see. This means you have to have high confidence in what you expect to see. Use a source or DUT you are familiar with to explore the new instrument.

4. *Before* you start a measurement, do some estimates with paper and pencil so you know what to expect. Anticipating what you expect to see in each measurement is Bogatin's Rule #9. If you don't know what to expect, how can you tell if the measurement is reasonable? Without a simple estimate, you cannot apply Rule #9.

5. When you perform a measurement, *always* do at least some preliminary data analysis or consistency tests with the measurements BEFORE you finish the measurements. *Do not* take a lot of data, and then the next day or the next week, look at the measurements and do an analysis. When you analyze the result, if you find the measurement is off by a factor of 10 from what you expect, only if the measurement system is in front of you will you have a chance to debug it and find the source of the error.

6. You can never do too many consistency tests. Whenever possible, ask yourself, if this is true, what else can I measure to test this idea?

1.10 Important Consistency Test: Rule #9

The first and most important consistency test to perform is to verify the measurement is consistent with what you expect. This is the motivation for:

> *Bogatin's Rule #9: Never perform a measurement or simulation without first anticipating what you expect to see.*

Is the measurement consistent with what you expect? If it is not, then you need to go back and check for problems in the measurement, or with the DUT, or reevaluate your understanding of what to expect. Either way, you will learn something valuable. When using any instrument, it is very easy to push a button and get a measurement either as a single number or a waveform on the screen. It is difficult to have confidence that the measurement is close to the reality of the actual signal on the DUT. Setting up the scope screen so you can see the pattern you care about and so you can read the figures of merit directly off the front screen is *always* important, no matter what else you want to do with the scope.

This way, you will have a rough idea of the signal waveform and the signal's features, like the period, the peak-to-peak value, or the signal rise time, for example. If you can't get a rough measure of these features from the front screen, how do you expect the scope's built-in functions to measure this?

Before you take a measurement or perform a simulation, it is important to apply Bogatin's Rule #9: Anticipate what you expect to see. It is not possible to ever be 100% sure your measurement is the actual, real voltage on your DUT. All you can do is perform consistency tests. The more consistency tests you can perform, the more confident you will be that the instrument, and your methodology using the instrument, provides some level of accuracy of the actual measurement of your DUT.

If you get in the habit of always thinking about the signal on the DUT before you turn on the scope, you will understand your system better, and you will gain confidence in your ability to analyze your system.

If the measurement you take is not what you expect, there is usually a reason for it. Don't proceed until you can reconcile the difference. Maybe the probe is not connected to the pin you think it is. Maybe the DUT is not powered on. Maybe your scope probe is introducing an artifact.

Sometimes, your understanding of your DUT is wrong, and this is a chance to move up the learning curve.

In order to apply Rule #9, you should get in the habit of adjusting the scales of the scope to make it easy to view the signal on the front screen so you can read valuable figures of merit directly off the screen. Once you have a reading, compare it to your expected value. If the scope scales are odd fractions or the signal trace is too small to read an amplitude, how do you know if the measured signal is consistent with rule #9? Before utilizing a cursor or a built-in measurement function, it's important to have a rough idea of the actual value being displayed.

A common error is applying the cursor or measurement function to the wrong channel on the scope. This is an easy problem to miss. You will always get a number displayed on the front screen, but it may not be about the measurement you expect. Applying Rule #9 will often catch this sort of problem. However experienced I am at best measurement practices, I sometimes make mistakes. Rule #9 helps me catch my errors early before they become serious errors.

1.11   Important Consistency Test: Measure a Known Signal

A second important consistency test is to measure something for which you know the answer. This can be a reference signal on a DUT that you have measured previously and have confidence in its value.

The compensation signal built into every scope is a signal for which you know the answer. A signal from a calibrated function generator is a signal for which you know the answer. Practice in measuring these sorts of signals will help you establish confidence in your instrument and your measurement methodology.

When in doubt about the quality of the measurement of a new DUT, you can always go back and measure a signal for which you know the answer. This is an important consistency test.

1.12   First Step in Any Measurement

Ken Johnson, product architect and director of marketing at Teledyne LeCroy, categorizes scope users into two groups: *viewers* and *analyzers*. A viewer looks at the voltage patterns on the front screen and extracts useful information just using their Mark 1 eyeball. This is basically what is done with all older analog scopes, which did not have any built-in functions.

An analyzer uses either built-in cursors or functions to extract numerical information from the measured data or exports the measurements into a file and uses an external tool like SPICE to transform the raw measurements into useful information.

The very first step in any measurement should be to extract useful information using your Mark 1 eyeball. This is what you will use to apply Rule #9 to compare to your expectation and to any other measurement you might extract using built-in measurement functions. This is a critically important consistency test. It is an important error trap to find potential measurement artifacts. While it is no guarantee, it is an important first step.

This means you want to set up the screen features to make it easy to get this information quickly. Use scales that expand the DUT's signal features you care about but are in units easy to read off the front screen, like in units of 1 V/div instead of a scale of 1.74 V/div.

For example, **Figure 1.12** shows two different scale settings for the same measurements. In one case, using simple, easy-to-read scales of 0.5 V/div and 200 nsec/div, and in the other case, with an odd value of 456 mV/div and 160 nsec/div with some arbitrary offset. Which scale setting makes it effortless to read the important figures of merit of the signal from the front screen?

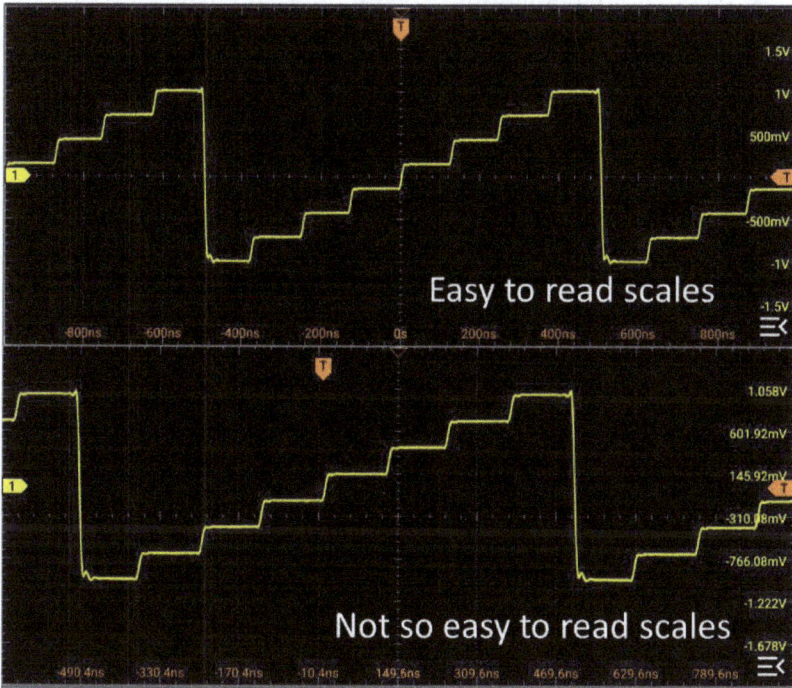

Figure 1.12 Examples of the same measurements with two different scale settings. On the top figure, it is easy to extract figures of merit from the front screen, while on the bottom screen, it is not so easy.

From the first screen, the following important figures of merit can be read off the front screen with little effort:

✓ peak-to-peak voltage is 4 div x 0.5 V/div = 2 V

✓ repeat period is 5 divisions x 200 nsec/div = 1 usec

✓ step interval is 0.6 div x 200 nsec/div = 120 nsec

✓ step voltage is 0.5 div x 500 mV/div = 250 mV.

It was effortless to extract these approximate values from the front screen using the screen units and our Mark 1 eyeball. This is the basis for applying Rule #9 and checking the consistency with any subsequent measurement functions.

Using the scales on the bottom screen requires some pencil and paper or calculator calculations. The raw measurements are exactly the same. The only difference is in how they are displayed. The more barriers you place in the way of reading quick estimates of the figures of merit from the front screen, the less likely you will do it.

This is why it is so important to always start with scale settings that make it easy to read approximate values of the important features of a signal directly off the front screen.

1.13   Second Step in Any Measurement

The second step in extracting useful figures of merit from the raw measurements is to leverage the built-in features of the scope. The simplest feature is using cursors to read the voltage value or time value of a point in a waveform based on the value stored in the acquisition buffer.

Using cursors does not require the scales to be adjusted to make it easy to read by eye. However, before cursors are used, it is important to apply Rule #9, which *does* require convenient scale settings. A common error is having the cursors on the wrong waveform. You will always measure a voltage or a time value as reported by the cursors, but it may be the wrong waveform. Unless you know what value to expect, you may not be able to catch this mistake.

The other built-in feature to extract figures of merit is measurement functions. These typically operate on all the data in the acquisition buffer and perform some calculations based on the assumptions of the features in the waveform. Many scopes have built-in functions to extract figures of merit for the entire buffer of data like the average, the peak to peak, the rise time, the period, and the RMS value. These figures of merit characterize the waveform. **Figure 1.13** is an example of a measured signal with some of the figures of merit options available in this mid-range scope.

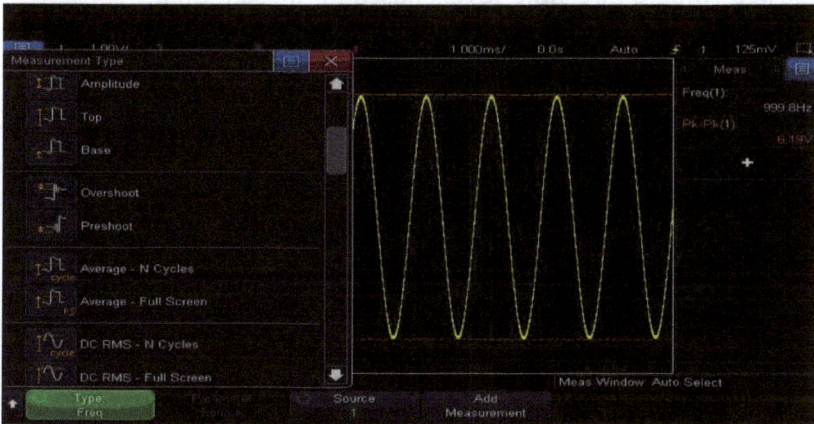

Figure 1.13 An example of the screen on a Keysight DSOX3024 scope shows some of the built-in measurement functions that can automatically display useful information from the measured waveform.

Always be aware that when the scope performs some calculations on the measured buffer of voltage over time data, assumptions are made in the calculation. Not always do these assumptions match what you would like them to be.

For example, the RMS value calculated usually includes the DC component of the signal. If you want the DC component subtracted, you want to select the ACRMS measurement or the standard deviation measurement.

The rise time is usually the 10-9 rise time, based on finding the 0% level and 100% level of the signal that is recorded in the current acquisition buffer. If the edge is not a linear ramp or a Gaussian shape, but has a long, graduated tail, the 10-90 rise time displayed may not be the term you care about. You will absolutely get a number, but it may not be the number you want.

For example, **Figure 1.14** shows the measured rising edge of a signal from a fast source. The built-in rise time measurement function follows a specific algorithm that measures the 0% level of the signal on the left edge of the screen, the 100% signal at the right edge of the screen, and finds the time intervals for the 10% and 90% voltage threshold levels based on these limits.

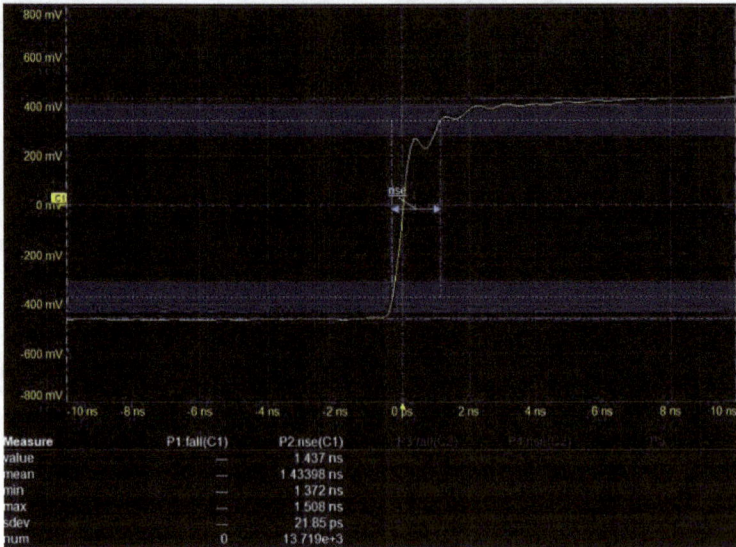

Figure 1.14 Measured 10-90 rise time using a built-in measurement feature. This value is misleadingly long compared to the rapid rise of the initial part of the signal. This was measured on a Teledyne LeCroy WavePro HD scope.

Is this one number returned, 1.43 nsec, an accurate measure of the rise time of the signal? In this example, a simple rise time is not a very precise figure of merit because this leading edge has structure to it.

If you want a rise time in order to estimate the bandwidth of the signal, the steep part of the leading edge may be more appropriate, which is closer to 0.4 sec. If you want to know when the ringing has settled, a longer value closer to 2.5 nsec may be a better figure of merit. This illustrates the danger of blindly accepting a measurement just because the function is labeled "rise time," "duty cycle," or "overshoot."

Higher-end scopes have a remarkable array of specific built-in functions to extract useful figures of merit for applications such as clock signals, power rail signals, motor drivers, data bus traffic, and even eye diagrams. All of these functions can be selected with just a few mouse clicks if you know where to find the menu item.

Before blindly accepting the value as read by the scope, always verify the number is consistent with what you expect and what you estimated using your Mark 1 eyeball.

The ultimate analysis of measured data is to use the scope as a very sophisticated data acquisition system and export the measured voltage over time data into an analysis tool such as SPICE or MATLAB or Python. If the scope you have does not have the built-in function to perform the specific analysis you need, using an external tool will give you the flexibility to get you the answer you need.

While this book focuses on applications using the built-in functions common to most scopes, a few examples are provided to show how the raw or processed V(t) measured voltage can be exported from the scope as a CSV file and then imported into common analysis tools such as LTSPICE, Keysight ADS, and QUCS.

The starting place for either a viewer or analyzer application is quality measurements.

1.14 The Bottom Line

1. Measurements are essential to open the window to the world of electrical signals that you cannot sense.

2. A DMM is an essential tool for any lab, but has significant limitations.

3. A scope is an essential tool for any lab. It will display the time dependence of a measured voltage.

4. A measurement artifact is a distortion of the true measurement of the DUT signal due to how the measurement is performed and the measurement system.

5. There exists an absolute reality of the actual signal on the DUT. The purpose of best measurement practices is to reduce measurement artifacts and reveal the true DUT signal.

6. All scopes today are digital storage scopes that translate the analog voltage signal that changes in time into digital measurements.

7. A figure of merit is a number extracted from many individual measurements that characterize some feature of the measurement.

8. A DSO should always be the first instrument of choice over a DMM when measuring any voltage signal to identify potential noise on the signal invisible to a DMM.

9. Always start out by measuring a signal for which you know the answer to get familiar with the best measurement practices for your instrument.

10. Always apply rule #9 to any measurement or simulation.

Chapter 2
Four Must-Have Instruments

No one instrument is going to meet all of the required needs for general measurements, including performance, ease of use, and shortest time to insight, with the fewest measurement artifacts. Instead, you want to have a few instruments available that will cover most of your needs. Each instrument will vary in price and capability.

As a general rule, the value of any instrument is related to the ratio of its performance to its price. A high-value instrument means it has a high enough performance you want at a low enough price you can afford.

Of course, since many applications are very specific, the performance you want may be different than the performance someone else wants. If your application is measuring biometric voltages from nerves, for example, low frequency, low noise, and high precision may be more important than higher bandwidth. Be aware of what features contribute to value in your specific application.

As general-purpose measurement tools, every engineer should have these four instruments available:

- A DMM for quick, DC-type measurements.
- A PC sound card scope. It's free and will enable you to practice with scope measurements and can be used for many audio signal applications.

- A low-cost multifunction scope with an integrated function generator. For the price point, it can be a workhorse instrument for many sophisticated stimulus-response measurements. However, it is limited in bandwidth, generally under 50 MHz.

- A general-purpose high-bandwidth scope. Its measurement bandwidth should be at least 3x your signal's bandwidth. This can range from 100 MHz all the way up to 100 GHz.

Each of these instruments is described below.

2.1   The DMM

A digital multimeter will display the numerical value it measures or computes on the front screen, which is updated about every second. Fundamentally, all a DMM measures is voltage. It measures a current by measuring the voltage the current generates across an internal, calibrated sense resistor. It measures resistance by measuring the voltage drop created by the resistor under test connected to an internally created, calibrated Thevenin circuit.

For AC signals, a DMM uses internal diode circuits to rectify and then average the signal into a DC voltage level, which is then measured and displayed. This is related to the root mean square (rms) voltage of the signal.

When a DMM reports an AC voltage, it is reporting a rms value but is calibrated in a limited frequency range and for a rms number of signal waveforms. For this reason, a scope should be the first choice if you want accurate AC voltage measurements.

The DMM should be used for simple measurements such as

- *DC voltage*

- *Continuity*

- *Resistance*

- *Isolated, floating voltages*

A handheld DMM is battery-powered. One chief value in a battery-powered DMM is the ability to measure the voltage in circuits that float and have no connection to earth-ground. It can also measure the true differential voltage across any two points in a circuit. While scopes can also do these measurements, they require special probes to connect to the DUT. A DMM does it natively right out of the box. The topic of measuring signals from sources that are grounded or floating is covered in a later chapter.

There is a wide range of DMM instruments available, with prices ranging from less than $10 to > $200. They differ in their features, but all have an absolute accuracy for measuring a DC voltage of better than 0.5% +/- 2 digits. Some are rated with a NIST traceable accuracy of less than 0.01%.

For example, **Figure 2.1** shows a selection of six different DMM units and models, all connected to the same voltage source. A high-end Keithley 196 DMM with NIST traceable accuracy of better than 0.01%, measured this same voltage source as 5.284 V +/- 0.001 V.

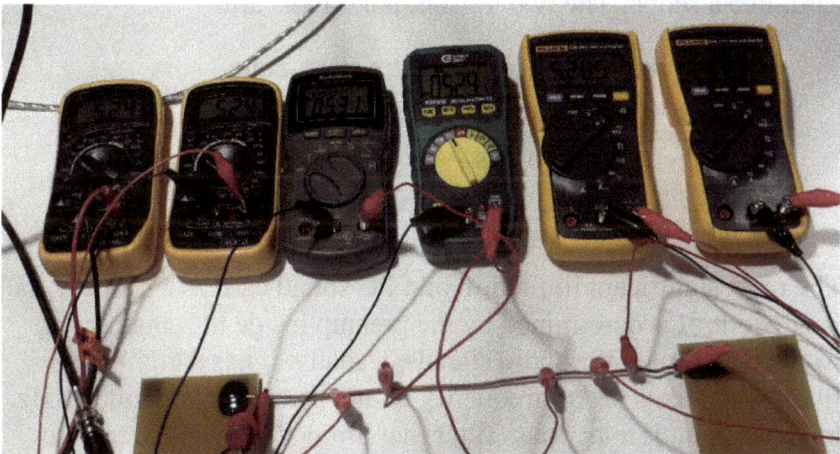

Figure 2.1 Examples of six different DMMs with a range of prices and capabilities, all measuring the same DC voltage source. Note that their measured values vary from 5.24 V to 5.31 V. This is a range of +/- 35 mV out of about 5 V or +/- 0.7%.

Each handheld DMM measured the same 5.284 V DC signal, and all agreed to within +/- 35 mV. This is about 0.035 V/5 V or ~ +/- 0.7% relative precision, close to their typical rated absolute accuracy error.

The absolute accuracy of even the $7 DMM, without any post-factory calibration, is 0.7%. With post-factory manual calibration, the absolute accuracy of this $7 DMM can be reduced to 0.1%.

A DMM is the ideal instrument for quick, simple DC measurements when the historical pattern of the signal is not important. For all other measurements, a scope is the right instrument to use when measuring a voltage.

2.2   Free PC Sound Card Scope

If you have never used a scope before, or you want an instrument to evaluate audio signals, a scope that uses your computer's built-in sound card is the right instrument for you. Best of all, with freely available software, it is basically a free scope anyone can access.

There are many free tools available that can access your sound card, like a scope. The two most popular tools are:

- ✓ PC sound card scope by Christian Zeitz (https://www.zeitnitz.eu/scope_en)
- ✓ Digilent Waveforms (https://digilent.com/shop/software/digilent-waveforms/download)

The most significant limitation of a PC sound card as a scope is its bandwidth. It is determined by the preamp inside the sound card, which is designed for audio signals. Usually, it has a high pass filter, which cuts off all frequency components below about 20 Hz. In series there is a low pass filter that limits the highest frequency it can record, roughly in the 10 kHz to 20 kHz frequency range. This limitation makes it ideal for audio frequency signals but not much else.

Its primary value as a scope is as a tool to practice using a scope and to experiment with some of the features common to all scopes, with virtually no barrier to entry. There is nothing stopping you from getting started with this scope right now. The details on how to download the recommended software and get started using your sound card as a scope are described in Chapter 6.

If you care about audio signals, both analyzing them and generating them, a free PC-based scope may be all you need to use. With its integrated voltage measurement and output signal from the function generator, which can synthesize and play any sort of signal over your PC speakers, a free sound card scope is really a free multifunction instrument.

As a scope, it will measure the voltage waveform at an analog input channel. These measurements can be displayed and processed using built-in functions in the software. The PC sound card scope I recommend, from Digilent, using the Waveforms software, can perform many common operations on measured data that most digital scopes can perform, such as:

- *Display of an analog input channel*

- *On-screen cursors*

- *More than 30 different automated measurement functions*

- *Built-in spectrum analyzer using an FFT algorithm*

- *Data logging to a CSV file with the raw measurements or figures of merit, such as the average value or rms value*

An example of the measured voltage from a sound card input using the Digilent Waveforms software, displaying the scope measurement of the microphone voltage signal and its real time spectrum, is shown in **Figure 2.2**.

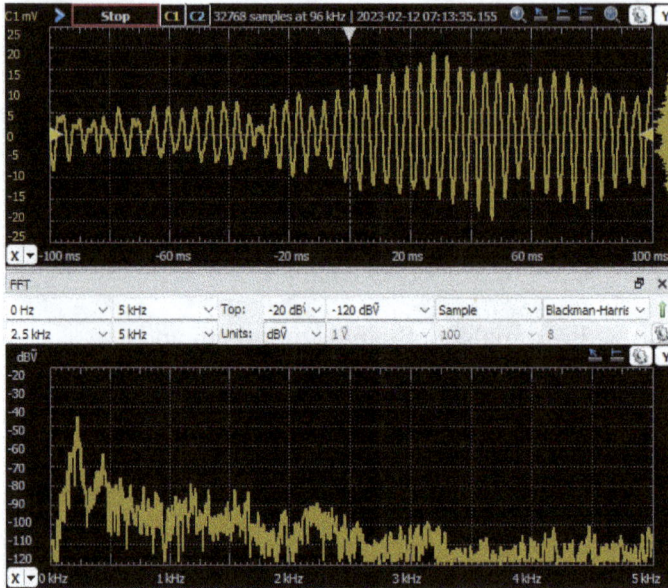

Figure 2.2 An example of the screen of the Waveforms application using my PC's sound card as the data acquisition front end of a scope. This shows the measured voltage waveform and its resulting spectrum in real time. The sound recorded was music.

The software can also generate sound by driving the speaker output using an output channel driven by a function generator.

The combination of the stimulus source and response measurements means the free sound card scope-multifunction instrument can also be used to measure the transfer function or mechanical vibration spectrum of physical objects or the acoustic shielding effectiveness of absorbers.

When the function generator output from the sound card is used as a stimulus voltage signal to a DUT and the input to the sound card measures the response, this free PCB sound card scope can also be used to:

- *Measure the transfer function or Bode plot of any passive or active circuit.*

- *Measure the impedance of any passive or active circuit.*

Just a caution: when connecting any external device to your PC's sound card other than a microphone or speaker, be very careful. If you apply an input voltage that is too large, you always run the risk of damaging the sound card inside your PC. To protect your computer, always use an external USB sound card in these applications.

2.3   Multifunction Instruments

A scope, which is a specialized form of a digital waveform recorder, is a powerful instrument to measure the signals from voltage sources. When combined with a stimulus source, it can measure the transfer function of an active device, the Bode plot of a filter system, and even the impedance profile of any electronic circuit.

The combination of a scope with a stimulus source, or function generator, and the software to integrate their functions together is referred to as a *multifunction instrument.*

It is a much more powerful instrument than just a scope. The scope and function generator are separate instruments. It's the software that combines them together. For this reason, these multifunction instruments are sometimes referred to as *software-defined instruments.* The principle of operation of a multifunction instrument that can create a stimulus and measure a response is shown in **Figure 2.3**.

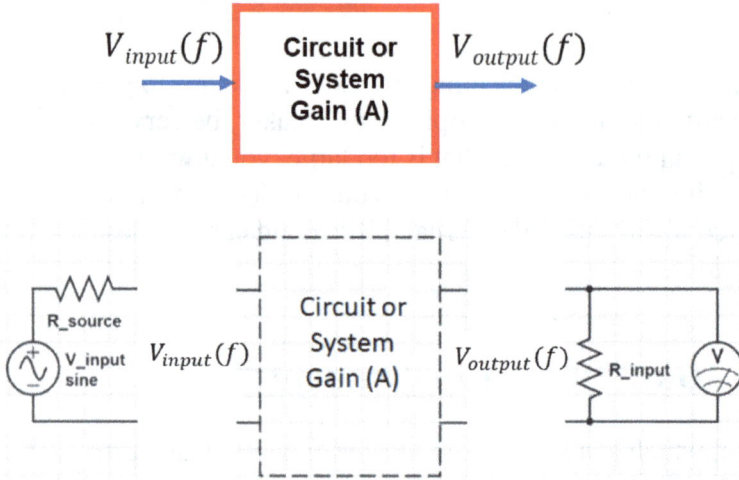

Figure 2.3 The principle behind stimulus-response measurements, specifically when the input and output signals are both voltages.

An example of a stimulus-response measurement of the Bode response of an TLV4110 opAmp using both a small signal amplitude of about 10 mV and a large signal of 1 V amplitude, showing the lower bandwidth for larger signal amplitudes, measured with a Digilent AD2 multifunction instrument is shown in **Figure 2.4**.

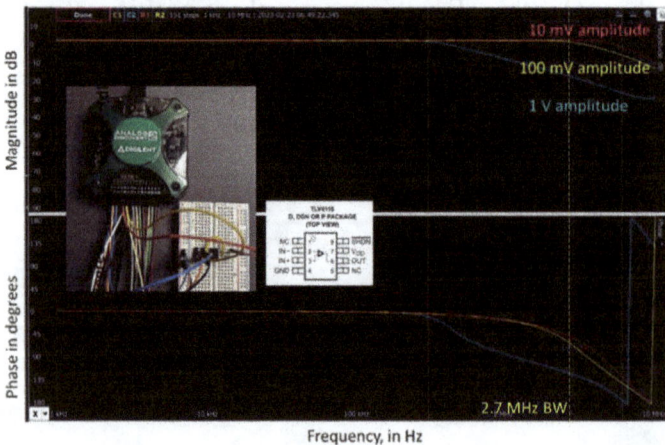

Figure 2.4 Measured Bode plot of an OpAmp measured with a multifunction scope, shown in the inset.

There are many commercially available instruments with an integrated scope and function generator that can act as a multifunction instrument. They differ in the bandwidth of the signal source or of the scope, the variety of functions available, the user interface, and their price point. The table below lists a few examples of commercial multifunction instruments available.

Vendor/URL	Instrument	BW Limits	Price
PicoTech https://www.picotech.com/	2204	10 MHz	$150
Analog Devices https://www.analog.com/en/design-center/evaluation-hardware-and-software/evaluation-boards-kits/adalm1000.html	ADALM1k	100 kHz	$80
Analog Devices https://www.analog.com/en/design-center/evaluation-hardware-and-software/evaluation-boards-kits/adalm2000.html	ADALM2k	10 MHz	$195
Liquid Instruments https://www.liquidinstruments.com/products/hardware-platforms/mokugo/	Moku:go	30 MHz	$599
Digilent https://digilent.com/shop/analog-discovery-2-100ms-s-usb-oscilloscope-logic-analyzer-and-variable-power-supply/	AD3	35 MHz	$399
Red Pitaya https://redpitaya.com/product/stemlab-125-14-board-for-oem-partners/	STEMlab 125-14	60 MHz	$377

Many scopes, such as the Keysight DSOX3024, the Rigol DHO924S, and the Rohde and Schwarz MXO5 series scopes, all have an integrated function generator and are capable of stimulus-response multifunction measurements. Though these scopes have bandwidths into the 1 GHz range, the function generators have bandwidths typically below 25 MHz and are comparable to the lower-end USB-based multifunction scopes for these applications.

The chief limitation of most multifunction instruments is their function generator bandwidth. If an application fits within the bandwidth limit of a multifunction instrument, generally, this is the right instrument to use. In addition to the scope functions, you get all the other instruments enabled by the software features, such as data logging, Bode plots, impedance profiles, and a three-terminal curve tracer.

2.4   A Higher Bandwidth Scope

The most important differentiator between all scopes is the highest sine wave frequency component they can measure. This metric is generally referred to as the bandwidth of the scope.

If a constant amplitude sine wave signal is input to the scope, the measured amplitude will be constant until the frequency approaches the limits of the scope. At some frequency, the amplitude will begin to decrease. The bandwidth of the scope is defined by the frequency at which an input sine wave will be measured with an amplitude that has decreased to 70% (-3 dB down) of the source.

The bandwidth of a scope is also a measure of the shortest rise time the scope can measure if the input is a step signal with 0 psec rise time. As a rough approximation, the shortest rise time a scope should be used to measure is when the rise time of the source = 1/bandwidth of the scope. If the bandwidth of the scope is 100 MHz, the shortest rise time the scope should be used to measure accurately is about 10 nsec. This is further discussed in Chapter 4.

The bandwidth of commercial scopes spans a wide range but generally falls into three families:

- ✓ Low-end with a bandwidth below 100 MHz (> 10 nsec signal rise times)

- ✓ Mid-range with a bandwidth from 100 MHz to 1 GHz (> 1 nsec signal rise times)

- ✓ High-end with a bandwidth greater than 1 GHz (< 1 nsec signal rise times)

Scopes with a measurement bandwidth below 100 MHz are incredibly valuable but are limited in application. Most multifunction scopes fall into this range. They are perfect for audio applications, analog electronic circuits, sensor-based systems, and embedded systems with a rise time longer than 10 sec. They generally have the lowest price.

Mid-range scopes have a bandwidth above multifunction scopes. This enables measuring rise times down to 1 sec, which is common in all microcontrollers, embedded systems, and IoT applications, for example. This is all that is needed for many microcontroller bus systems such as UART, I2C, I3C, and SPI. Serial decode of these digital channels is a common built-in function in mid-range scopes. This is also a perfect range for power integrity measurements where the power rail noise spectrum rarely extends above 1 GHz.

Generally, high-end scopes are used in applications where their higher bandwidth is required and can justify their higher price. This is the case for all high-speed serial links such as USB, HDMI, PCIe, and Ethernet. Many RF applications use real time DSO instruments to view the modulated or pulse shapes of the RF signals such as for Wi-Fi, BLE, communications, and radar. These applications often require additional software such as eye diagrams, vector signal analysis (VSA), and jitter analysis.

An example of an 8 GHz bandwidth Teledyne LeCroy WavePro HD scope measuring a PRBS source is shown in **Figure 2.5**. This sort of high bandwidth measurement is not possible with a low- or mid-range scope.

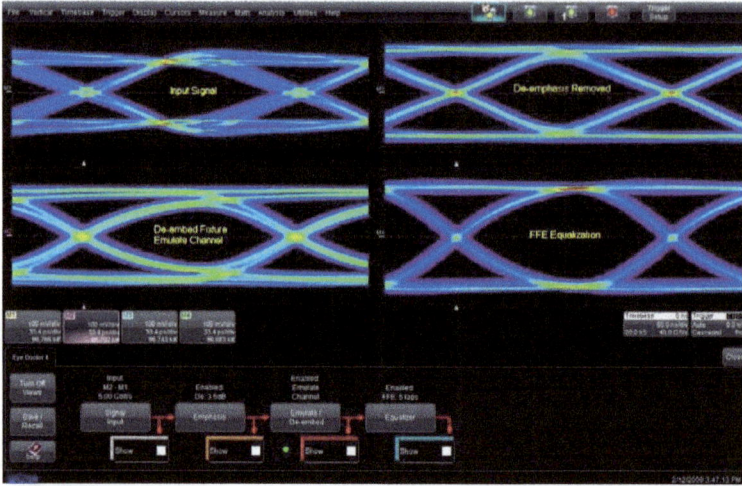

Figure 2.5 Examples of a measured PRBS signal with equalization features added to the signal. This is measured with a 5 Gbps PRBS data pattern measured with an 8 GHz bandwidth scope.

In this book, I use a variety of low- and mid-range scopes to illustrate best practices used in general and specific applications. The specific features and user interfaces may be different, but they all do the same thing: display the pattern generated by voltage signals and provide a short time to insight.

2.5   The Bottom Line

1. Every lab should have four different instruments: a DMM, a free PC sound card scope, a low bandwidth multifunction scope, and a high bandwidth scope.

2. While a DMM is easy to use, it should only be used to measure a voltage or current that is already known to be

constant. If the signal is actually changing, you will get a measurement from the DMM, but it will be misleading.

3. A free PC sound card-based scope is a powerful learning tool.

4. The simplest and most powerful free PC sound card scope is the one available from Digilent. It uses the same user interface as is used for their multifunction scope.

5. A multifunction scope can be used to measure the transfer function of a system either as a voltage transfer and displayed as a Bode plot or as a current-voltage transfer function displayed as an impedance.

6. All multifunction instruments have a bandwidth limited by the function generator to below 50 MHz.

7. The bandwidth of a scope is the highest sine wave frequency that can be measured with an amplitude that has dropped no lower than 70% of the input amplitude.

8. Choose a scope bandwidth that is larger than the 1/rise time of your signal.

9. If your applications have signal bandwidth above 100 MHz, a higher bandwidth scope should be in your future. This is their key differentiating feature.

10. A high bandwidth scope is generally much more expensive than a mid-range scope but also includes features such as eye diagrams and jitter analysis.

Chapter 3
Mastering the DMM

The DMM should be your first choice to make continuity tests, resistance tests, and DC voltage or current measurements when your DUT is floating or when you need a true differential DC voltage measurement.

3.1   Introducing the DMM

A digital multimeter will display the numerical value it measures or computes on the front screen, which is updated about every second. While it is called a multimeter, fundamentally, all it measures is voltage.

It converts a current into a voltage by measuring the voltage the current generates across an internal, calibrated sense resistor. It measures resistance by measuring the voltage drop created by the resistor under test connected to an internally created, calibrated Thevenin circuit. This turns the resistor under test into a voltage divider circuit with a known voltage source and known resistance.

AC signals are measured by using an internal diode circuit to rectify and then average the signal into a DC voltage, which is then measured and displayed.

The DMM should be the first choice when used for simple measurements such as

- DC voltage
- Continuity
- Resistance

- DC current

- Isolated, floating DC voltages

There is a wide range of DMM instruments available, with prices ranging from less than $10 to more than $200. They differ in their features, but all have an absolute accuracy for measuring a DC voltage specified at better than 0.5%. Some DMMs have a specification of 0.01% absolute accuracy. This is traceable to a National Institute for Standards and Technology (NIST) absolute voltage reference source.

A DMM is the ideal instrument for fast, simple DC measurements when the historical pattern of the signal is not important. For all other measurements, a scope is the right instrument to use.

Many DMMs are battery-powered with a plastic enclosure. This means they float relative to earth-ground, and they measure the true differential voltage between their two terminals.

3.2   How a DMM Measures Voltage

All modern DMMs use a high-resolution analog-to-digital converter for their input. This ranges from a 12-bit to a 24-bit resolution ADC.

With a 12-bit ADC, the dynamic range is only 1 part in 4,095 levels or counts. However, if 100 measurements are averaged together, it is possible to increase this dynamic range by a factor of sqrt(100) = 10, to 40,950. This is sometimes referred to as over-sampling and is a common method of increasing the effective number of bits (ENOB). If the measured noise is random or Gaussian, the rms amplitude of the digitizing noise will decrease with the square root of the number of samples averaged together.

Typically, an ADC is limited to an input voltage range of 0 V to about 5 V. A preamp and voltage divider in the front will scale the signal and offset it so that bipolar signals in the 1 mV to 100 V range can be measured. These components and circuits are usually referred to as the analog front end (AFE). All the electronics from the DUT to the input to the ADC are lumped in the general category of the AFE.

Since they are designed to display a numerical value on the front screen, each measurement of voltage is typically updated at a rate of about one reading a second. During the displayed time interval, a DMM displays the average voltage over this time period.

This means that if it is measuring a changing signal, all that is displayed is the average value over the roughly 1-second averaging time interval.

For example, if the signal from a DUT is a pulse width modulated (PWM) signal, the DMM will only show the average or DC component. **Figure 3.1** shows an example of a DMM measuring an EKG signal with an average value of about 19 mV. The simultaneous scope measurement shows a voltage changing with a well-defined pattern with 4 V peaks. However, the DMM only shows the average value, with no indication of the real voltage pattern from the device under test.

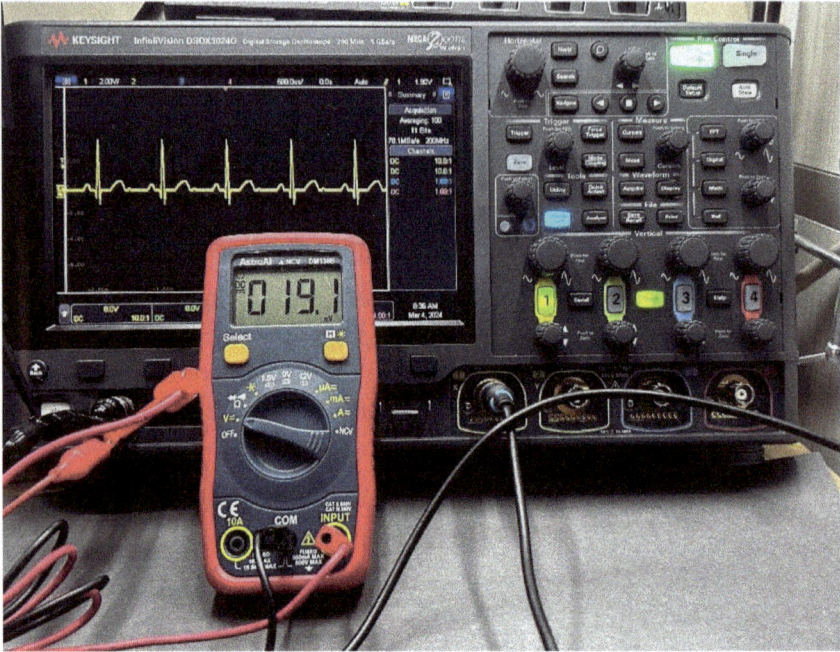

Figure 3.1 A DMM displays only a DC average value over a second interval, even when the underlying voltage signal is a complex waveform.

If the DMM is connected to a sine wave signal with an amplitude of 1 V, a period of 1 msec, and an offset value of 0 V, the DMM will measure a 0 V signal. The DMM tells you nothing about the voltage pattern present, only the voltage averaged over a 1-second time interval.

> *A DMM is not suitable to measure the voltage of a signal that changes much on a time scale of 1 sec.*

This is an important source of artifacts. Part of situational awareness is to consider the long integration time of the DMM when you interpret measured voltages that are displayed on the DMM.

Most DMMs will measure multiple functions, such as current, resistance, diodes, and even test transistors. They have to convert every other type of measurement into a voltage and measure this voltage, then display it back in terms of the original type of signal.

3.3   Situational Awareness in DMM Measurements

There is a difference between an ideal voltmeter and a real voltmeter. For example:

- ✓ An ideal voltmeter has an infinite input impedance. A real voltmeter has some finite input impedance.

- ✓ An ideal voltmeter reads the instantaneous voltage at the input. A real voltmeter reads the average voltage over some finite period of time.

- ✓ An ideal voltmeter has an absolute accuracy that is perfect. A real voltmeter has some accuracy error based on its traceability to a NIST standard, which is typically 0.5% +/- the last 2 digits.

- ✓ An ideal voltmeter has no random noise. A real voltmeter has some random noise from its amplifier.

The purpose of an instrument is to measure the real, actual voltage between the terminals of the DUT, over time with some expected accuracy, precision, and time response.

We refer to the actual, real voltage on the DUT as the signal. Anything else that is measured is noise.

Any actual voltage coming from the DUT and its environment is considered a signal. When the DUT is a power rail, for example, we want to measure the voltage on the power rail. If we expect the power rail to be a constant 5 V, for example, any variation other than 5 V is noise on the power rail. Some of this noise comes from the DUT and is what should be measured. Some of this noise is a measurement artifact and should be reduced or at least measured.

When we connect the instrument to the DUT, we create a system. If the process of connecting the instrument to the DUT causes the system to distort the actual, real signal from the DUT, we refer to this distortion as a measurement artifact. A measurement artifact is introduced because of the electrical properties of the DUT, the interconnect, and the instrument. **Figure 3.2** shows an example of the equivalent circuit of the DUT and the DMM, illustrating where an important measurement artifact would arise.

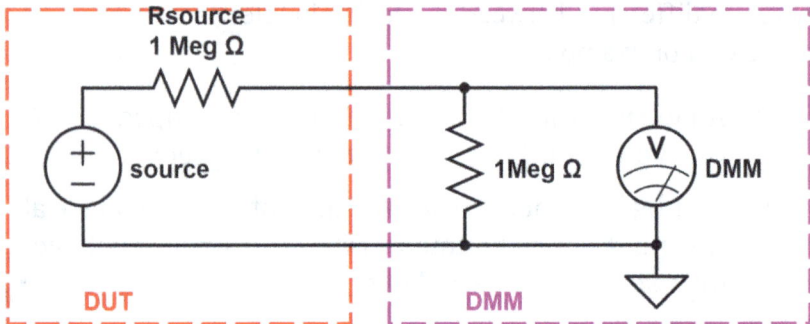

Figure 3.2 The equivalent circuit model of the DUT and the DMM identifying a voltage divider between the DUT output resistance and the DMM input resistance as an important measurement artifact.

The DMM adds a measurement artifact to the DUT voltage. If we were not aware of the voltage divider created by the system, we would not know the measurement had a potential artifact when measuring a high-impedance source.

Measurement artifacts arise when the electrical properties of the measurement instrument interfere with the measurement and distort it in some way. The way to avoid these measurement artifacts is by paying attention to situational awareness. This means thinking of the equivalent circuit model of the entire system of the DUT; the interconnects to the instrument, which is sometimes called the fixture, the probe, or the cables, and the instrument itself.

Situational awareness applies to all measurements. With a DMM, situational awareness forces you to think about the equivalent circuit of the DMM in each of its measurement modes, as well as the limitations of the ADC. It will only display average voltages over a one-second interval. This means one artifact is that the measurement is an average voltage, not the instantaneous voltage. If the voltage is changing faster than about once a second, the DMM will measure an artifact.

Due to the input impedance of the DMM when used as a voltmeter, if the DUT source impedance is above about 10k ohms, the DMM will measure a lower voltage by more than 1% from the actual voltage on the DUT due to the voltage divider created by the measurement system. These are measurement artifacts.

The higher the bandwidth of the signals, the more impact there will be from the interactions of the DUT, the interconnects, and the instrument, and the more likely the measurement system will introduce artifacts.

This is why situational awareness is an important best measurement practice. Always be aware of the equivalent circuit model of the DUT-interconnect-instrument system and possible sources of artifacts.

The first step in applying situational awareness is to understand the equivalent circuit model of the interconnects to the instrument and the instrument itself and consider possible artifacts.

3.4   Reverse Engineering the Input Resistance of a DMM

An important principle of situational awareness is to think of the combination of the DUT and the instrument as a system and how the equivalent circuit model of the instrument interacts with the equivalent circuit model of the DUT.

Measurement artifacts arise when the equivalent circuit of the instrument, which is an approximation of its real behavior, distorts the signal from the DUT. In this case, what is measured by the instrument is not the true signal from the DUT.

This means that knowing the equivalent circuit model of an instrument is important for situational awareness. When the spec sheet of the instrument does not spell out the equivalent circuit elements, which are important to know, the best measurement practice is to extract these values by reverse engineering the instrument.

The general principle of reverse engineering is to create or synthesize your best guess of the equivalent circuit model of the DUT or system of interest and then use best measurement practices and every other tool at your disposal to extract these circuit element parameters. The first step is to create the equivalent circuit topology of the instrument.

The simplest, first-order model of a DMM is an ideal voltmeter. The input impedance of an ideal voltmeter is infinite, an open. A second-order model adds a shunt resistance, which is the input resistance of the voltmeter. This second-order model is a good starting model for any instrument that measures a voltage at its input. The input resistance of the instrument is an important figure of merit that characterizes the instrument. This model is shown in **Figure 3.3**.

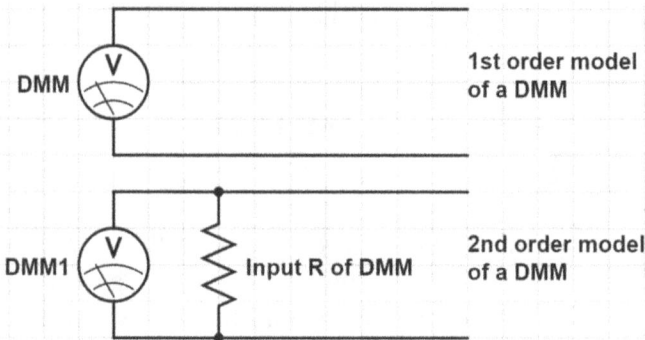

Figure 3.3 Example of the equivalent circuit model of an ideal DMM showing the first-order model and second-order model.

A third-order model would add input capacitance to the input of the voltmeter. This complex model would add the equivalent series resistance and inductance of the interconnect wires from the DUT to the DMM. This third-order model is shown in **Figure 3.4**.

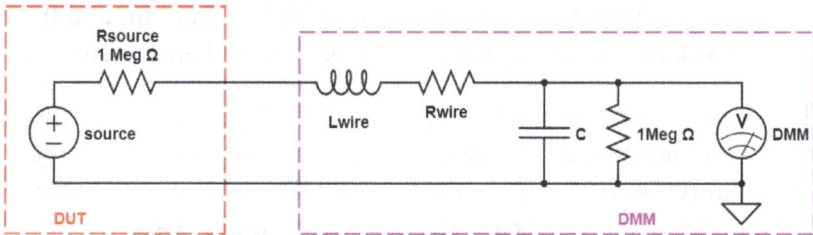

Figure 3.4 A third-order model of the DUT and the DMM identifies multiple circuit elements in the DMM. This is still an ideal model, just more complex.

This third-order model, while complex, is still an ideal model for a DMM. It is composed of ideal circuit elements that behave with mathematical precision. It is not correct to call this a nonideal model of the DMM. It is just a more complex model than the first-order model of a simple ideal voltmeter. Its purpose is to interpret the measurement from the DUT and evaluate how much of the measurement might be an artifact due to the measurement system.

In most applications of a DMM as a voltmeter, the voltage drop across the loop inductance of the DMM leads is negligible and can be ignored. The input capacitance has little effect on the slowly changing signals that could be measured and can be ignored.

Likewise, because of the high input resistance of the voltmeter, the current through the lead resistance is so tiny as to contribute negligible voltage drop artifacts in voltage measurements and can be ignored.

However, the input resistance of the DMM can introduce an artifact in some cases. This is the most important equivalent circuit model of a real voltmeter that should be used in situational awareness. The most important figure of merit to extract that characterizes the instrument is the input resistance of the voltmeter.

There are two simple ways of measuring the input resistance of any voltmeter. The simplest is to literally connect a second Ohmmeter to the front terminals of the DMM and use this to directly measure the input resistance of the DMM. The limitation is that some Ohmmeters only measure up to 1 megaohm, and the input impedance of the voltmeter may be higher than 1 megaohm.

The second way is to provide a precision voltage source that the instrument has measured. Then, a large resistor, such as a 1 megaohm resistor, is added in series, and the voltage of the source is measured. This creates a voltage divider circuit, as shown in **Figure 3.5**.

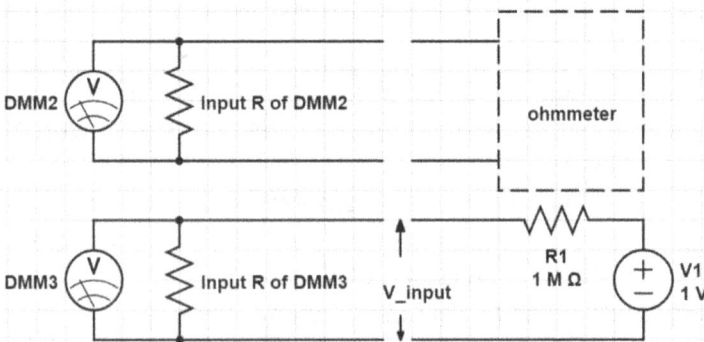

Figure 3.5 Two methods to reverse engineer the input resistance of any voltage-sensitive device under test.

The voltage source of the voltage divider, V1, can be measured directly by the voltmeter, without the 1 megaohm, R1, series resistor. We assume the output resistance of the voltage source by itself is much less than 1 megaohm.

With the 1 Meg series resistor inserted between the DMM and the 1 V source, the V_input voltage of the divider voltage can be measured directly by the voltmeter. From the source voltage, the external series resistance, R1, and the measured input voltage, the unknown input resistance of the DMM can be calculated.

The first method of just connecting an ohmmeter capable of measuring a high resistance is shown in **Figure 3.6** where the input resistance of an ANENG AN8008 DMM is measured as about 11 megaohms by a UNI-T UI6E DMM.

Figure 3.6 Example of the direct measurement of the input resistance of a DMM using another DMM as an ohmmeter. The input resistance of the ANENG DMM, on the right, is measured as 10.95 megaohm by the UNI-T UI6E DMM on the left.

The input impedance of any voltmeter is important to know to avoid potential artifacts. Any voltage measurement from any DUT creates a voltage divider with the input resistance of the voltage meter. What the DMM or other voltmeter ever measures is the voltage divider voltage across its input resistance.

When the source impedance of the DUT is larger than 1% of the input impedance of the voltmeter instrument, the voltage divider voltage is more than 1% lower than the source voltage. This means the voltmeter will report a lower voltage than the actual voltage on the DUT.

As part of situational awareness and watching out for potential measurement artifacts, always be aware of the source impedance of the DUT and the input impedance of the voltmeter. When they are within 1% of each other, they recognize that there is a potential measurement artifact.

This applies to any measurement, whether with a DMM, multifunction scope, or other scope.

3.5   Reverse Engineering How a DMM Measures Current

A DMM only measures voltage. To measure a current, the current is turned into a voltage using a sense resistor in series with the current. The second-order equivalent circuit model of a DMM set to measure current is shown in **Figure 3.7**.

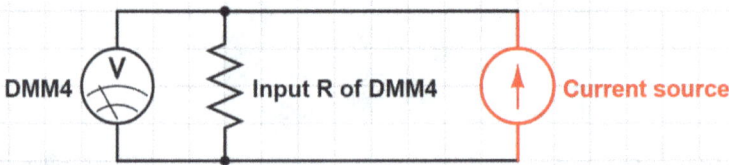

Figure 3.7 An equivalent circuit model of a DMM set to measure current.

When on the current setting, the DMM connects a calibrated shunt resistor through which the current being measured flows. The voltage across this calibrated resistor is a direct measure of the current through the circuit in which the DMM is connected.

To implement this circuit, the DMM will add a series resistor and a series voltage drop in the circuit to which it is inserted. We refer to the extra voltage drop created by the DMM in the circuit, the *voltage burden*. The series resistor and the voltage burden introduce measurement artifacts. As part of situational awareness, this series resistance and the voltage burden should always be estimated.

In principle, it should be possible to measure the sense resistance of the ammeter. In practice, since this is usually on the order of 1 ohm, it is sometimes difficult to measure with a 2-terminal DMM due to the contributions from the lead resistance and wire lead resistances of the DMM and its fixture.

An alternative and simple way to measure the sense resistor in a DMM measurement is to set up the DMM to measure a current and simultaneously use a second DMM to measure the voltage burden. From the measured current and voltage burden, the sense resistance can be calculated. This is illustrated in **Figure 3.8**.

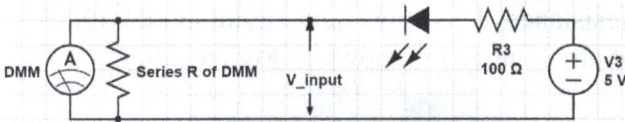

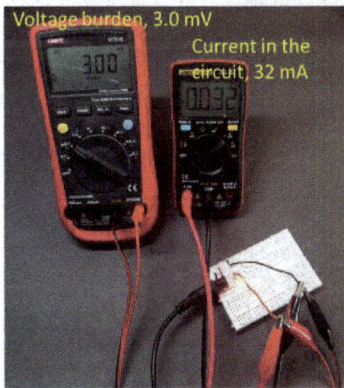

$$R_{sense} = \frac{V_{burden}}{I_{measured}} = \frac{3.0mV}{32mA} = 0.094 ohms$$

Figure 3.8 Example of the measurement setup and the actual measurement and calculation of the series resistor as 0.094 ohms.

In this example, the measured sense resistor is 0.094 ohms. Due to the uncertainty in the last digit of the current, this sense resistor is probably closer to 0.1 ohms. This same resistance was measured for all the current scales of this DMM in the ammeter setting.

This means that whenever this ammeter is inserted in series in a circuit under test, it adds a 0.1-ohm series resistor in the circuit. When the other resistances in the circuit are of this order, the

DMM series resistance will affect the circuit. Likewise, if the current were 1 A, for example, the DMM would add a voltage burden to the circuit of as much as 1 A x 0.1 ohm = 0.1 V.

3.6   Reverse Engineering How a DMM Measures Resistance

There is no such thing as an ohmmeter. Instead, every ohmmeter creates a voltage divider circuit and measures the voltage across the unknown resistor, given the precision voltage of the source and the source series resistance. This equivalent circuit model of the inside workings of an ohmmeter is shown in **Figure 3.9**.

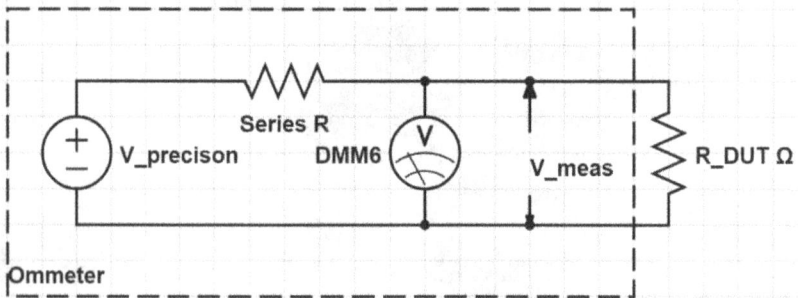

Figure 3.9 Equivalent circuit model of how an ohmmeter actually measures resistance using a voltameter and a precision voltage source.

The two important figures of merit that describe the behavior of the DMM when measuring resistance, the precision voltage source and the series resistance, can be reverse-engineered very easily.

First, the internal precision voltage source can be measured with another voltmeter when the ohmmeter is set for a specific ohms scale. This measurement is shown in **Figure 3.10**.

Figure 3.10 An example of using the UNI-T DMM as a voltmeter (left), to measure the internal precision voltage of the ANENG AN8008 DMM, set as an ohmmeter (right), on the kOhm scale.

Be aware of the input resistance of the voltmeter when measuring a voltage. The input impedance of the UNI-T DMM used as a voltameter is 10 megaohms. If the impedance of the ohmmeter is smaller than about 100 kOhms (1% of 10 megaohms), the input resistance of the voltmeter does not affect the voltage reading. In the ANENG DMM used as an ohmmeter, its precision voltage reference is 1 V.

The second term to extract for the ohmmeter is its series resistance. Unfortunately, this cannot be measured directly with an external ohmmeter. An external ohmmeter set up to measure the input resistance of the ohmmeter under test will apply its own precision voltage, and this voltage will interfere with the internal voltage of the ohmmeter being tested.

This is an example of why you should never use an ohmmeter to measure the resistance of a component that is part of a powered-on, active circuit. Your ohmmeter may display a resistance value, but this value reported may have no relevance to reality because the voltage from the circuit may create a voltage across the resistor under test that changes the voltage across the resistor under test and is misinterpreted by the DMM. This is another example of a potential measurement artifact.

Even when the circuit power is turned off, all you really measure is the equivalent resistance of the resistor in the circuit, which includes the resistor under test and anything else connected to it in parallel. If it has a 1 ohm resistor in parallel somewhere else in the circuit, all you may measure is the 1 ohm parallel resistor.

This is another example of situational awareness. Be aware of how your instrument is measuring what it displays so you can identify and avoid possible measurement artifacts.

This is why the general rule for using a DMM to measure the resistance of a resistor is,

Only measure the resistance of a DUT when it is pulled out of and isolated from the circuit to which it is part. Never measure the resistance of a component when the resistor is part of a powered-on circuit. When connected to a circuit, ALL you measure is the equivalent parallel resistance.

To measure the internal resistor that forms part of the voltage divider circuit requires a little more analysis. The voltage divider circuit when an external R_DUT is attached is simply

$$V\_meas = V\_precision \frac{R\_DUT}{R\_series + R\_DUT}$$

The process to extract the R_series resistance inside the ohmmeter is first to measure the internal precision voltage. Then, add a known resistor across the input to the ohmmeter, R_DUT. A second voltmeter is used to measure the voltage divider voltage across this external resistor, V_meas. The internal series resistance can be extracted from:

$$R\_series = R\_DUT\left(\frac{V\_precision}{V\_meas} - 1\right) = 1kOhm\left(\frac{1V}{0.0096V} - 1\right) = 103kohms$$

This circuit was set up with a 1 kOhm external resistor as the DUT. **Figure 3.11** shows an example of this setup with the 1 ohm resistor and ANENG AN8008 DMM set to measure ohms on the 1 ohms scale. A second DMM, the UNI-T DMM measured the voltage across this external resistor.

Figure 3.11 The measured voltage across a 1 kOhm resistor (left DMM) is measured as 0.0096 V by the UNI-T DMM when the ohmmeter (right) measured the resistance of the DUT as 0.983 kOhms on the 1 kOhm scale.

In this example, the V_precision was measured as 1 V. The R_DUT was measured as 1 kOhm. The V_meas was measured by the external DMM as 9.62 mV. This means the R_series is:

$$R\_series = R\_DUT\left(\frac{V\_precision}{V\_meas} - 1\right) = 1kOhm\left(\frac{1V}{0.0096V} - 1\right) = 103kohms$$

The internal series resistance of the DMM, when set to the ohms scale, is 103k ohms. This is within 3% of 100k ohms. Since the tolerance rating of the 1k external resistor was only 5%, it is likely the internal resistance of the DMM was 100k ohms.

In this way, the internal equivalent circuit model of the DMM as an ohmmeter can be extracted.

Knowing the equivalent circuit model of the instrument under test is important to understand potential measurement artifacts. It is also an important principle in engineering that there are no black boxes. Always try to reverse engineer what goes on inside of any measurement instrument. This is an important situational awareness principle.

3.7   How a DMM Measures Continuity

Continuity is another important function of a DMM. Continuity is nothing more than a low resistance between the leads of the DMM. It is a measure of the connectivity between two nodes in a circuit. They are connected if there is continuity, a low resistance, between them. Most DMMs can be set up to sound a buzzer when the measured resistance drops below a value that ranges between 50 ohms and 200 ohms. This is what continuity means.

The buzzer is useful so that continuity can be tested while using two hands to hold the probes without looking at the DMM. This way, many connections can be tested quickly.

It is easy to reverse engineer the threshold of what resistance causes the buzzer to go off. When the DMM is in continuity mode, the resistance it measures is also displayed on the screen. A few resistor values can be measured to see the threshold when the buzzer is activated.

For the ANENG AN8008, the buzzer activates at a resistance of less than 99 ohms. This level or lower is considered a connection. In other DMMs, continuity is when the resistance is no higher than 50 ohms.

3.8 Application: Continuity Testing of a Solderless Breadboard

The advantage of using a DMM as a continuity meter is that it is very simple and very fast to return valuable information.

A solderless breadboard is a structure with small holes on 100 mil centers to connect through-hole components quickly. It works particularly well with 22 AWG wire, which has a snug fit in the holes. It is a workhorse component for quickly prototyping circuits. **Figure 3.12** is an example of a prototype circuit built with a solderless breadboard.

Figure 3.12 An example of a quick and simple circuit prototype created with a solderless breadboard.

To use it effectively, it is important to know the continuity between the holes. **Figure 3.13** shows the connections created by metal strips inside the plastic housing.

Figure 3.13 The internal connectivity pattern of metal strips inside the plastic housing of a solderless breadboard.

The five adjacent holes in a horizontal row are all connected together. The two rows across a center span are not connected together. The two vertical columns on the solderless breadboard's left and right sides are all connected up and down the column, but adjacent columns are not connected.

These simple observations can be verified quickly and easily using a DMM set for continuity.

One of the subtle features of a full-size SBB is that in some units, the top and bottom halves of the columns are not connected together. If you assume they are connected and use the SBB as though they were, you may find your circuit does not work.

This is easily verified using the continuity setting of the DMM. In the example shown in **Figure 3.14**, the same column is not connected between the top half and bottom half of the SBB.

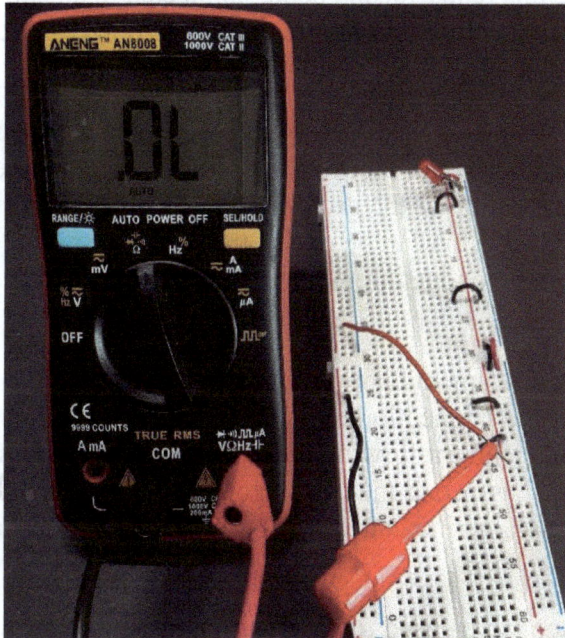

Figure 3.14 The connectivity meter shows an open between the top and bottom half of the left-most column. To use this SBB effectively, jumper wires, as shown on the right, need to connect the top and bottom half of the columns.

As a general rule, if you don't measure the connectivity of the SBB, assume the top and bottom halves are not connected together and use jumper wires to connect them.

3.9 Application: Is Earth-Ground Connected to Chassis Ground

A useful application of a DMM is to measure the resistance between the chassis of an instrument and earth-ground. This helps to identify which connections on an instrument are actually earth-ground connected.

Knowing the earth-ground connections to a device helps in identifying good connection points to reduce ESD (electrostatic damage). When it is recommended to "ground" yourself to prevent ESD events, this really means being connected to an earth-ground conductor. Connecting to a floating circuit-ground provides no ESD protection.

In a house or building that passes US civil codes for wiring, the (usually) round prong of a power socket is connected to earth-ground. The location of the ground prong in other countries' power plugs may differ. This can be one of the reference nodes between which any other conductor can be tested for connectivity or low resistance.

Whether the instrument is powered on or not, it should not change the earth-ground connection to the chassis. There should not be any intentional current flowing in the earth-ground path from an instrument's chassis and the earth-ground plug in the wall socket. If there is, this is an indication of a significant safety hazard with the instrument.

An ohmmeter can be used to measure the resistance between the chassis of an instrument to a known earth-ground conductor. If this is on the order of a few ohms or less, the surfaces are all connected together and to earth-ground.

Once it is established that the chassis-ground of an instrument is connected to earth-ground, it can be used as a reference surface. When a rack of instruments use 3-prong power plugs and are all plugged into wall power, generally, all their chassis-grounds are connected together through the common earth-ground. While this can be a small DC resistance, this path may have a large, frequency-dependent impedance at higher frequencies. This is sometimes a source of EMI certification test failures. A DMM will not be sensitive to this effect.

Many 3-prong power supplies that plug into the wall, especially used in laptop supplies, have only 2 wire connections as their output to the device they are powering. Using an ohmmeter, the resistance between the earth-ground connection of the power plug and the outer cable shield of the laptop power plug, the circuit-ground connection, can be measured.

Generally, this resistance varies between 1 megaohm to 1 kOhms. Its purpose is to provide protection from the possibility of any static charge buildup on the circuit-ground connection to your laptop. If there is any metal surface on your laptop, this is the chassis. Generally, the chassis will be connected to the circuit-ground of your laptop. When plugged into the power socket, the circuit-ground is connected to earth-ground by a 1 k resistor. It is not meant to carry large currents, just leakage or static buildup currents.

3.10 Reverse Engineering What a DMM Measures as AC Voltage

Most DMMs can measure DC voltages as long as they do not change faster than about 0.5 Hz. Any faster than this and either the displayed voltage is not updated fast enough, or the displayed voltage is a low-pass filtered version of the actual voltage.

In addition, many DMMs have a setting to measure AC, or changing voltages. But what exactly is this a measure of? Is it the peak-to-peak value, the amplitude, or the rms value?

There are two different RMS values, which are sometimes confused unless they are specifically called out. The DC or average value of the voltage can be included in the RMS value. This is sometimes referred to as the DCRMS value. The DCRMS value of a waveform will vary as the DC voltage offset changes, even if the peak-to-peak value is constant. It is defined as

$$DCRMS = \sqrt{\frac{1}{T}\int_0^T V^2(t)dt}$$

The second type of RMS value is with the DC or average component first subtracted off of each voltage measurement. This is referred to as the ACRMS value. It is equivalent to the standard deviation calculation. A DC component will have no impact on the ACRMS value. It is given by

$$ACRMS = \sqrt{\frac{1}{T}\int_0^T (V(t) - \langle V(t)\rangle)^2 dt}$$

In most DMMs, when it measures an AC voltage, it displays the AC root mean square, ACRMS, value of the signal, not the DCRMS value. For example, the ACRMS value of a sine wave, averaged over an integral number of cycles, is 0.707 x the amplitude.

If the waveform is a 50% duty cycle square wave, the ACRMS value is equal to the ½ x the peak-to-peak value. For any other duty cycle, the ACRMS value decreases compared to the peak-to-peak value.

If it is important to measure an AC signal, a far better instrument to use is a scope. This will display the waveform shape and allow direct measurement of many figures of merit.

However, the advantage of using a DMM for an AC signal is that the DMM floats and can measure a voltage from a source that is also floating. It can also measure a differential signal.

The RMS value read by a DMM varies between DMM models depending on what exactly they are doing. If this measurement is important, use a DMM, which is rated as a "true RMS" meter.

The AN8008 is rated as a true RMS meter. To reverse engineer this instrument and test its limitations in measuring an RMS value, the wave generator of an AD2 scope was used as the signal source. This signal was measured by both the AD2 scope and the DMM set to AC voltage. **Figure 3.15** shows the setup and an example of measuring a 1 kHz frequency, 1 V amplitude sine wave with the scope and the AN8008.

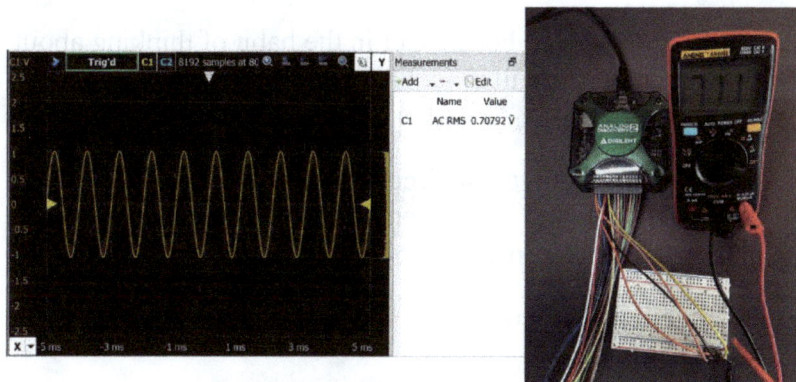

Figure 3.15 An example of measuring a sine wave with the AD2 scope and the AN8008 DMM set on AC voltage.

For this sine wave, the expected RMS value was 0.707 V. What is measured by the AD2 scope is 0.708 V. The AN8008 DMM measured an RMS value of 0.711 V. This difference is less than 0.5%.

Using a swept frequency sine wave source, the frequency range over which the AN8008 measured an ACRMS value that was within 10% of the AD2 scope's RMS measured value was 20 Hz to 5 kHz. Beyond this frequency range, the displayed RMS value on the DMM was off by more than 10%. This was for amplitudes of 100 mV up to the 5 V maximum range of the function generator.

This was also the case for a square wave and other waveforms, even a random noise waveform. The AN8008 reported the true RMS value of the waveform as being within 10% of the true RMS value.

If the voltage to be measured is a single-ended signal, in an earth-grounded circuit, a scope is a better instrument. There is a broader frequency range for the measurement, and more information about the actual waveform is available.

3.11   The Bottom Line

1. There are no black boxes. Get in the habit of thinking about the equivalent circuit model for all instruments and the DUT.

2. Based on the equivalent circuit model of the instrument, think about how to reverse engineer the important parameters of the instrument.

3. The input impedance of a DMM is typically about 10 megaohms.

4. When used as an ammeter, a DMM typically has 0.1 ohm series resistance in its path.

5. The voltage drop across the DMM when used as an ammeter is called the voltage burden. Knowing the internal series resistance and the current, the voltage burden can be estimated.

6. Situational awareness is about thinking about the impact the equivalent circuit model of the instrument has on the circuit of the DUT.

7. Never use a DMM as an ohmmeter when your DUT is powered on. The voltage in the circuit will interfere with the voltage generated internally by the ohmmeter.

8. When you measure the resistance of a resistor while it is in the circuit, be sure to power the circuit off, and then you will measure the equivalent circuit of all the resistance in parallel.

9. When you are not sure if the signal you are measuring is really constant, use a scope to measure the signal instead of a DMM.

10. A true RMS DMM will report the ACRMS value of the waveform for frequencies roughly between 20 Hz and 5 kHz.

Chapter 4
Most Important Scope
Differentiator: Bandwidth

The bandwidth of a scope is its most important differentiating feature. It is the first criterion used to select the appropriate scope for your application. Bandwidth is fundamentally about the highest sine wave frequency that can be measured with an amplitude close to (no lower than 71% of) the input signal's amplitude. Higher-frequency sine waves would be measured with an amplitude less than 71% of their actual amplitude.

The bandwidth in the frequency domain translates into the time-domain as a measure of the shortest rise time the scope can measure and display. To measure a shorter rise time signal requires a higher bandwidth scope.

The very first question you should answer when selecting a scope for your application is: what is the bandwidth of your signal? Then select a scope with a bandwidth at least three times this value. Choosing a scope with adequate bandwidth will help you avoid a potentially very important measurement artifact.

If the bandwidth of your scope is too low, even if the scope is free, it will be worthless in your application. In fact, it may be even less than worthless. You will always get a measurement, but the measurement may be wrong or misleading, and you may not be aware of this.

Of course, the downside of selecting a scope with too high a bandwidth is that it will be more expensive than a lower bandwidth scope. Unless you are investing in the future, you may be paying for more value than you need. This leads to the general practice of purchasing a scope with at least as much bandwidth as required for your application and then as much bandwidth as you can afford.

4.1   Consequences of Too Low a Scope Bandwidth

The bandwidth of a scope affects at least three qualities of the measured signal: the rise time of the signal, the magnitude of short pulses, such as from switching noise, and the potential artifact of ringing noise from Gibbs ringing.

For example, the rise time of the signal from a digital pin of a popular Arduino board, which uses a Microchip Atmega 328 microcontroller chip, is about 4.7 nsec when measured with a scope with a 200 MHz bandwidth or higher. If the scope bandwidth is less than 200 MHz, the rise time will appear longer. Features in the signal that change on the time scale of 5 nsec or shorter may not be visible with a lower bandwidth measurement. This is a measurement artifact.

Figure 4.1 shows the measured rise time of a digital I/O signal as measured using a 10x probe having a bandwidth of 500 MHz and a scope measurement bandwidth of 50 MHz. The rise time of the signal is much longer with the lower bandwidth measurement.

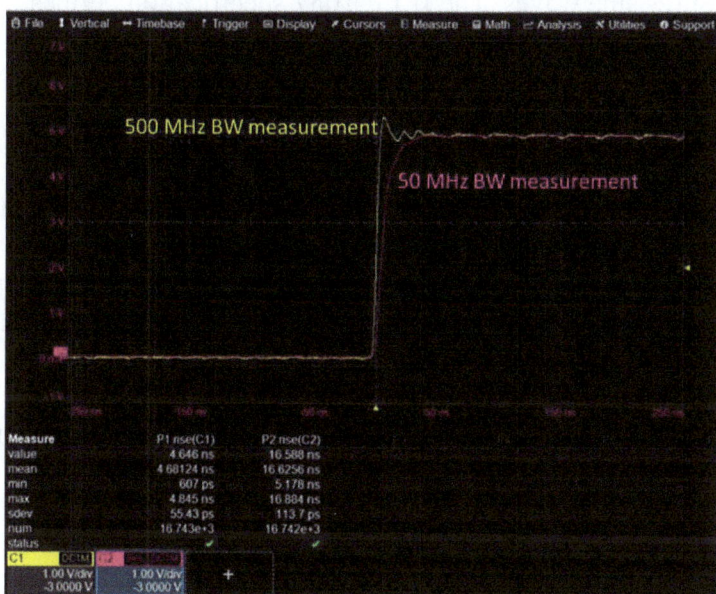

Figure 4.1 The same measured digital signal with two different bandwidth settings. The rise time is longer, and the ripple noise is less in the low bandwidth measurement.

Switching noise is driven by the dI/dt of the signal's rising edge and the inductance in the signal-return paths. This noise signature mostly has high-frequency components.

If the scope bandwidth is too low, you will measure this switching noise, but it may be misleadingly low. You may be fooled into thinking there is no problem when, in fact, your switching noise is too large, and your measurement hides a design flaw that could cause your product to fail.

4.2 The Bandwidth of a Scope

The bandwidth of a scope is the highest sine wave frequency component the scope can accurately measure. After all, the purpose of a scope is to show how a signal changes over time. How fast a change in voltage can be measured is an important figure of merit for the scope.

The simplest definition of the bandwidth of a scope relates to how it measures sine waves. The IEEE 1057 specification defines electrical bandwidth as the frequency at which the measured amplitude of an input sine wave is reduced by -3 dB (i.e., attenuated to 70.7% of the actual value of the signal, a fall of approximately 30%) relative to its amplitude at the input.

Many scopes behave like an n-pole filter, with n between 1 and 4. This means the transfer function between the input signal and the measured signal looks like a low-pass filter response. The point where the transfer function drops by -3 dB compared to the pass-through region is considered its bandwidth. **Figure 4.2** is an example of the measured transfer function of a Keysight 4024 scope. It is rated at a 200 MHz bandwidth, and its -3 dB frequency is measured at 240 MHz, which easily meets this spec.

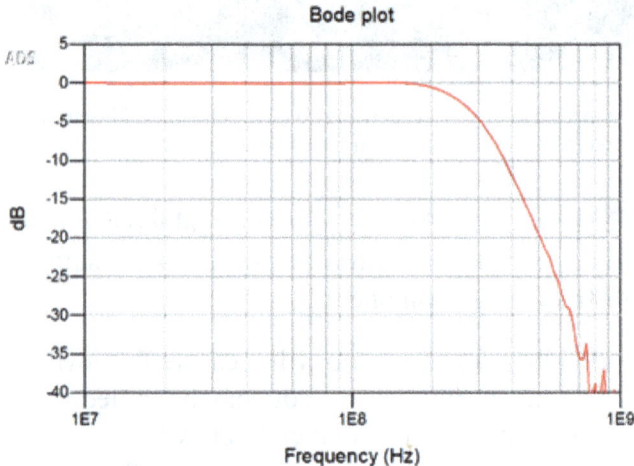

Figure 4.2 The measured transfer function of a Keysight 4024 scope is rated for a 200 MHz bandwidth. The drop off in the high pass region is about 80 dB/decade, which is a fourth-order filter response.

This type of low pass filter's transfer function has a step response in the time domain. An input step signal with a very short rise time would be filtered by this transfer function and appear as a step response with a longer rise time.

There is a connection between the -3 dB bandwidth of a filter and the 10% to 90% rise time of its step response. This is in the general form of

$$RTBW = BW \times RT$$

Where

RTBW = the rise time bandwidth product, a figure of merit of all filters

BW = the filter's -3 dB bandwidth

RT = the filter's step response's 10% to 90% rise time

Using a simple numerical simulation, the empirical relationship between the bandwidth of a filter and its step response's rise time can be calculated for different order filters. This relationship is shown in **Figure 4.3**.

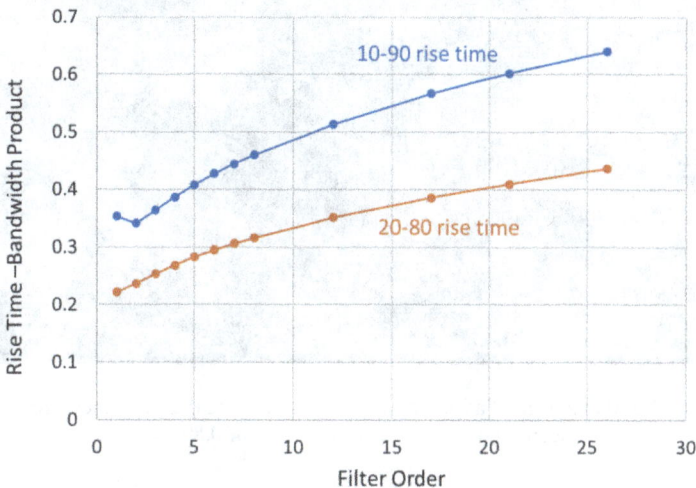

Figure 4.3 Numerical experiment showing the extracted 10-90 rise time-bandwidth product for the step response of different ideal filters. The pole frequencies were the same for each filter. The 10-90 and 20-80 rise times show different Rise Time-Bandwidth products but similar trends. Courtesy of Aditya Rao, MS Thesis, University of Colorado, Boulder, 2020.

For the special case of a fourth-order filter, the rise time-bandwidth product is about 0.4. This is the relationship for a typical mid-range scope's DSP.

For example, if the bandwidth of the scope is 500 MHz, the step response 10-90 rise time of the scope is

$$RT_{scope} = \frac{0.4}{BW_{scope}} = \frac{0.4}{0.5 \ GHz} = 0.8 \ \text{nsec}$$

This means if a 1 psec rise time step signal were the input, the scope would measure a 0.8 nsec 10-90 rise time. As an example, a source with a 0.05 nsec 10-90 rise time was measured with a Keysight MSOX3054G scope with a 500 MHz rated bandwidth. The signal, with 8 averages, is shown in **Figure 4.4**. The estimated rise time based on an eyeball estimate from this screen is about 1 nsec, compared to a Rule #9 estimate of 0.8 nsec. The built-in measurement function for the 10-90 rise time recorded a value of 0.89 nsec, very close to both estimates.

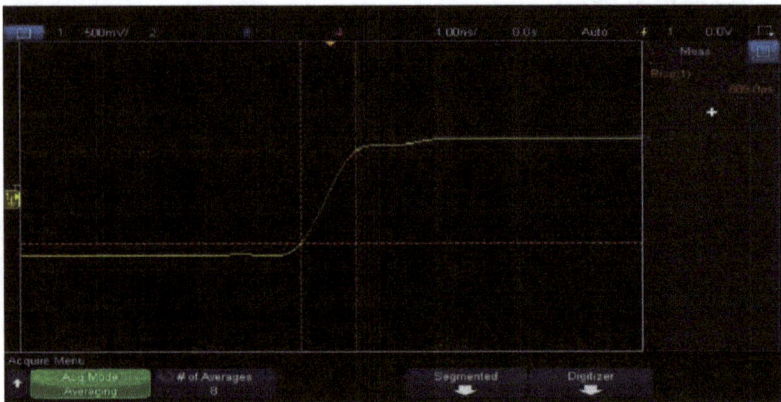

Figure 4.4 The rising edge of a square wave with a 0.05 sec rise time, measured with a 500 MHz bandwidth scope shows a measured rise time of 0.89 sec.

When the input signal has a rise time comparable to the scope's intrinsic step response rise time, the measured signal's rise time will be distorted due to the finite step response of the scope.

4.3   Scopes with a DSP Filter and Gibbs Ringing

Most professional-level scopes with bandwidths of 200 MHz or higher have significant digital signal processing (DSP) front ends, which help to flatten or equalize the frequency response of the amplifier. This adds an n-pole digital filter after the ADC to increase the bandwidth slightly. It means that the step response of the scope is that of an n-pole filter, with n on the order of 4 for many high-end scopes.

The step response of a fourth-order filter is not the same as that of a first-order or RC filter response. It has some artificial ringing due to the sharp cutoff in the frequency response compared to the higher frequency components in the input step. This effect is called Gibbs ringing or Gibbs ears. It is an artifact in the step response due to the band-limited response of the fourth-order filter. This comparison between the step response of different filters is shown in **Figure 4.5**.

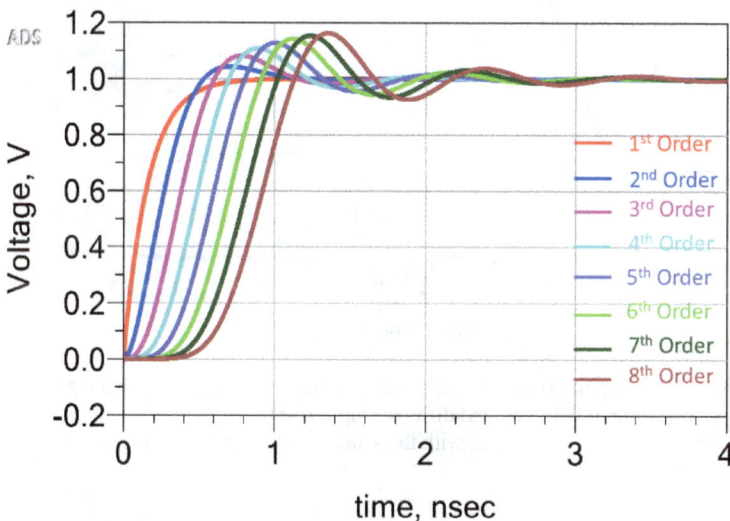

Figure 4.5 The step response of different Butterworth filters all with the same pole frequency, but different orders. Note the Gibbs ringing on the step response. Courtesy of Aditya Rao.

This means that when a scope with a DSP front end measures a signal with a bandwidth higher than the scope's cutoff frequency or pole frequency, the step response will show some ringing that is not present in the input signal. It is a measurement artifact.

If you are not aware of this measurement artifact, you might suspect this ringing is due to some sort of reflection noise and head down a rat hole trying to uncover the source of this ringing, which has nothing to do with your DUT.

Figure 4.6 shows an example of the response of a fourth-order filter with a pole frequency of 1 GHz. The intrinsic step response of the filter is a 10-90 rise time of 0.4 nsec. When the input signal's bandwidth is higher than the filter's response, so that the filter's response is "band-limited" by the fourth-order filter, we see Gibbs ringing. When the bandwidth of the signal is below the filter's pole frequency, we see a smooth rise time response.

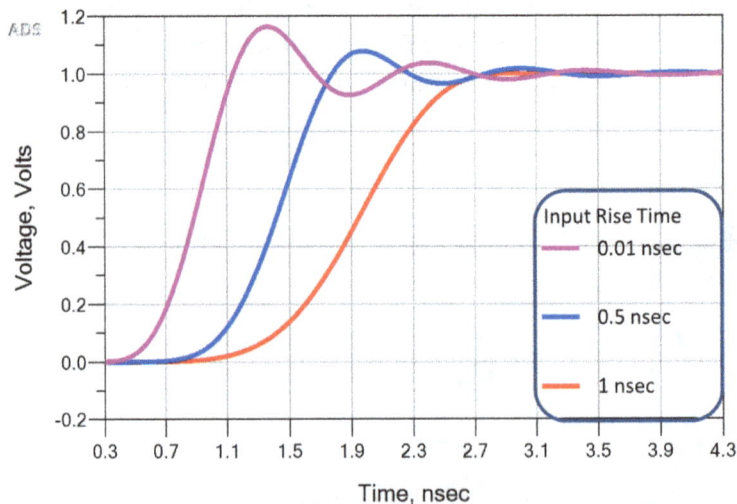

Figure 4.6 Simulated step response of a filter with a 1 GHz pole frequency, having a step response 10-90 rise time of 0.4 nsec, and different input rise times. When the input signal's rise time is too short, compared with the scope bandwidth, there is artificial ringing called Gibbs ringing.

An example of the measured step response of a DSO with a 4 GHz and 8 GHz bandwidth, having an internal fourth-order DSP filter,

showing the measured and simulated Gibbs ringing, is shown in
Figure 4.7. Note that the actual signal's edge has no ringing, but
the overshoot is all measurement artifact from the bandwidth
limitation of the scope.

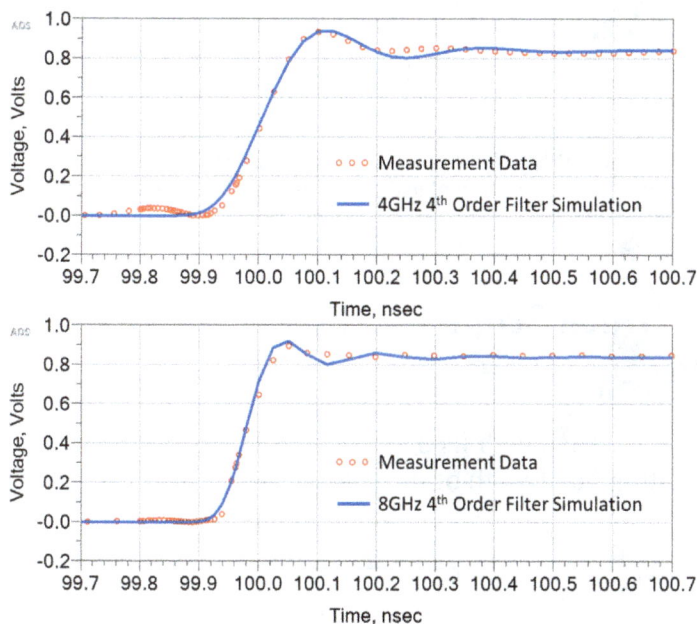

Figure 4.7 Measured step response of a 4 GHz and 8 GHz bandwidth scope showing the
measured response and the simulated response using a fourth-order model. Note the
Gibbs ringing in each case.

A pseudo-random bit sequence (PRBS) signal with a bandwidth
higher than the scope's bandwidth will show Gibbs ringing in its
eye diagram. This ringing looks like a real feature in the DUT's
signal when, in fact, it is just a measurement artifact. When a
higher bandwidth scope is used to measure exactly the same signal,
this ringing artifact goes away.

An example of the measured eye from a 5 Gbps PRBS signal
source is shown in **Figure 4.8** using a 4 GHz and 8 GHz
bandwidth scope. The ringing is a measurement artifact.

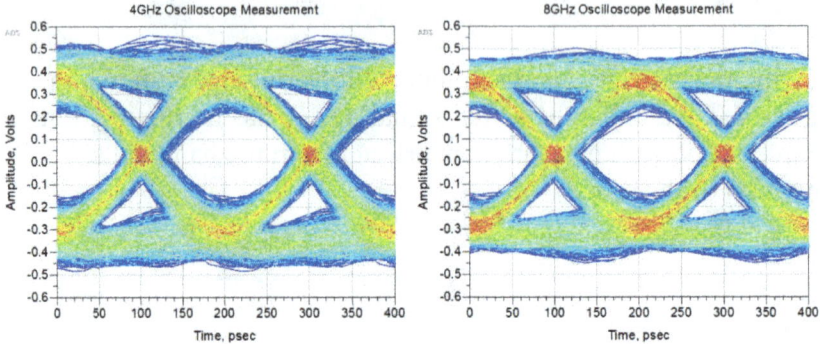

Figure 4.8 Measured 5 Gbps PRBS signal in a 4 GHz bandwidth scope (left) and an 8 GHz bandwidth scope (right). The overshoot in the lower bandwidth scope is Gibbs ringing, a measurement artifact.

4.4 How Much Scope Bandwidth Do You Need?

When two systems, each with its own bandwidth, are cascaded in series, the resulting bandwidth of the concatenated system is related by:

$$\frac{1}{BW_{system}} = \sqrt{\left(\frac{1}{BW_1}\right)^2 + \left(\frac{1}{BW_2}\right)^2}$$

To first order, if each system has the same relationship between its bandwidth and step response rise time, this connects the rise times of each subsystem and the final system as:

$$RT_{system} = \sqrt{RT^2_1 + RT^2_2}$$

For example, if the scope bandwidth is 200 MHz, or 0.2 GHz, its intrinsic rise time is 0.4/0.2 GHz = 2 nsec. If the input signal rise time is 10 nsec, the measured rise time would be

$$RT_{measured} = \sqrt{10^2 + 2^2} = 10.2 \text{ nsec}$$

The measured rise time would be within 2% of the signal's actual rise time. But, if the signal rise time were just 2x the scope's step response rise time, 4 nsec in this example, the measured rise time by the scope would be

$$RT_{measured} = \sqrt{4^2 + 2^2} = 4.47 \text{ nsec}$$

This is a measured rise time that would be about 12% greater than the actual rise time. If the signal's rise time were any shorter, the error in the scope's measured rise time of this signal would show a larger error than about 12%.

When the scope's bandwidth is included in this analysis, the relationship between the signal's actual rise time and its rise time measured by the scope is

$$RT_{measured} = \sqrt{RT^2{}_{signal} + \left(\frac{0.4}{BW_{scope}}\right)^2}$$

Using 10% as the threshold for maximum acceptable measurement error sets a limit to the minimum scope bandwidth required to see a signal's undistorted rise time as:

$$1.1 \times RT_{signal} > \sqrt{RT^2{}_{signal} + \left(\frac{0.4}{BW_{scope}}\right)^2}$$

or

$$BW_{scope} > \frac{0.96}{RT_{signal}} \sim \frac{1}{RT_{signal}}$$

This is a very simple and easy-to-remember rule of thumb. In order to measure no more than a 10% error in the rise time of a signal, use a scope bandwidth higher than 1/ rise time of the signal. To measure a signal with a 10-90 rise time of 1 nsec, to within 10% of its actual rise time, requires a scope bandwidth of at least 1/1 nsec = 1 GHz.

This also suggests that if the scope bandwidth is 200 MHz, the shortest rise time you should use the scope to reliably measure for 10% accuracy of the rise time is

$$RT_{signal} > \frac{1}{BW_{scope}} = \frac{1}{200 \text{ MHz}} = 5 \text{ nsec}$$

Even though the scope's step response rise time is 0.4/200 MHz = 2 nsec, when the signal's rise time is shorter than 5 nsec, the measured rise time will be more than 10% higher than the actual rise time of the signal.

For many typical signals, the bandwidth of the signal is roughly

$$BW_{signal} = \frac{0.35}{RT_{signal}}$$

This suggests a simple relationship between the minimum bandwidth of a scope, compared to the bandwidth of the signal to be

$$BW_{scope} > \frac{1}{RT_{signal}} = \frac{1}{0.35} BW_{signal} = 2.85 \times BW_{signal}$$

Since this is really just a rule of thumb, we approximate this to be $BW_{scope} > 3 \times BW_{signal}$.

This is the origin of the guideline that the minimum bandwidth of the scope should be at least 3x the bandwidth of the signal to maintain the accuracy of a measured rise time and to reduce any potential Gibbs ringing artifact.

Another way of stating this is that the bandwidth of the scope should be at least 3x the bandwidth of the signal, so the rise time measurement is within 10% of the actual signal's rise time.

In a scope with a well-behaved system response, the bandwidth of the scope is unambiguous using the -3 dB point definition. However, the bandwidth of a signal is more ambiguous. Some assumptions have to be made about the shape of the signal's edges and what features in the spectrum of the signal constitute its highest significant value.

Due to this ambiguity, these relationships between the signal's features and the recommended bandwidth of the scope should only be used as a guideline. If you worry if the relationship should really be scope bandwidth > 2.5 x signal bandwidth, you are using this rule of thumb incorrectly. It is only a rule of thumb, providing only a guideline estimate.

4.5   Scope Bandwidth as a Differentiator

The price point of a scope scales with its bandwidth, the vertical resolution, and the number of input channels. This is because of the complexity and cost of the preamp, the ADC, and all the associated electronics that stream the digital data into local memory and then process the data in real time. The higher the bandwidth, the faster the electronics needed. Faster electronics and memory are always more expensive than slower electronics.

While low-end scopes generally have a bandwidth of less than 100 MHz, this is the start of the bandwidth range for the higher-end or mid-range scopes. Commercial scopes vary in price from the low end of about $200 to the high end of over $1,000,000. They can differ in their functionality, special features, number of channels, and display size, but the most important differentiator and figure of merit of a scope is its bandwidth.

When selecting a suitable scope for your application, the first question to answer is what bandwidth is needed. Identify the bandwidth of the shortest rise time signals you may have and select a scope bandwidth at least three times this frequency.

If you use a lower bandwidth scope, you will always measure a signal with your scope, but its measured rise time features will be distorted from those of the DUT's. This is a type of measurement artifact.

Even if the scope you use is free, if the bandwidth is too low for your application, it will be worthless.

Deciding on a scope purchase is about not just your current signal bandwidth requirements but what you expect to measure over the next 2-5 years. This is the typical time between generations of released scope advances.

At the very least, always be aware of your signal's bandwidth and your scope's bandwidth. When the signal bandwidth exceeds 1/3x scope's bandwidth, a caution flag should be raised to be careful in the interpretation of your signal.

Once you establish the scope bandwidth required for your applications, the next criteria in selecting a scope are the specific functions required, ease of use, and compatibility with other scopes you use.

There are many vendors of scopes. While their general features are the same, where the controls are located on the instrument, the way various special functions work and even the color scheme of the traces on the screen vary from vendor to vendor.

Within a vendor's series of scopes, the user interface is generally the same. If you are familiar with one vendor's user interface, it will make using another scope in that vendor's family easier to use than switching to a different vendor.

4.6   Measuring the Bandwidth of a Scope

A simple way of measuring the bandwidth of a scope, or at least the shortest rise time of the step response of the scope, is measuring a very fast edge signal that has a rise time shorter than the scope's intrinsic step response.

One useful fast-edge signal source is available from Leo Bodnar Electronics. The LBE-1320 fast pulse generator has an SMA connector and generates a 10 MHz square wave signal with a rise time of about 30 psec.

Using the guidelines above, the bandwidth of a scope to accurately measure this rise time should be 1/0.03 nsec = 33 GHz. For a scope with a bandwidth of 200 MHz and having a 1-pole response, the measured rise time of this pulse generator would be RT = 0.35/0.2 = 1.75 nsec. **Figure 4.9** shows the measured 10-90 rise time of 1.82 nsec, which is very close to this Rule #9 estimate. This signal really had a rise time of 30 psec but was measured as 1.82 nsec.

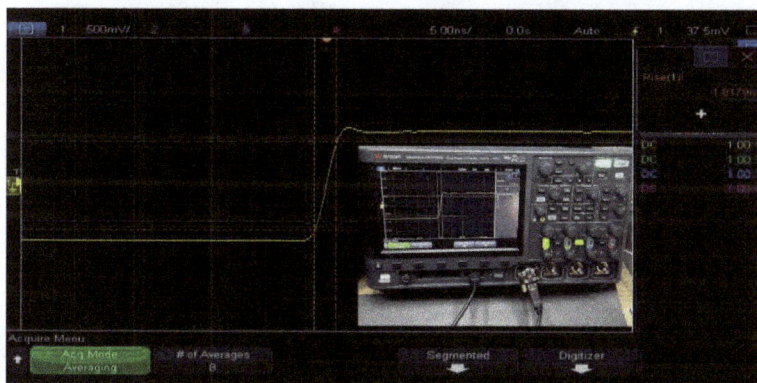

Figure 4.9 Example of the measured 10-90 rise time of a 30 psec rise time signal source using a Keysight DSOX3024, 200 MHz bandwidth scope. The scope with the pulse generator powered by USB is shown in the inset.

4.7   Low-End and Mid-Range Scopes

In this book, we make the arbitrary distinction of a *low-end scope* having a bandwidth below 100 MHz. At or above 100 MHz, the scope is considered a mid-range scope. Mid-range scopes are usually part of a family with higher bandwidth scopes in the series. Below 100 MHz, the scope is either stand-alone or a multifunction scope.

One exception to this is the PicoTech family of scopes. Their entry model, the 2204, has a bandwidth of 10 MHz and their high-end 9404-16 scope has a bandwidth of 16 GHz. They all use the same USB form factor and the same software. This means an engineer starting out can become familiar with the software using a $150, 10 MHz scope and use the same interface to which they are familiar with on the highest bandwidth scope. This is one way of capturing mindshare.

Generally, a mid-range scope also has a deep acquisition memory, typically larger than 1 M samples acquired in one continuous stream.

A mid-range scope is generally more expensive than a low-end scope. This is because of the technology of the ADC, and the memory and processing capability required. Many mid-range scopes are part of a family with a range of higher performance. The 100 MHz version is usually the entry-level version of the family.

Some families of scopes have an entry-level version that starts at 50 MHz. For example, the InfiniVision family of scopes from Keysight, shown in **Figure 4.10**, starts at 50-MHz bandwidth.

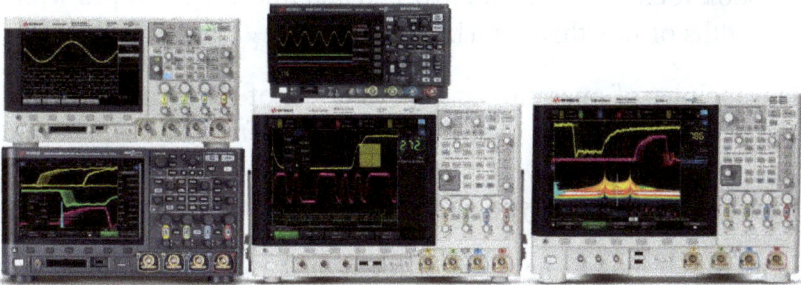

Figure 4.10 The InfiniVision series of scopes from Keysight has bandwidths from 50 MHz to 6 GHz.

The higher price point of mid-range scopes compared with low-end scopes is partly because of the intended market and also because of the technology that goes into the mid-range scopes. The higher bandwidth and deeper memory mean faster electronics and more memory chips. This will always be more expensive. Often, the same technology is used in the lower-end as in the higher-end scopes in the same family, so the high-end scopes sometimes help to subsidize the development costs of the entire family.

A scope bandwidth of 1 GHz is suitable to measure a signal with a rise time of 1/1 GHz = 1 nsec or longer. If the rise time of the signals of interest has a shorter rise time than 1 nsec, an even higher bandwidth scope would be needed.

This is the arbitrary threshold to separate mid-range scopes from high-end scopes. The higher bandwidth scopes require higher-performance electronics and higher bandwidth cable-connector-probe interfaces to the DUT. The entire ecosystem needs to be reconsidered at bandwidths above 1 GHz.

Many applications, especially in analog circuits or sensors or even Internet of Things (IoT) applications, have signal rise times on the order of 20 nsec or longer. This means a scope with a bandwidth of 50 MHz would be perfectly adequate.

However, many digital signal edges are typically shorter than 10 nsec, so a mid-range scope becomes an essential tool in any lab. This book focuses mostly on low-end and mid-range scopes with bandwidths of less than 1 GHz.

The principles of operation and the analysis of the measured results are essentially the same over the entire range from low-end to mid-range. The differences will appear in the bandwidth of the measurements and the type of applications that can be analyzed.

The main differences between different mid-range scope vendors are the specific functions available and the user interface to access these functions.

This means the skills gained in using a low-end scope or an entry-level mid-range scope will transfer to other, higher bandwidth scopes in the family.

4.8   Low-End Scopes

Below a 100 MHz bandwidth, there are many scopes available, ranging in price from < $100 to $500, with varying features and capabilities.

This is not to imply that low-end scopes are toys. Many with price points near $400 have all the features and capabilities of mid-range scopes. They all have built-in measurement functions and even FFT features. Their main limitations are a lower bandwidth and much smaller acquisition buffer sizes, typically < 10k samples.

If your application is for a bandwidth under 35 MHz, scopes such as the Digilent Analog Discovery 3 scope, at $400, may be the right tool for your application.

See, for example, https://digilent.com/shop/analog-discovery-2-100ms-s-usb-oscilloscope-logic-analyzer-and-variable-power-supply/

Or the Red Pitaya: https://redpitaya.com/

In addition, many low-end scopes also have built-in function generators, which, using the provided software, can be used as a stimulus source and the scope as the response measurement to do automated Bode plots or even impedance analysis of active or passive components.

Even though these are referred to as low-end instruments, they are still powerful lab instruments, just with low bandwidth and small acquisition buffer sizes.

If you are only interested in audio signals or bandwidths under 100 kHz, there are a variety of free software tools that can turn the sound card in your PC into a scope with all the standard features of a mid-range scope. One example is the sound card oscilloscope from Christian Zeitnitz. This turns your internal or external sound card into a 2-channel scope and 2-channel function generator.

It can be downloaded from here: https://www.zeitnitz.eu/scope_en

Another valuable free software tool that turns your sound card into a scope is built into Waveforms from Digilent. This is the same user interface used with their external USB scopes. It has many of the advanced features of a pro-level scope. Used with a sound card, it has a bandwidth of about 100 kHz and is limited by the input limitations of your sound card.

It can be downloaded here
https://digilent.com/shop/software/digilent-waveforms/

4.9   The Bottom Line

1. Always be aware of the bandwidth of your signal and the bandwidth of your scope. When they are close, be on watch for artifacts such as Gibbs ringing.

2. The bandwidth of a scope is the frequency at which the measured sine wave amplitude would be reduced to about 70% of the actual sine wave's amplitude.

3. Bandwidth is the most important differentiator for a scope.

4. The bandwidth of your scope should be at least three times the bandwidth of your intended signal. Then, purchase the highest bandwidth scope you can afford above this.

5. As a rough rule of thumb, to measure a signal with a 10-90 rise time to within 10% of the actual rise time, select a scope with a bandwidth = 1/rise time.

6. The consequence of using a scope with too low a bandwidth is the possibility of introducing measurement artifacts.

7. Many mid-range scope families start with a bandwidth of about 100 MHz and go up well above 5 GHz.

8. Consider other specialized features required for your application.

9. The initial rise time of your scope is the rise time of its step response, which can be directly measured using a source with a rise time much shorter than the step response.

Chapter 5
Common Scope Features

All modern scopes in the low-end and mid-range families have many similar features and some important differentiators. Any scope in either family is a good vehicle to explore the common features. The most important process to accelerate up the learning curve in using a scope with best measurement practices is to practice. This means hands-on experience with a functioning scope.

Most of the examples in this book can be run on whatever scope you have available. If you do not have your own scope available, anyone can use the free sound card-based scope described in Chapter 2 and amplified in this chapter. In addition, if you want to try the look and feel of a mid-range scope, a free scope emulator tool is introduced later in this chapter.

5.1   Thirteen Differentiators Between Scopes

Every low, and mid-range scope looks slightly different from one another and has different performance metrics, but many have the same common features.

There are 13 specific features in most modern DSOs for which you should become familiar. When evaluating a scope or using it for a specific application, you should be aware of the performance limitations for each feature. These will influence how you interpret any measurement. When a measurement is close to one of these limits, watch out for measurement artifacts.

Keep a list of these features and their value for your scope. When you use a new scope, always be aware of the following features and where to find their values on your scope:

1. **Maximum bandwidth of each channel**. This is the most important differentiator. The scope bandwidth should be 3x your signal's bandwidth. In terms of the signal's rise time, the scope bandwidth should be at least 1/rise time of the shortest signal's 10-90 rise time you want to measure.

2. **Maximum sample rate of one channel**. This sets the ultimate time resolution. The Nyquist frequency (1/2 x sample rate) should be at least 2.5x the scope bandwidth to avoid aliasing. This generally means the scope's maximum sample rate will be at least 5x the scope's bandwidth and should be at least 15x the signal's bandwidth.

3. **Maximum sample rate per channel with all channels on at the same time**. If multiple channels share the same ADC, this will cut down the sample rate per channel. This is called multiplexing. Some scopes will share two pairs of channels with one ADC. Identify which channels are on separate ADCs so that you can use them when the highest sample rate is needed.

4. **Vertical bit resolution of the ADC at the highest sample rate**. This is usually 8-bit, 10-bit, or 12-bit. In rare cases, it is as high as 14-bit or even 16-bit. Using averaging, the effective number of bits (ENOB) can be increased but with a sacrifice in sample rate, time resolution, and maximum signal bandwidth.

5. **Voltage noise floor on the most sensitive scale and highest bandwidth**. This is related to the preamplifier, the bit resolution of the ADC, and the smallest voltage signal that can be measured practically. The noise is measured as the RMS value and is usually on the order of 0.1 mV to 1 mV rms.

6. **Number of channels**. This is usually 2 to 4 and, in some scopes, as many as 8.

7. **Maximum acquisition buffer size for one channel**. In low-end scopes, this will be 1k to 10k samples. In mid-range scopes, this will extend to > 100M samples. The actual buffer length for a specific acquisition will be related to the sample rate and total acquisition time. In some scopes, the entire acquired buffer may not be exported to a CSV file. It will sometimes be decimated to reduce the file size.

8. **Maximum acquisition buffer size with all channels on**. This may be reduced as multiple channels share the same memory space.

9. **Integrated function generator**. This will have its own set of controls offering various signals such as sine, square, pulse, and sometimes arbitrary wave generator. The maximum generated frequency is usually limited to 25 MHz. The output impedance is generally 50 ohms so cannot drive currents above about 30 mA.

10. **Built-in multifunction stimulus response functions**. This should include at a minimum Bode plot or network analysis and impedance. With multiple function generators, it can also include transistor curve tracing.

11. **Options for specialized probes**. These include current sensor probes, active differential probes, rail probes for power rail measurements, high voltage probes, DC isolated probes, and specialized probes such as higher bandwidth active probes. These will usually be proprietary to each scope vendor's family and are not generally interchangeable.

12. **Advanced triggering options**. These functions, such as specific byte patterns in serial buses, are constantly evolving as specific customers suggest new patterns, which

become integrated as software upgrades to existing instruments.

13. **Advanced measurement functions**. These are constantly evolving. Some scopes allow users to write scripts to create their own functions. Some of the more advanced mid-range scopes treat the measurements as scope traces to plot them over time or display their statistics from acquisition to acquisition. This is called trends and tracks.

5.2   Ten Common Features to All Scopes

Regardless of the advanced features of your scope, *all* scopes in the low- to mid-range bandwidth families all have a set of common features that behave exactly the same on all scopes. The only difference may be the location of the controls and the range you can select.

If you learn the best practices for using these features on one scope, they will apply directly to all other scopes you encounter. This section lists these features, and they are each reviewed in detail in the following sections. Some of the specific features, such as measurement functions and FFT, are covered in detail in later chapters.

These features common to all scopes and described in later chapters are:

1. Triggering

2. Vertical scale and offset control

3. Horizontal scale and offset control

4. Additional display modes such as X-Y plots, roll, and strip chart mode

5. Saving traces on screen or externally

6. Cursors

7. Built-in measurement functions

8. Waveform math

9. Serial decode

10. FFT

5.3   No Substitute for Hands-On Experience

As with any tool, there are generally two different types of users:

- The casual user

- The expert user

When you encounter a new instrument, decide which type of user you are. Your approach and enthusiasm in using the instrument will be decided by the type of user you are and your goal.

If you have never touched a scope before, you will probably take the same journey to get up to speed if you intend to be a casual user or an expert user. If you are a casual user, you probably don't want to invest the extra effort to become an expert user. You just want to learn enough to use the instrument to solve the problem at hand. Your challenge is remembering where all the hidden menus that you discovered are until the next time you use the scope.

However, if you are excited about measurements, want to get the most out of your measurements, or want to enhance your value in the lab, you will want to invest the effort to become an expert user.

Regardless of your end goal, the secret to success in using an instrument effectively is *practice*. The essence of any measurement is that the device under test is a real component and there is a real instrument that will be doing the measuring. The key ingredient to success in any measurement is hands-on practice.

While it is important to review the principles and the best measurement practices as outlined in this book, the only way to become proficient using any instrument is by getting your hands on a scope and working through some of the measurement examples shown in this book so you can replicate them and learn how to set up the instrument, avoid artifacts, and interpret the results.

With this guiding principle in mind, two free versions of a scope are presented in this book.

One uses the sound card in your computer and allows you to practice measuring real time signals picked up by the microphone from your built-in or external sound card.

The second tool allows you to experience the look and feel of a mid-range and high-end scope using a software emulator. With the exception of triggering, many of the features of a scope can be explored and analyzed with this software scope emulator.

Both of these software tools will give you hands-on experience using a scope if you do not have one readily available. These options are covered in detail in the following chapter.

Chapter 6
Setting Up Two Free Scopes

If you have never used a scope before, or you want an instrument to evaluate audio signals, the Digilent Waveform's software scope that uses your computer's built-in sound card is the right instrument for you. Best of all, with software that can be downloaded online, it is basically a free scope anyone can access.

Its primary value as a scope is as a tool to practice using a scope and to experiment with some of the features common to all scopes, with virtually no barrier to entry. There is nothing stopping you from getting started with this scope right now.

A key feature of this tool is that it uses the same user interface as with the Digilent external USB scope with a 35-MHz bandwidth.

There are some limitations to the free PC sound card. Since the PC sound card ADC is AC-coupled, the DC offset of the signal never changes, and the vertical offset control has no real value.

The ADC built into all sound cards is fixed in its full-scale range and sample rate. This means that changing the vertical and horizontal controls does not change the amplifier settings.

Likewise, the acquisition buffer size and the sample rate are fixed and cannot be changed. Otherwise, all the common features found in a higher-end scope can be explored in the free sound card scope.

To exercise some of these more advanced features found in a higher-end scope, another free tool is offered in this chapter if you do not have your own scope available to get valuable hands-on experience. This software-based scope emulator is available for free from Teledyne LeCroy. It runs under Windows and gives you the look and feel of a mid- to high-end scope with all of the features you can explore in real time on your computer, except for adjusting the trigger settings. The trigger is internally set and not available in this emulator tool. Otherwise, it allows you to explore all the settings using a built-in function generator as the synthesized source.

All the important, common features of all scopes can be explored between these two free scope options. This chapter describes how to access and install these two free scope tools.

6.1   Turn Your Computer into a Scope

The sound card in your personal computer (PC) that records from an internal or external microphone uses an analog-to-digital converter (ADC) to turn the voltage generated by the microphone into a digital signal. With the appropriate software, this ADC can be turned into a scope.

After all, this is exactly what a modern, digital scope is. An analog signal present at the front end goes through a buffer amplifier to adjust its voltage level to be more compatible with the ADC. Since this buffer at the input to the scope is the first amplifier the signal sees, it is often referred to as a preamp.

The preamp conditions the signal into the ADC to change its voltage range, its bandwidth, and its source impedance. The ADC converts this conditioned signal into a digital signal.

The digital signal is in the form of an integer description of the input voltage level, transforming the input range of the ADC, such as -1 V to +1 V, into some number of discrete levels. For example, in a 12-bit scope, there are 2^{12} - 1 discrete levels, or 4095 different integer values and 0. The ADC at the front of the PC sound card is usually a 16-bit ADC. This has 2^{16} - 1 levels, or 65,535 integer levels.

These integer values for each measurement point are stored in your computer's memory. The intelligent part of the scope, either a microcontroller, microprocessor, FPGA, or ASIC then manipulates or processes this digital information and displays it on the screen as a plot of the measured voltage on the vertical scale versus time on the horizontal scale. An example of the display of a voltage signal from the sound card of my PC is shown in **Figure 6.1**. This signal was recorded as I listened to piano music.

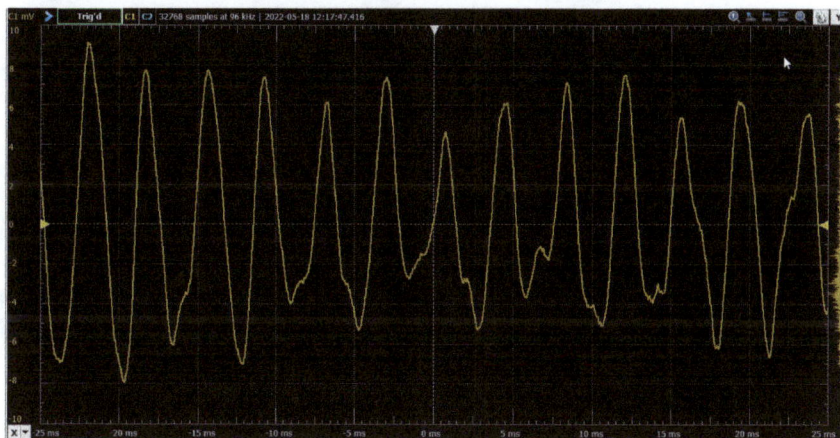

Figure 6.1 A simple example of a voltage signal displayed on the vertical axis in 2 mV/div as a function of time on the horizontal axis as msec/div. Following the directions in this chapter, you will see a similar display on your computer.

The sound card in every PC is basically a preamp, an ADC, and the interface to stream the ADC measurements into memory. The digitized information is available through some digital interface, such as a UART or a universal serial bus (USB), to the microprocessor in your computer.

The combination of the analog front end (AFE), ADC, and digital interface in your sound card, with the appropriate software, can turn your sound card into a simple DSO. However, given the acceptable voltage levels and bandwidth limitations of the preamp and ADC in your sound card, the performance of this sound card-based DSO is limited.

Its value is in learning all the important features found in any scope, and best of all, it is free and available to anyone. This means there is no excuse not to use it to learn best scope practices.

6.2   Options for PC Sound Card Scopes

There are many free tools available to use your sound card like a scope. The two most popular tools are:

- ✓ PC sound card scope by Christian Zeitz (https://www.zeitnitz.eu/scope_en)

- ✓ Digilent Waveforms (https://digilent.com/shop/software/digilent-waveforms/download)

The tool from Digilent is much more versatile and powerful and has the same interface used in their multifunction scope. This is the tool we use in this book to illustrate the principles of PC sound card scopes and multifunction scopes.

Many free versions of software are available that use a simple circuit plugged into your sound card to combine the microphone input as the response and the speaker output as the stimulus to create many instruments. An example of an RLC meter is described here.
https://icom.hsr.ch/fileadmin/user_upload/icom.hsr.ch/publikatione
n/mathis/trade_publications/MAH_Klaper_2008_RLC_Meter_EN
_elektro0608.pdf

Another example of a multifunction instrument that uses a PC sound card and simple external circuits with free software is Daqarta. https://www.daqarta.com/index.htm.

When the sound card is used as just a scope, it will measure the voltage waveform at an analog input channel. These measurements can be displayed and processed using built-in functions.

All the free sound card scope software tools also have a function generator built in. They use this signal to drive the speaker output and generate sounds based on the function generator waveform settings.

The combination of the stimulus to a speaker and response measured from the microphone input means the free sound card scope is a multifunction instrument. In principle, it can be used to measure the transfer function or mechanical vibration spectrum of physical objects or the acoustic shielding effectiveness of absorbers.

In principle, the audio interface of your PC can be used to interact with external voltage signals. An external voltage can be measured by the microphone input plug. The voltage range is very low, typically less than +/- 0.5 V, and is AC coupled. The speaker jack can be used as a voltage source, though the output voltage range is small and the output impedance is high.

Just a caution: When connecting any external device to your PC's sound card, other than a microphone or speaker, be very careful. You always run the risk of damaging the sound card inside your PC. To protect your computer, always use an external USB sound card in these applications.

If you want to use an external sound card, the Sabrent USB sound card is a low-cost ($8) alternative to consider. (https://www.sabrent.com/product/AU-MMSA/usb-external-stereo-3d-sound-adapter-black/#description). It has an internal 16-bit analog-to-digital converter (ADC) that can sample up to 196 kS/sec but has a limited input frequency range from about 100 Hz to 20 kHz. Many free software scope tools can drive this USB sound card. An example of this USB sound card is shown in **Figure 6.2**.

Figure 6.2 An example of a simple, low-cost USB-based sound card that can plug into any computer.

6.2.1   The Frequency Response of the Sabrent PC Sound Card

The most significant limitation to a PC sound card as a scope is its bandwidth. The bandwidth is limited by the preamp inside the sound card, which is designed for audio signals. Usually, the input is capacitively coupled, which acts as a high-pass filter, cutting off all frequency components below about 20 Hz.

In series, there is a low-pass filter that limits the highest frequency it can record, roughly in the 10 Hz to 20 kHz frequency range. This limitation makes it ideal for audio frequency signals but not much else.

The frequency response of the PC sound card can be measured in a relatively straightforward process. In principle, we need to generate a sine wave input to the scope. The input amplitude to the scope is measured along with the scope-measured amplitude. The ratio of the scope-measured amplitude to the input amplitude is the frequency response or transfer function of the scope, often described as H(f).

As the frequency of the source is swept, the frequency response is measured and plotted on a log-log scale.

To implement this measurement of the frequency response of the Sabrent USB sound card, I used an external sine wave signal source. The amplitude was constant from DC to 10 MHz. I measured this sine wave signal with the Sabrent sound card using the Waveforms interface. **Figure 6.3** is an example of the measured sine wave signal set for 1 kHz.

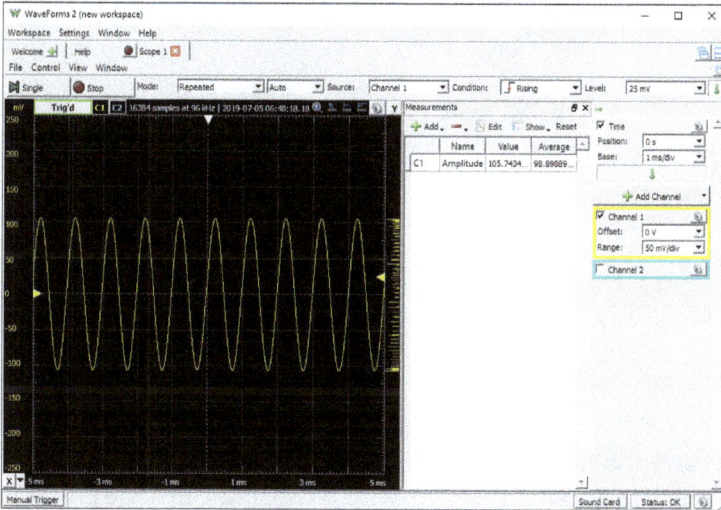

Figure 6.3 Measured input to the sound card with a 1 kHz sine wave as the source.

In this experiment, the sine wave frequency was changed from 1 Hz to 100 kHz, and the amplitude was measured by the sound card as it was displayed in the Waveforms software. The ratio of the measured response of the sound card scope to the input amplitude was plotted on a log scale with the log of the sine wave frequency.

When plotted this way, the ratio of the response (the measured sine wave amplitude) to the stimulus (the known input amplitude of the sine wave) is called a Bode Plot. It is a direct measure of the *frequency response* or *transfer function* of the scope for sine wave signals. The measured transfer function of this specific sound card scope is shown in **Figure 6.4**.

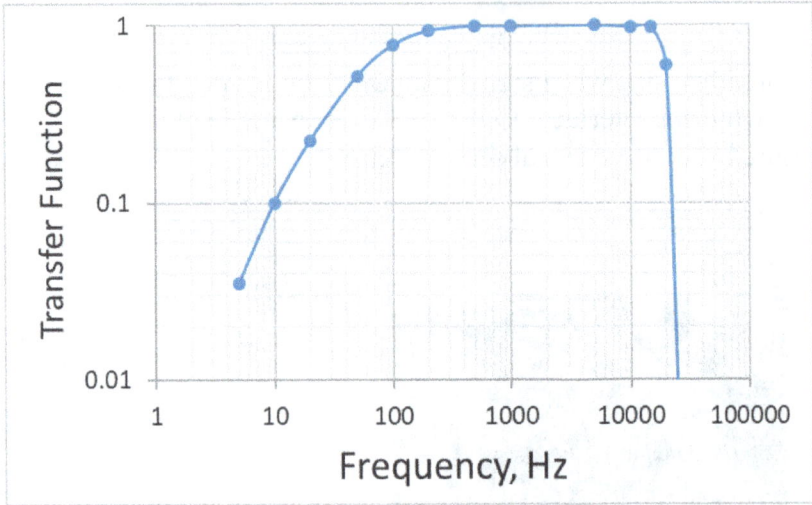

Figure 6.4 Frequency response of the ADC built into the Sabrent sound card to sine waves.

At the low-frequency end, the sound card behaves like a high-pass filter because it is AC coupled with a capacitor in series. The pole frequency, where the response is 70% of the pass band region in the specific sound card, is about 100 Hz.

At the high-frequency end, the sound card has a faster roll-off with a pole frequency of about 20 kHz in this case. These features limit the frequency response of this sound card to a range of about 100 Hz to 20 kHz. While this limits the use of a sound card as a general-purpose scope, it is an excellent scope for audio signals.

Other sound cards may have slightly different transfer functions, but all will have low-frequency and high-frequency limits on this order.

If you want to experiment with the features of a DSO and open up a window into the invisible world of time-varying voltage signals in the audio range, a sound card-based DSO can be a powerful introductory tool to explore the features and capabilities of a scope. Best of all, it is free.

6.3   Download Waveforms Software from Digilent

All of the common features found in every scope can be found in the PC sound card scope created using the Waveforms software from Digilent.

This interface is exactly the same as used in their higher performance multifunction scope, the Analog Discovery 2 (AD2) or AD3, which we explore in more detail in a later part of this book. All the experience you will gain using Waveforms with your sound card will apply when using any higher-performance scope.

The first step is to download Waveforms from the Digilent web site: https://digilent.com/shop/software/digilent-waveforms/download

While the latest version of the Waveforms software will access any internal or external audio sound card, the later versions after v3.19 will not allow you to use a trigger function. The latest version of Waveforms for which you can explore the trigger function is v3.18.1. This is the version you should download and install.

If you purchase and use an Analog Discovery 3 scope, you will need to install v3.2x, which will access all the features of the AD3. The latest versions allow you to access the trigger function in the USB scope so you will not need an earlier rev.

If you use your sound card, download, install, and run v3.18.1. You will be given a dialog box to select the source of the ADC. If you do not have an AD2 scope connected, you can select the sound card option.

If this is the first time you are using this feature, you may be asked to register, but it is completely free.

Your setup screen will look like what you see in **Figure 6.5**.

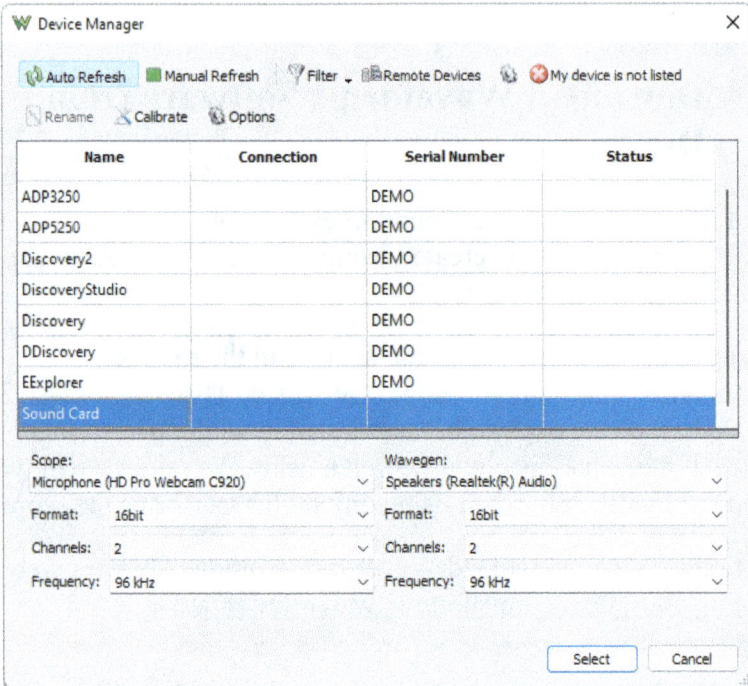

Figure 6.5 Setup screen to use Waveforms with your sound card. Select and highlight the Sound Card option.

Once installed and selected, your sound card will be the source of the ADC and you will see the scope interface of Waveforms with the input from the microphone of your sound card displayed as a scope trace. You will be able to select any of the microphones plugged into your computer as input to the scope instrument and any of the speaker options plugged into your computer as Wavegen output options.

The main selection screen will open with all of your instrument options. This is shown in **Figure 6.6**.

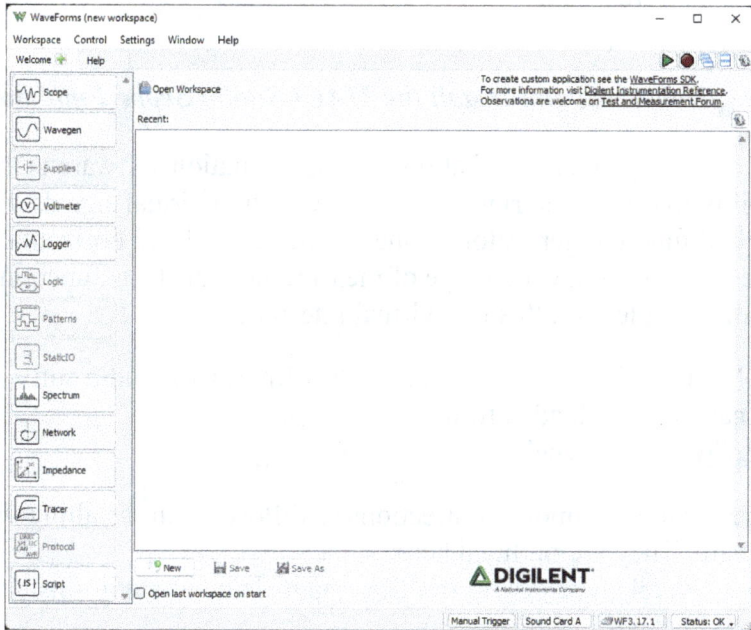

Figure 6.6 Main instrument selection screen.

In this chapter, we will only be using the scope instrument. This instrument will allow you to explore all the common features used in every scope. From this experience, the learning curve to use another scope is much lower than starting from scratch.

After these common features are reviewed, the best measurement practices will be applied to specific measurement applications using higher-performance scopes.

6.4 The MAUI Studio Scope Emulator

If you do not have hands-on access to a scope and want to explore the features found in a higher-end scope, Teledyne LeCroy offers a free software emulator that performs like their mid-range scopes but uses synthesized measurements. This software tool, running under Windows, will allow anyone to recreate the display and analysis of some common waveforms, as though they were measured in real time.

6.4.1 *Download and Install the MAUI Studio Scope Emulator*

One of the significant limitations of this emulation software is the inability to emulate a trigger function. Creating virtual signals with a virtual function generator, changing the vertical and horizontal scales, and using a wide range of measurement analysis functions are all possible with this free virtual interface.

The MAUI (Manual or Automated User Interface) Studio software tool can be downloaded from https://www.teledynelecroy.com/mauistudio/ .

There are a few important directions to follow when installing this software. They are outlined here: https://cdn.teledynelecroy.com/files/manuals/maui-studio-install-instructions.pdf.

If you had installed an earlier version of MAUI Studio before 2021, you will have to manually uninstall the .lic file. Uninstall this file: **C:\LeCroy\XStream\Licenses\HostId.lic**

The next step is to download the MAUI Studio software. Go to this https://www.teledynelecroy.com/support/user/ and register. You will receive a link in your email to download the MAUI Studio software and installation instructions.

After installing the free version, you will need to register the host ID information of your computer back to the MAUI Studio web site.

A screen will open with the three pieces of information needed to register for a free node locked license. This page is shown in **Figure 6.7**.

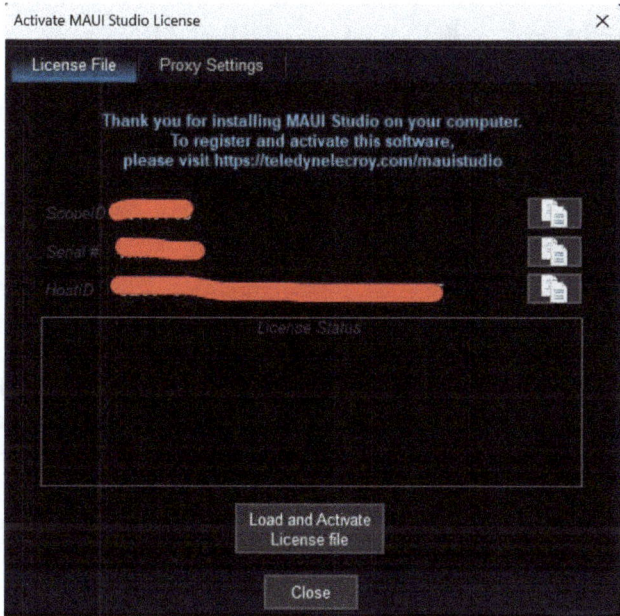

Figure 6.7 The screen on your computer after installing the MAUI software to extract the registration information of your computer.

There are three pieces of ID information about your computer you will have to provide back to the Teledyne LeCroy website to receive your MAUI Studio license. When you return to the registration website, https://www.teledynelecroy.com/mauistudioregister/, select the New MAUI Studio Registration as shown in **Figure 6.8**.

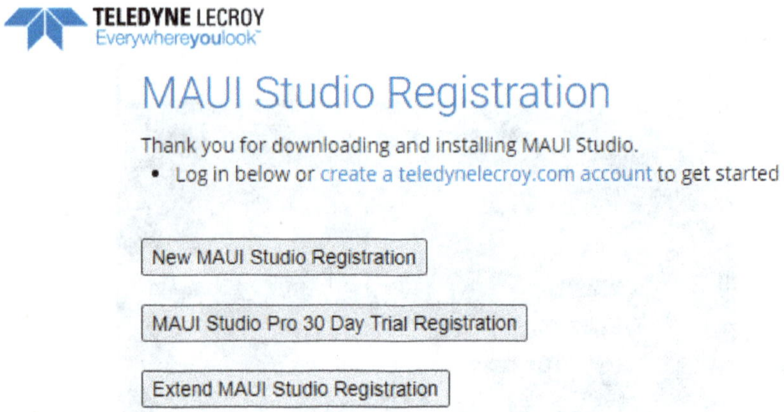

TELEDYNE LECROY
Everywhere**you**look"

MAUI Studio Registration

Thank you for downloading and installing MAUI Studio.
* Log in below or create a teledynelecroy.com account to get started

New MAUI Studio Registration

MAUI Studio Pro 30 Day Trial Registration

Extend MAUI Studio Registration

Figure 6.8 Once the MAUI software is installed, return to this web page and click the New MAUI Studio Registration to post your computer's information to get a node-locked license generated.

In the new registration page that opens, you can individually copy and paste the files from your computer into this form, shown in **Figure 6.9**.

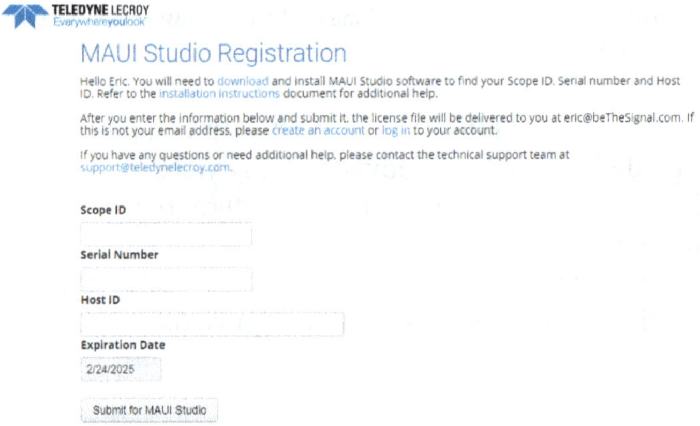

TELEDYNE LECROY
Everywhere**you**look"

MAUI Studio Registration

Hello Eric. You will need to download and install MAUI Studio software to find your Scope ID. Serial number and Host ID. Refer to the installation instructions document for additional help.

After you enter the information below and submit it, the license file will be delivered to you at eric@beTheSignal.com. If this is not your email address, please create an account or log in to your account.

If you have any questions or need additional help, please contact the technical support team at support@teledynelecroy.com.

Scope ID

Serial Number

Host ID

Expiration Date
2/24/2025

Submit for MAUI Studio

Figure 6.9 Final page to copy and paste your computer's information to have the node-locked license automatically generated.

An email with the license file will be sent to your account. While the website says within 24 hours, it usually takes about 1 minute to receive the license file in an email.

Once you obtain this license file, place it on your desktop, and then click the "load and activate license file" button on your MAUI Studio installation page on your computer. Then, point to this license file. The next time you open up MAUI Studio, your scope emulator will appear.

When you run MAUI Studio for the first time, it will come up emulating the HDO W4000 scope. This is the entry-level version of the HDO mid-range scope family.

Normally, a blank scope screen comes up. By selecting the Trigger/auto menu, you can turn on an emulated live scope screen as though you were measuring an actual signal. This is shown in **Figure 6.10**.

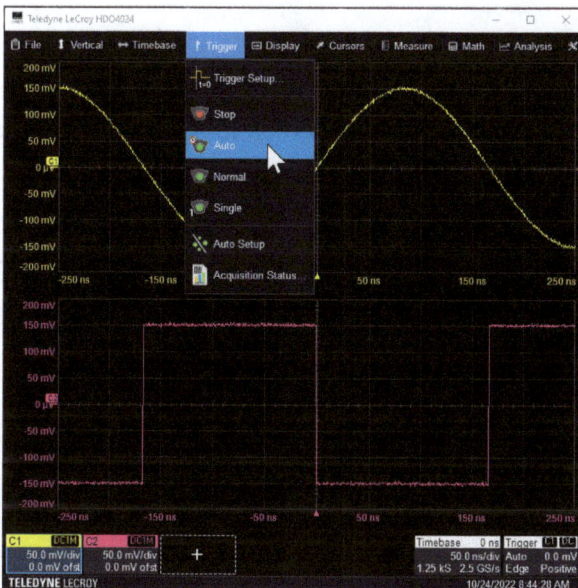

Figure 6.10 Selecting trigger/auto will turn your MAUI Studio software tool into an emulator of a live scope.

When you have this screen set up, you are ready to pretend you have a live scope on your PC and can explore all the important scope features except the trigger.

6.4.2   Quick Start Guide with the Scope Emulator

If you want to change the number of screens displayed on the front of the scope, select the Display menu and try either the single grid or the tandem grid. This selection menu is shown in **Figure 6.11**.

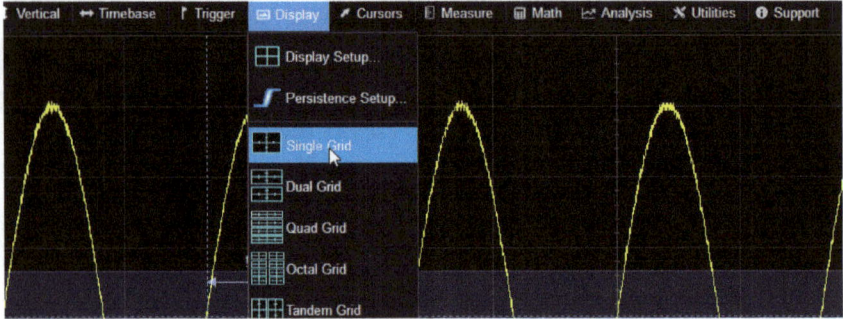

Figure 6.11 Under the Display menu, change the look and feel of the screen to display multiple grids.

To access the vertical controls for any channel, click on the box in the lower left corner below the screen labeled with the channel numbers C1, C2, C3, or C4.

To access the horizontal controls, click on the small icon box on the lower right corner of the screen that says Tbase, shown in **Figure 6.12**.

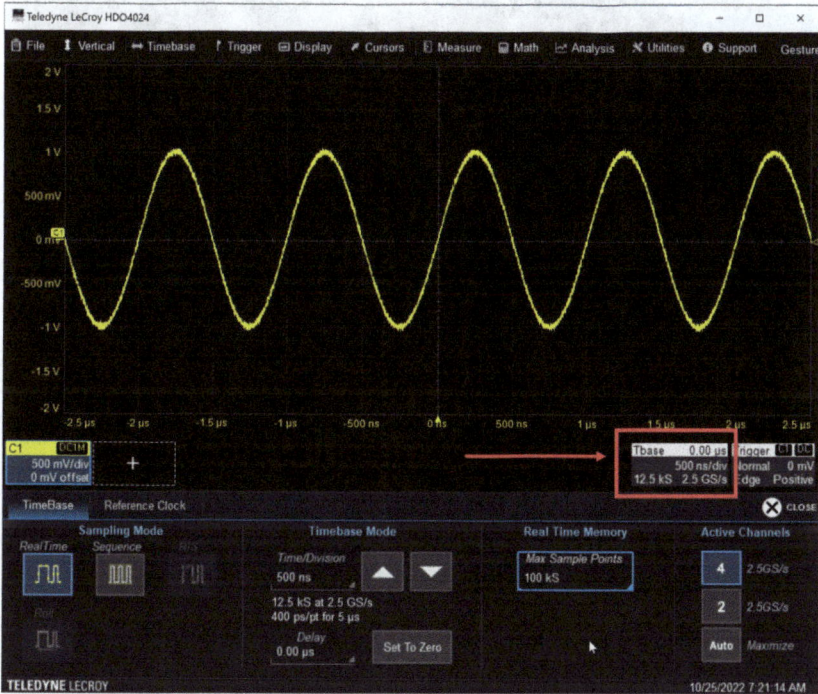

Figure 6.12 Open the controls for the time base by clicking on the Tbase icon on the lower right of the display screen.

Note that in the MAUI Studio emulator, while the trigger functions are available to select, they do nothing to the operation of the scope emulator. The trigger behaves as though it were in Auto mode with a rising edge and 0 V threshold. To exercise the trigger features of a scope, and see them in action, use the PC sound card scope.

6.4.3   The Function Generator

An actual HDO4000 scope has a built-in function generator, The MAUI Studio emulator has a virtual one built in as well. It can be accessed under the file menu. Once selected, the control panel for the function generator opens. This is shown in **Figure 6.13**.

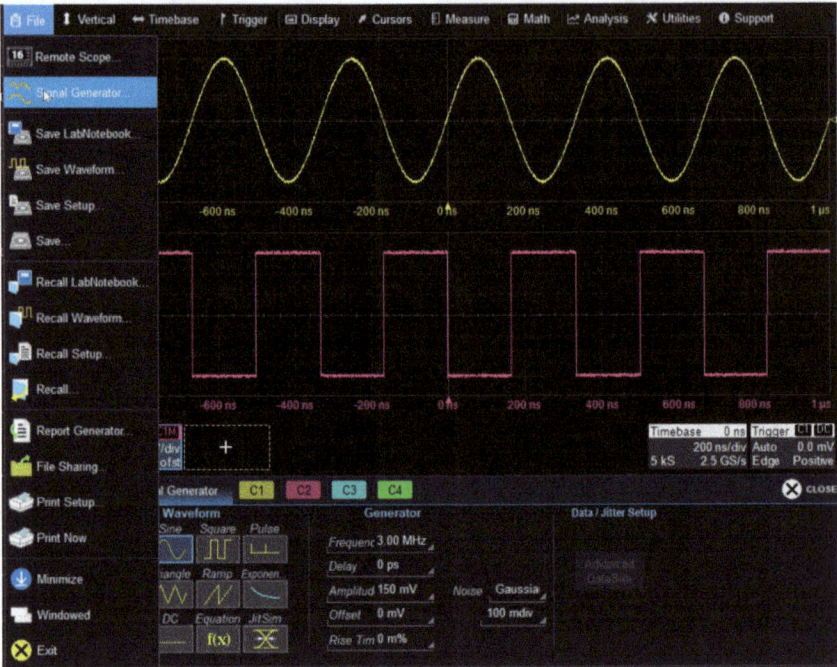

Figure 6.13 Open up the controls for the signal generator under the file menu.

This signal generator can create a variety of waveforms that can be used as simulated signals to exercise the basic functions of the scope.

When the signal generator is activated and a dynamic sine wave signal is displayed on the screen, of course, no actual waveforms are produced, and no actual measurements are taken. This software emulates what the scope would display. It is a chance to practice basic scope measurement functions and the controls a real scope would use.

The first signal we will explore is a sine wave. There are three figures of merit that describe an ideal sine wave: its frequency, amplitude, and phase. In principle, the offset of an ideal sine wave is 0 V. A signal generator can add an offset or average value to a real sine wave.

These four figures of merit that define the "real" generated sine wave signal are available in the signal generator tab setting. For other waveforms, the rise time is also a figure of merit that can be selected for the synthesized waveform. If it ever gets hidden, it can be opened by clicking on the channel icon box, labeled as C1. This will open up the channel controls tab, and, next to it, the signal generator tab, located directly under the screen with the traces. This is shown in **Figure 6.14**.

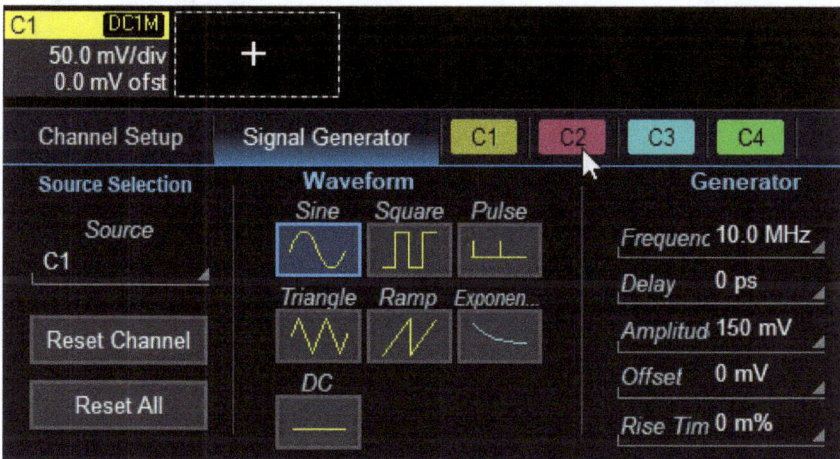

Figure 6.14 The signal generator control tab is next to the channel setup tab. It is found by clicking on the channel select. Note the controls for the five figures of merit for a sine wave.

As an experiment, try changing each of the parameter values in the selection window and see the impact on the live, displayed waveform. This waveform can be used to exercise the vertical and horizontal controls for the scope.

This emulation software is for a 4-channel scope. Each channel is designated as C1, C2, C3, and C4. They are color-coded. As a starting place, uncheck the Trace-on box to turn off all the channels except C1.

6.4.4   *Experiments to Try*

1. Install the MAUI Studio emulator or another emulator of another scope.

2. Synthesize a sine wave waveform with about 1 kHz period and 1 V amplitude.

3. Adjust the time base and vertical scale to display this signal so that the figures of merit are easily visible by eye.

4. Make the waveform a 5 V offset and a 10 mV amplitude. Adjust the scales to read this amplitude from the front screen directly.

5. Explore each of the waveforms from the function generator.

6. Expand the time base to the shortest time. Can you see the sample interval on the scope trace of a square wave?

7. Experiment with different trigger settings. Even though the emulator does not give any control of the trigger function, you can try the stop trigger to see one trace frozen on the screen and the jitter in consecutive traces.

8. Select a sine wave at 1 GHz. The highest sample rate is 2.5 GSps. How many samples are measured in one cycle? What does this waveform look like? Try a lower frequency so there are 10 samples per period.

9. Create a sine wave of 1 MHz and 1 V amplitude. What are the optimized scale settings to easily measure the important figures of merit directly from the front screen?

10. What is the noise floor of a DC measurement? When you select a DC offset, what is the most sensitive scale, and what is the variation in the DC value on different time bases? Does it matter if the DC value is 1 V, 10 V, or -1 V? What offset would you use in each case to view this DC voltage on the most sensitive voltage scale?

11. Export an example waveform into a text file and import this file into your favorite circuit simulator.

Chapter 7
Mastering Basic Scope
Functions

Everything you need to know about using a scope can be explored, demonstrated, and practiced using the free sound card scope and the MAUI Studio scope emulator described in the previous chapter. They will also give you some insight into the behaviors of signals.

The basic function of a scope is to display the V(t) waveform over some time interval to gain insight into the behavior of a signal.

7.1   Shorter Time to Insight

The real value of a scope is to open up a window into a world we cannot see with our senses. Ultimately, the value of any scope measurement is to gain insight about the signal. Our goal in setting up and using a scope is to engineer a shorter time to this insight.

There are generally three ways of using a scope depending on the level of insight we seek:

- To view the waveform on the front screen, look for patterns and extract a few figures of merit using your mark 1 eyeball.

- To characterize a waveform and extract a few figures of merit using built-in analysis tools.

135

- To use the scope as an advanced data acquisition system and acquire waveforms that will be analyzed with external software tools.

All of these features are demonstrated with the PC sound card scope and MAUI Studio scope emulator.

7.2   Quick Start Guide to a PC Sound Card Scope

The Digilent Waveforms software, which turns your PC sound card into a scope, can be downloaded and installed from here:

- ✓ Digilent Waveforms (https://digilent.com/shop/software/digilent-waveforms/download)

After installation, set up the PC sound card scope instrument by first selecting running Waveforms and then selecting the correct microphone and speaker that are available and active in your computer. This can also be selected from the initial setup screen by selecting the Settings/Device Manager menu option, as shown in **Figure 7.1**, and select a microphone option that is available.

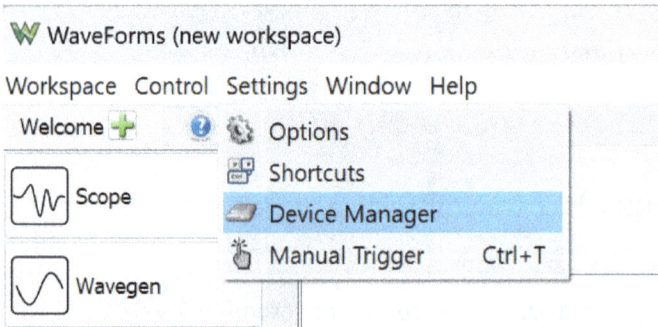

Figure 7.1 To select a different microphone input, go to the Settings Menu and Device Manager and pull down the list of microphone options. Select an active one.

Once you have selected the specific sound card to use as your scope's *analog front end* (AFE), and selected the scope as the instrument to start, the scope screen will open. Click the green start arrow and Waveforms will show you a standard scope window interface. You should see a waveform on the screen corresponding to the voltage measured from your computer's microphone.

To see your first measurement of a signal from the microphone input, click the green run arrow in the upper left part of the screen. This button is highlighted with a red arrow in **Figure 7.2**.

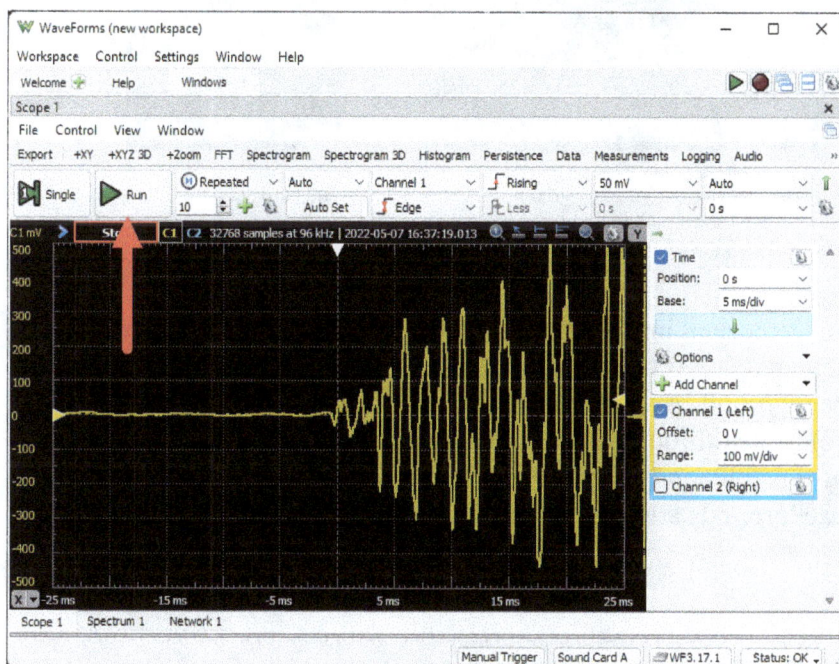

Figure 7.2 The first instrument is the scope. Click the run button to start an acquisition. Keep all the other features at their default values.

While the scope can be adjusted to control the horizontal time base, the vertical voltage scale, and the trigger level, using only the default values will allow you to see your first measured voltage waveform on the screen. The run button starts the acquisition.

There should be a yellow trace displayed on the screen, but it may look like it is not active. To convince yourself it is working, make some noise near your microphone input, and you will see something change on the screen. **Figure 7.3** is an example of the screen I saw using the standard default conditions when I whistled.

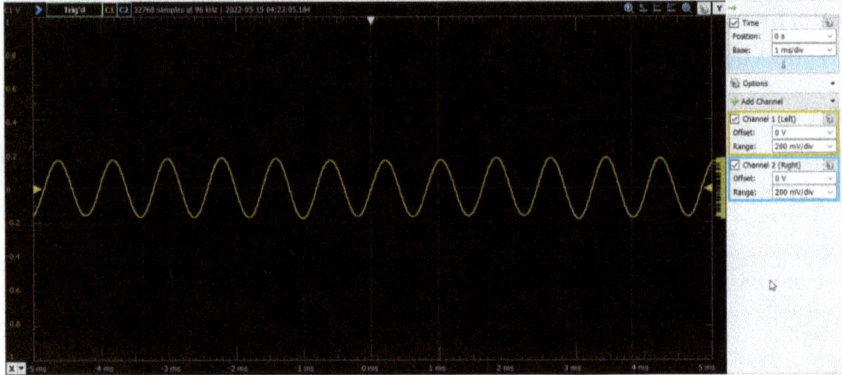

Figure 7.3 After clicking the green run arrow, your sound card scope will display the voltage waveform picked up by the microphone. This is the measured voltage when I whistled into the microphone.

An important initial option to select is your preference for the screen color. You can select either a black background or a white background. As with all the optional settings, this feature is controlled through a menu hidden behind a gear icon. The location of this selection is shown in **Figure 7.4**. The white and black backgrounds are used interchangeably in this book.

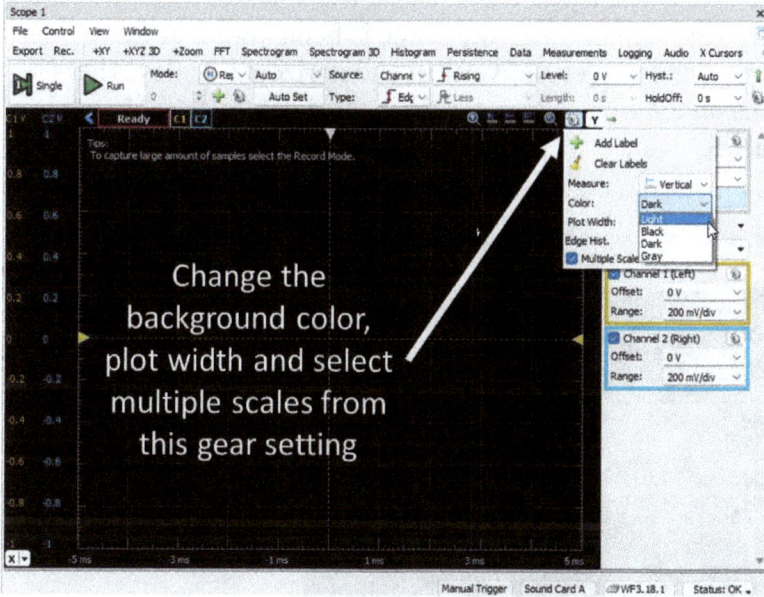

Figure 7.4 Select both the background color and the plot width from this menu. In addition, the check box for multiple scales should be checked.

In the same menu, two other important settings can be set. The width of a trace on the screen in pixels adjusts the width of a trace on the screen. Experiment with the size of the trace that shows up most clearly but still shows good resolution on your screen.

The check box for multiple scales should be checked. This will turn on the vertical scale labels for both channel 1 and channel 2 on the left edge of the screen. This printed scale simplifies taking a reading directly from the front screen. Experiment with these controls to find your preference.

You now have measured your first scope waveform. From this signal, you can explore the three basic features of every scope: the trigger, the vertical scale adjust, and the horizontal scale adjust.

7.3   The Standard DSO Functions

The very first step in using any scope is being aware of the
locations of the trigger and vertical and horizontal scales. The
locations on the front screen for these three controls are shown in
Figure 7.5.

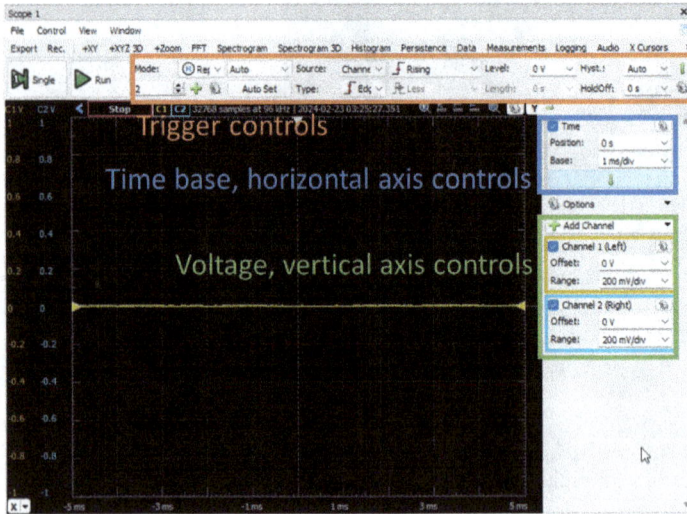

Figure 7.5 The location on the screen for the three most important controls is shown
color-coded for the trigger, time, and voltage scales.

The scope's display screen is divided into squares, which we refer
to as *divisions*. This is a source of confusion. The squares are
further divided into small marks called tick marks. There are five
tick marks per division. On many scopes, for historical reasons,
there are eight vertical divisions and 10 horizontal divisions. Even
in scopes that use an LCD display, which is basically a graphical
interface and can create arbitrary scales, this tradition is still
followed.

However, in the Digilent AD3 scope interface, there are 10 vertical
divisions and 10 horizontal divisions. Be aware of the number of
horizontal and vertical divisions on the scope you use. What you
set for any scale is the value per division; knowing the number of
divisions in full scale helps you identify the full-scale voltage and
time values.

Time is displayed across the horizontal axis. In almost all scopes, the *full scale* (fs) horizontal axis or time base is divided into 10 divisions. When we set the horizontal scale, we are setting the time per horizontal division. In the example above, the horizontal scale is 1 msec/div. This is the default value of Waveforms. With a total of ten divisions in the horizontal direction, this is 10 div x 1 msec/div = 10 msec full scale.

In most scopes, the vertical scale is divided into eight divisions. This is based on historical reasons. The very first scopes developed more than 70 years ago used eight vertical divisions and 10 horizontal divisions, and this has become a de facto standard.

However, in the Waveforms software, 10 divisions are used in the vertical scale. This makes the screen divided into 10 x 10 divisions.

In the measured example above, the vertical scale is 200 mV/div. With a total of 10 vertical divisions, this is 10 div x 0.2 V/div = 2 V full scale.

These two scales can be independently adjusted. Never hesitate to adjust the scales to show the features in the signal most important in your application. Keep in mind that there are 10 vertical and horizontal divisions on the AD3. Why keep the scale set so your signal is within less than 1 division when you have the rest of the screen into which to expand the signal?

Each control region has a gear icon with settings that can be adjusted. You cannot break anything if you experiment with the settings. If you introduce a setting combination you can't get out of just close the Waveforms application and do not save the settings, when asked. This way, Waveforms will open with its default settings when you start up again.

Alternatively, if you find a combination of settings for all the options you prefer, save this with a unique file and it will open up with these settings the next time you open Waveforms.

7.4   Triggering

The trigger is one of the most confusing and most useful features of any scope. The trigger settings define a *trigger event*, which sets the time when $t = 0$ occurs. Once a trigger event occurs, all the data in the acquired buffer is assigned a time stamp based on the point at which $t = 0$.

The trigger event is defined based on a few features of the incoming signal, such as the voltage level equal to 1.15 V and the signal increasing in value on a rising edge.

When the measurements are displayed on the screen with the $t = 0$ location in the center of the screen, every other part of the waveform is positioned relative to this same feature. When the signal is repetitive, every feature of the signal is always positioned at the same location on the screen in every acquisition. It looks like the signal is stationary.

The purpose of the trigger is to make a repetitive signal appear stationary on the screen. If the repetitive signal appears to move across the screen, the trigger is not set correctly.

Even if the scope is not showing a waveform on the screen, it is still acquiring measurements. It's just not displaying them. The trigger setting also defines when the data in the acquisition buffer is displayed on the screen.

Without this function, the scope will just acquire a full buffer of data and display it whenever it is full. This may be suitable if all you want to see are snapshots of the measured waveforms, but it does not provide valuable information if you are searching for a specific transient signal or want to see how stable a periodic waveform is.

Instead, you want to be able to see the measured data displayed on the screen based on specific features in the waveform. This is what triggering does.

7.4.1   A Trigger Event

Every scope has a trigger circuit implemented either in hardware or software, which monitors the incoming analog signal and looks for a specific criterion to identify a trigger event.

The simplest criterion is when the signal passes through a voltage level, either as a rising or falling signal. This is usually the most important criterion to use to specify a trigger event.

The time instant at which a trigger event has been identified defines the t = 0 instant. Every other time value in the measured acquisition buffer is defined with respect to this instant.

In the Waveforms software, the t = 0 instant is identified with a small downward arrow, referred to as a *caret*, pointing at the start time location when the trigger event was received.

Figure 7.6 shows an example of a scope measurement with a rising edge trigger level. Note the incoming voltage at t = 0, where the vertical caret is at the upper center of the display, is the trigger level, 200 mV.

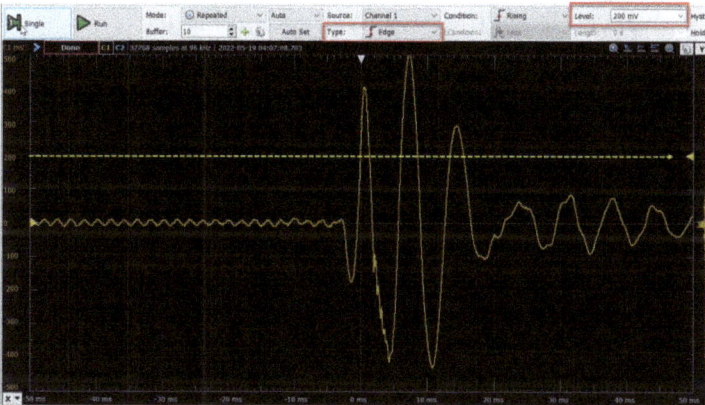

Figure 7.6 The scope display has the trigger level set for 200 mV, with a rising edge. These settings are identified in the red rectangles in the upper part of the screen.

In Waveforms, the trigger condition is set by either a rising or falling edge and a voltage level. These are the most common features of any trigger event. The exact data point for which the input signal on the channel specified meets these conditions is set as the trigger event, and the point at which it meets these conditions is defined as $t = 0$.

Normally, once the trigger event is received and the current acquisition buffer is displayed on the screen, the scope resets or is "armed" to start searching for the next trigger event. There is a deadtime during which the scope is resetting its controls and is not looking for a trigger. In higher-end scopes, this dead time can be microseconds or less.

If there are other features in your signal that would meet the trigger criteria that occur right after the event you care about, and you do not want the scope to respond to them, you can set the scope to increase the deadtime, during which it will not arm and look for or respond to a trigger event. This time you set is referred to as the *HoldOff time* and is set in the menu item in the trigger section to the far right of the screen. All scopes have this trigger feature. It is useful when looking at a data pattern, and you want to wait for the data pattern to complete before arming the trigger again.

Other, more advanced triggering events found on higher performance scopes might be a specific pattern of 0s or 1s, or looking at the input signal with a low-pass or high-pass filter.

There is usually a long list of conditions that can be set up automatically to define a trigger event. **Figure 7.7** is an example of part of the menu list of patterns that can be used as a trigger event in a specific scope.

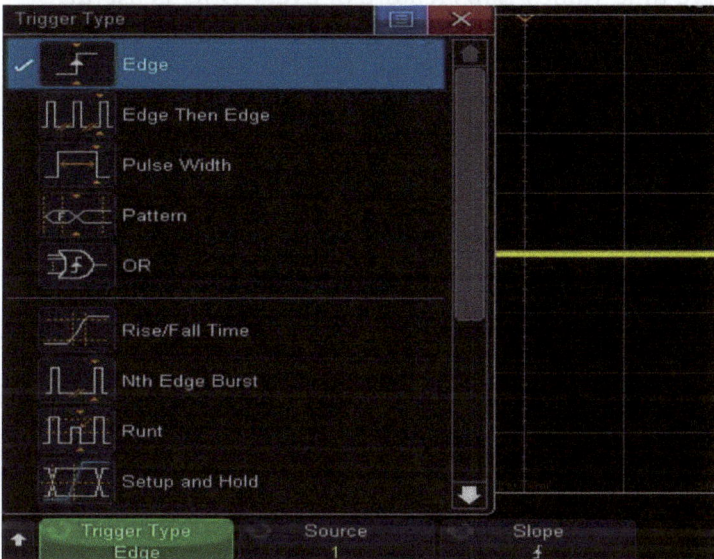

Figure 7.7 Some of the specific patterns in the signal that can be selected as trigger events for the Keysight 3042 scope.

7.4.2   Looking Back or Forward in Time

In a DSO, data is constantly being measured and written into a buffer. This buffer acts like a shift register or circular buffer. With each new measured voltage, the data in the buffer is shifted down until the earliest data point reaches the end of the buffer. With the next acquisition point, the first point is shifted out of the buffer and discarded.

In this way, there is a limited historical record of the measurements before the trigger event is identified.

When a trigger event is detected, and the time t = 0 is established, each measured data point in the acquisition buffer before t = 0 is assigned a time value, counting backward in time. This is how a DSO is able to look back in time, before the trigger event. After the trigger event, each successive measured point is given an incrementally higher time value.

The time base of a DSO can display measured data back in time, with negative time values. This feature of every DSO is useful when you want to see what might have led up to the trigger event.

Look at any scope trace shown previously, which had the t = 0 positioned near the center of the screen. All of the data located to the left of the t = 0 position is labeled as negative time.

There is a limit to how far back in time the measurements can be displayed. If the last acquired point in the buffer was the trigger event, then the farthest back in time the measured data in the buffer extends is the number of points x the time interval between points.

For example, if there are 1000 points in the acquisition buffer, and the sample rate is 100 kSamples/sec (Sps), with a time interval between points of 1/sample rate, or 10 usec, in this example, then the first acquired point was 1000 points x 10 usec = 10 msec back in time from the trigger event.

The horizontal offset position shifts the location of the t = 0 location and the entire waveform to the left or right. As the waveform moves, the caret identifying t = 0 shifts with the waveform, making it easy to always see the t = 0 location on the screen.

For example, **Figure 7.8** shows the measured sound of a finger snap triggered on the rising edge passing through 200 mV. Before the trigger event, there was a minor precursor sound that quickly rose up to the threshold level. This appears at a time before t = 0.

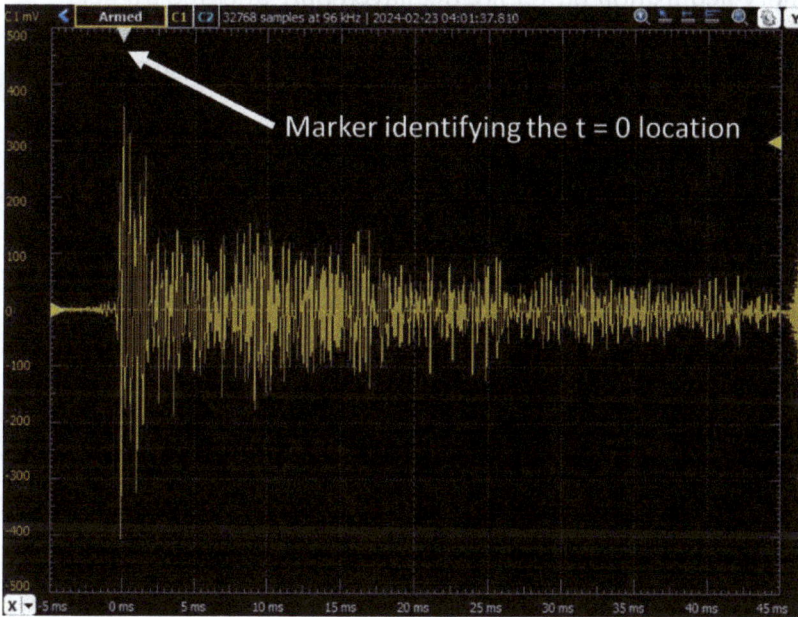

Marker identifying the t = 0 location

Figure 7.8 An example of seeing a measurement before t = 0, looking back in time. This is a feature of all DSOs.

The t = 0 location of the buffer of measured data can be moved anywhere on the screen, and even off the screen. For example, it is possible to set up a trigger event and then look at the buffer of measured data that occurs even 1 second after the trigger event using the horizontal time offset.

Most DSO scopes can look forward in time from the trigger event much farther than back in time. This is a useful feature when looking at the arrival time of a signal's feature compared to some other feature.

7.4.3   Normal and Auto Mode

There are two important ways for the trigger event to control the display of the scope measurements: *normal mode* and *auto mode*.

In normal mode, the measurements acquired and placed in the memory acquisition buffer are displayed *only* after the trigger event has been detected. This event defines t = 0 and the measured data is displayed as V(t) based on the selected scale settings.

If the signal is repetitive, and the trigger event level is chosen as a value that occurs only once in a cycle, the displayed data will always show the same stationary pattern on the screen. The same start of a cycle will be at t = 0 and multiple successive waveforms will appear on the screen in the same position. This will make a repetitive signal appear stationary on the screen.

A well-chosen trigger event makes repetitive signals appear to be stationary on the screen. This makes it possible to see the pattern in detail, especially any changes in the pattern from acquisition to acquisition.

If the signal is not repetitive, either aperiodic or transient, the display of the signal will only be refreshed on the screen after the trigger event has been detected. The scope will sit there idle, displaying the previous screen until a valid trigger event occurs.

Using normal mode makes it easy to detect and study transient events. The display is refreshed only after a new trigger event is reached. If the trigger level is set above the noise level but below the peak level expected, the scope will only trigger and display the new measurement for each new transient event.

For example, the sound of a snap of my fingers is shown in **Figure 7.9**. In this example, the trigger level was set high enough so that only the sound of my finger snap would create a large enough voltage on which to trigger. In normal mode, the scope sat there displaying the previous screen until my finger snap occurred.

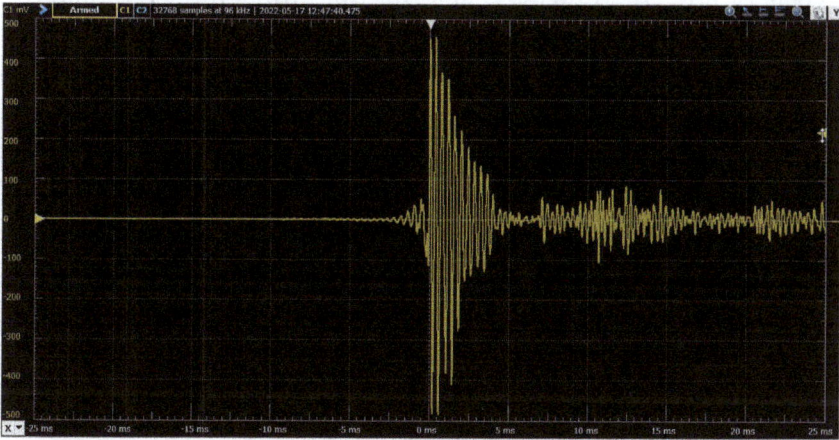

Figure 7.9 The signature of a finger snap recorded using the scope trigger on normal mode. The trigger level was adjusted high enough to only trigger on the sound of my snap.

The long decay of the sound, lasting beyond the 25 msec of the recording, is easily seen.

The downside of using the normal mode of triggering is that if the trigger level was set incorrectly and the input signal never meets the trigger conditions, you would never see the display refresh and you would have no idea if the trigger level was off, or if there were just no valid signal.

When you don't know what trigger level to set, it is sometimes useful to see the incoming signal displayed so that you have some idea of what is going on and then adjust the trigger level. This is when the **auto mode** should be used.

The auto mode behaves exactly the same as the normal mode, except that if no valid trigger event is detected after some period of time, typically between 50 msec and 200 msec, depending on the scope, a trigger is automatically created regardless of the incoming signal. This way, the current signal waveform is displayed, asynchronous with any pattern.

When you start out measuring a signal and have no idea what to expect or how to set up the scales, the auto-trigger mode should be used. This will show the current measurement display and gives you the chance to set up the scales and then adjust the trigger levels.

Of course, if a valid trigger event is present, the auto mode will behave exactly like the normal mode. A repetitive signal waveform will look stationary on the screen as long as it repeats within about 50 msec of the last trigger event.

As a best practice, the auto mode should always be the default setting to use. This will give you the chance to see the measured signal pattern and adjust the scales, while triggering on specific features for periodic signals. Only use the normal mode after the trigger levels are set and you only want to see a signal that specifically meets the trigger criteria.

7.4.4   Run and Stop Modes

In addition to the trigger mode, how often the scope updates the displayed measurements is also important. This is specified in the Run Control. There are two modes: *continuous* and *run and stop*, or *single shot*.

In continuous mode, the scope will refresh the displayed measurements every time a new trigger event is detected, deleting the previous buffer. In the single shot or stop mode, the scope will pause after a trigger event is detected and the measurement is displayed. You have to manually push the single shot button to re-arm the trigger and wait for another trigger event. This allows an opportunity to study or save a displayed waveform.

In principle, you can see any signal displayed on the front screen as a stationary signal by just setting the display for stop or single shot. This will freeze the display with the last acquisition. The displayed waveform will remain on the screen until the signal shot button is pressed again.

Single or single shot mode is the mode to use when the signal is not repetitive. When each acquisition after a trigger event is different from the last, it may be hard to see the details in each successive acquisition buffer unless the display remains frozen until you want it to refresh. Keeping the screen from changing and holding it fixed using the single shot mode is one way of giving yourself time to see these details.

If you are displaying a waveform and want to save the image or the data to a csv file, using the stop mode may be appropriate. If your signal is not periodic but changes with each cycle, a single shot will reveal the details of the signal patterns for that specific acquisition time period. It will change from cycle to cycle.

However, you are only looking at a single acquisition, not able to see if successive acquisitions are the same or different. Single shot mode should rarely be used on repetitive signals. It is not a substitute for correctly adjusting the trigger event settings.

Single shot mode is always the operation of last resort. Unless you have a strong compelling reason, do not use the single shot mode. You may not be capturing the event you want or you may be leaving valuable information about subtle changes of your signal happening during the very long time you are not viewing a live signal. In single shot, you lose the option of sniffing for clues about your signal for which you might not be aware.

Only after you have exhausted other triggering options on the continuous triggering mode should you consider using the stop mode.

7.4.5   *Types of Signals and Their Trigger Settings*

Different combinations of trigger settings should be used depending on whether the signal is *periodic, aperiodic, transient,* or *synchronous* with some other signal.

A periodic signal is repetitive and repeats itself over and over again with exactly the same pattern. You want to be able to see this pattern displayed on the scope as a stationary waveform. A signal from a function generator, a clock, or a pulse width modulated (PWM) signal is periodic. Normal or auto mode and continuous are the right trigger settings to use for these signals.

An *aperiodic* signal has a basic repetitive frequency, but each cycle may be different. A digital bus with data encoded in the signal is mostly periodic with the clock, but each packet of data will look different. The switching noise from a switch-mode power supply (SMPS) may have a period that changes slightly as the load changes. This means that successive cycles of the waveform do not always overlap each other.

For an aperiodic signal, a normal trigger mode would be the right trigger to use. The trigger settings can be adjusted while in auto mode and the trigger conditions set. If the data packet extends into the next triggered event cycle, it may be necessary to use a HoldOff value to trigger on the next packet. If the pattern changes from acquisition to acquisition, this might be a compelling reason to use the single shot mode so that a specific data pattern can be frozen on the display.

A transient signal is one that occurs at a random frequency with no predictable interval. It may be different from pulse to pulse. A sensor recording the impact of a hammer hitting a nail is transient. A reset signal pulling a digital pin from high to low for 1 msec is a transient event. A normal trigger mode on continuous triggering is the right trigger for this sort of event. It will enable the scope to wait for the trigger event and then trigger on it, even if the scope is armed and waiting for hours until the specific transient event occurs.

A synchronous signal is a stimulus to some other signal event. Cross talk from one channel to another is synchronous with the aggressor signal. A response signal is synchronous with the stimulus signal. Transients on the power rail may be synchronous with an I/O switching.

The scope should be triggered on a feature of the stimulus signal, and the response signal should be measured on another channel. This way the trigger event is based on the stimulus and the timing of the stimulus and the response are both measured at the same time, using the same time base.

Even if there is an external trigger input socket on the scope available, it is better to use one of the signal input channels to measure the external trigger signal rather than plug it into the scope's external trigger input. This way you can be sure of the signal quality of the external trigger signal and adjust the trigger based on the feature that is important, rather than let the scope's external trigger circuitry decide for you.

On a scope with multiple channels, there is only one-time base. Every ADC in the scope is triggered and records measurements using the same internal sampling clock and the measurements are displayed on the screen with the same time scale.

7.4.6   Roll or Strip Chart Mode

The normal operation of the scope is to display the V(t) on the screen. An acquisition buffer of measurements is displayed on the screen with the trigger event in the signal defining the t = 0 point. The buffer is displayed. Each time the trigger event is received, the screen is wiped and then refreshed with new measurements. Each successive measurement is displayed on top of where the previous measurement was displayed. This is sometimes referred to as the repeated signal mode.

When the time base is slowed down, typically to 100 msec/div or longer, or a full-scale time displayed of 1 second or longer, most scopes switch to a different mode to display the data. This is the roll or shift mode.

In this mode, the data is taken continuously with no trigger event needed and displayed on the screen as a first-in, first-out buffer.

The newest data is plotted on the far-right edge of the screen and with each new measurement, the entire measured trace is shifted to the left by one point. In this way, the measurements will appear to scroll from the right to left of the screen, reminiscent of a strip chart recorder. This is sometimes called the strip chart mode.

Figure 7.10 shows an example of the screen of the PC sound card with 1 sec/div as the time base. This mode is useful for looking at longer-term patterns in the signal.

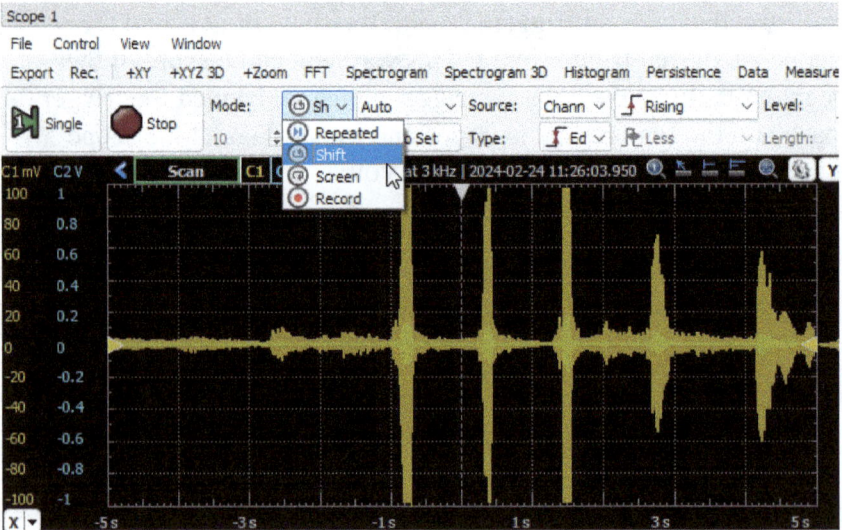

Figure 7.10 The shift mode of the PC sound card shows the longer-term behavior of the sound in my lab.

For the PC sound card scope, the largest acquisition buffer is 30,000 samples. The latest 30,000 samples are displayed to fill the time span. This means the sample rate of measurements is slowed down. For 30,000 samples measured and displayed over 10 sec, the sample rate is 30,000 S/10 sec = 3kSps.

7.4.7   *Summary of the Recommended Best Practices for Triggering a Scope*

✓ Use the auto-trigger mode as your default setup and adjust the scales to fit the measured trace on the screen.

✓ If the signal is repetitive, you should see a stationary pattern on the screen.

✓ If the signal is transient, use normal mode triggering to show only a valid triggered event, after setting up the scales and trigger using auto mode.

✓ When the signal is transient and you want to see the details of each acquisition, use the normal mode and single shot mode.

✓ When the signal is aperiodic, but changes in parts of the pattern in different parts of the buffer, use the auto and run mode to see the changes, but single to see the details in each acquisition.

✓ Use the holdoff setting when you want to delay the arming of the trigger to wait for a pattern to complete.

7.5   Vertical Scale Adjust

The first and most important purpose of a scope is to display the patterns of the signal on the front screen. The goal in adjusting the scope's scales is to display the signal in such a way as to help you get to insight as quickly as possible. Just because the signal is on the front screen does not mean you can easily and quickly read the important figures of merit from the screen. Setting up the scales correctly is a best practice that should become a habit.

Regardless of the scope, the guiding principle when adjusting the display settings is to use a setting that helps you get to the answer you are looking for quickly and easily.

7.5.1 *Vertical Scale Adjust and Control of the ADC*

The channel controls change how the signal is displayed on the front screen. In the Waveforms software using the PC sound card, the ADC input voltage range and sample rate are fixed and are not changed by the controls.

Even when Waveforms is used with the USB AD3 scope, there are only limited changes the scale adjustment does to the internal ADC. All it can do is add an attenuator before the ADC to increase the highest voltage range to +/- 50 V and it can slow down the sample rate on a longer time base.

In most high-end scopes, the offset voltage and scale setting do more than just change the display of the measured trace. Changing the scale and the offset also changes the analog front end (AFE) electronics.

An example of the equivalent circuit model of the AFE of a scope channel is shown in **Figure 7.11**. The input signal is summed with an offset voltage in an inverting summing amplifier. This is followed by another inverting amplifier that provides gain to the difference signal. The offset voltage can be provided by a physical resistor pot or, in most scopes, by an encoder that controls a DAC. This enables the offset voltage to be software-controllable.

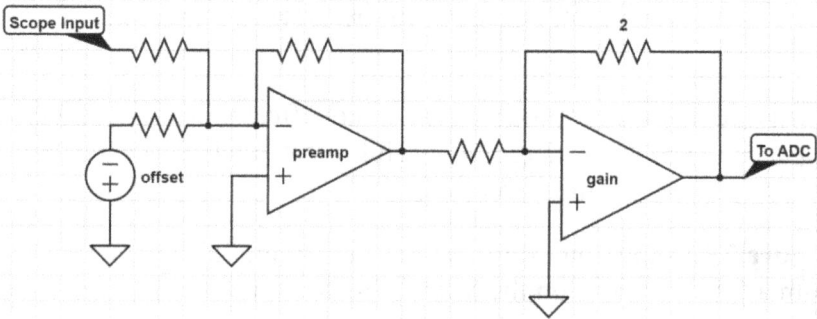

Figure 7.11 An example of an equivalent circuit of the offset added to the signal and then amplified before being digitized by the ADC.

This means that adjusting the offset and scale settings will often change the preamp settings and take advantage of the full dynamic range of the ADC. This is a good reason to adjust the vertical scales to make the signal expand to fit most of the screen, consistent with its figures of merit being easy to read off the front screen.

The vertical scale control sets the voltage per division. This controls the gain of the preamp. Adjust the vertical scale to expand the signal to cover most of the screen. This way, the limited dynamic range of the ADC can be used for the highest vertical resolution.

If there is a substantial DC component to the signal, the signal can be shifted up and down to offset the DC component. This is using the offset control. This subtracts from the signal that is input to the preamp.

7.5.2 Expanding the Vertical Scale About the Center of the Screen

When the vertical scale is expanded, there are generally two different options that can be selected about which the scale is expanded. The most useful setting is when the scale is expanded about the center of the screen, regardless of the voltage at which this is offset. With this setting, if you use the offset control to center your waveform, the scale will be expanded about the waveform.

The DC component of the signal will not affect how the signal is expanded. This is an incredibly useful feature and eliminates the need to AC couple the signal.

Figure 7.12 shows an example of a small amplitude sine wave with a 5 V DC offset on the MAUI Studio scope emulator. The first step is to adjust the offset to -5 V. The second step is to expand the vertical scale to display the signal over a larger fraction of the screen.

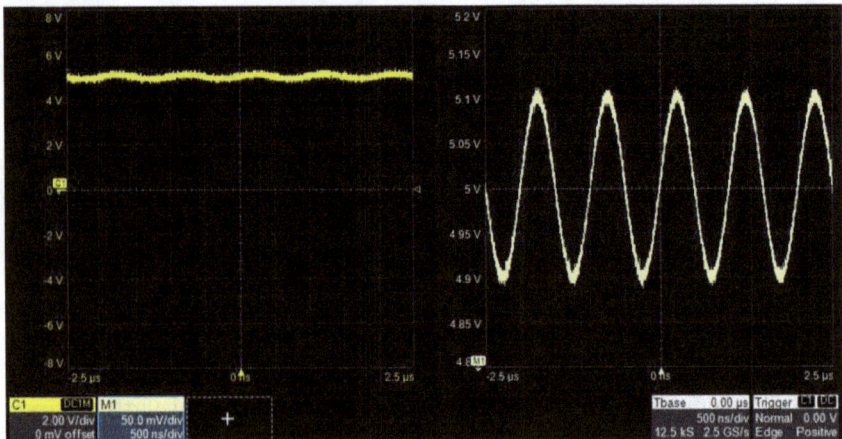

Figure 7.12 Starting with a signal with a large DC offset and a small signal, displayed on the left, change the offset setting to center the signal then expand the vertical scale to display the signal on the right.

All scopes have the option to expand the vertical scale about the center of the screen. Unfortunately, the default setting is different on different scopes, and this feature is called different things in different scope families.

In the MAUI Studio scope emulator, this feature is found under Preferences and then under the Acquisition tab. Selecting the Volts box in **Figure 7.13** will enable the scale expansion to expand about the center of the screen. Likewise, selecting the time box will allow the time expansion to be about the center of the screen.

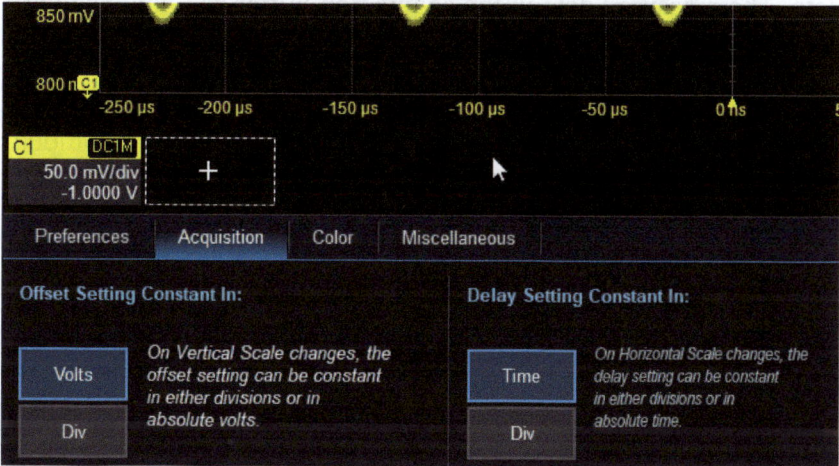

Figure 7.13 Under the utility menu and then selecting the preferences set up option brings up this dialog box. Selecting the Acquisition tab and the Volts box will set the scale to expand about the center of the screen.

If your scope does not expand about the center of the screen, it is worth finding this preference setting so that DC offsets can easily be compensated. Otherwise, each time the scale is changed, you have to continually adjust the offset value.

If all we did was change the displayed scale on previously measured data in the acquisition buffer, all we would be doing is expanding the digitized data stored in the acquisition buffer, and we would quickly be limited by the digitizing noise of the ADC.

In normal operation, the scope is dynamically adjusting the preamplifier settings and the measured signal at the ADC, as we change the scale and offset. This changes the next acquisition measurements but does not change the measurements in the previously stored and displayed buffer. This is an important consideration.

For example, **Figure 7.14** shows the same signal measured at 10 V/div, recorded in single shot mode in the MAUI Studio scope emulator. This acquisition buffer of data was then displayed after adjusting the scale to 1 V/div. The scope was not acquiring new data. The measured data in the acquisition buffer had not changed, only how it is displayed.

This expanded scale shows the digitizing noise in the saved data. This is compared with a live measurement in continuous run mode, with the offset and scale dynamically adjusted to 1 V/div. It is the same 12-bit ADC in the emulated scope, but the signal was adjusted going into the ADC to take advantage of the full scale and 4095 levels of the ADC.

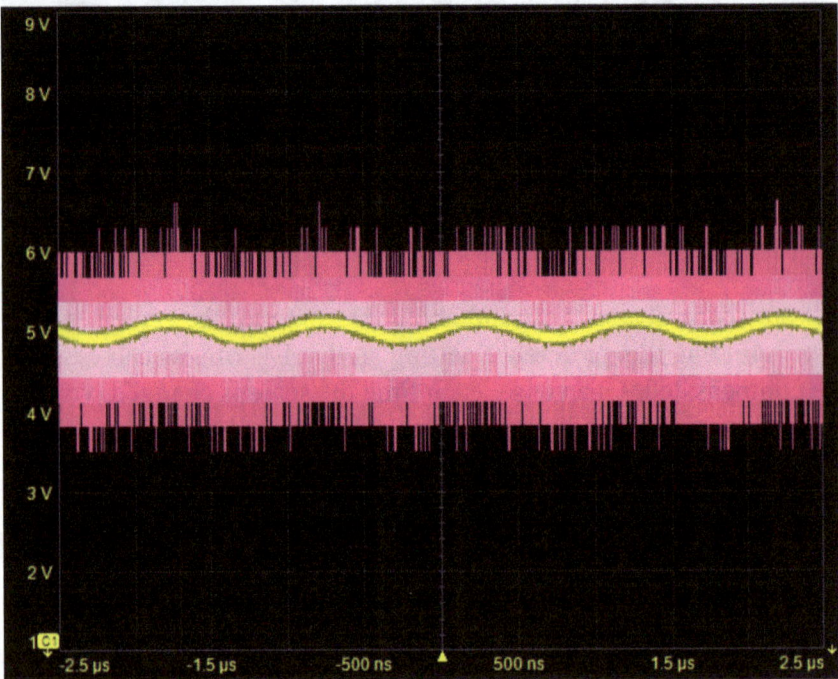

Figure 7.14 An example of the same signal measured two different ways. In the case of the trace with digitizing noise (red), the measurement was taken at 10 V/div and then just the scale of the stored measurement changed to 1 V/div. For the other trace (yellow), the scale and offset was adjusted on the live data, changing the preamp settings and taking full advantage of the limited vertical resolution of the ADC. Both measurements are displayed on the same vertical scale. Digitizing noise is very clear.

This supports another best measurement practice: to take full advantage of the limited vertical resolution of the ADC, adjust the scales and offset of the signal on the live measured data. Except for very rare situations, always display the live data and optimize the scales on the live data. This will optimize the preamplifier settings of the scope for each successive buffer acquisition.

Do not save a waveform, and then optimize the scales on the saved buffer of measurements. You will be increasingly limited by the digitizing noise of the ADC. The same best practice applies to the time base.

7.5.3   Adjusting the Vertical Scales in Waveforms

In the case of Waveforms, the scale-adjust controls just change how the signal is displayed on the front screen.

The vertical scale setting is the voltage corresponding to each division. The full-scale voltage range on Waveforms is 10x the volts/div setting. If the scale is 1 mV/div, the full-scale voltage range is 10 x 1 mV/div or 10 mV full scale.

Once the scale is adjusted to display the signal on the screen, the offset control moves the scale vertically. When using waveforms with a PC sound card, the offset control is not very important. The signal is AC coupled and always has a zero DC component. This means it will always be centered at V = 0. There is never a need to use the offset control with the PC sound card. However, it is immensely useful in other scopes.

The vertical scale adjustment is controlled by typing in a value in the displayed range box, or it can be selected from a pull-down menu. The options are based on the standard scope multiples of 1x, 2x, and 5x. This is shown in **Figure 7.15**.

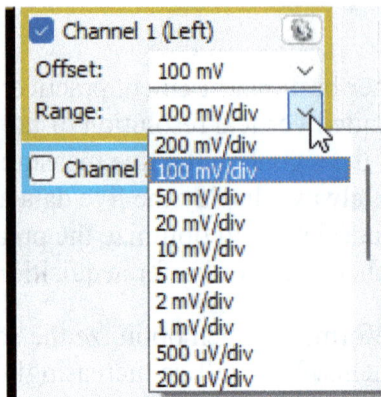

Figure 7.15 The vertical scale ranges can be selected as multiples of 1x, 2x, and 5x. This is typical of all scopes.

Using one of these scale settings makes reading a voltage from the front screen much easier than an arbitrary value like 12.7 mV/div.

7.5.4   Rescaling Vertical Axes into Other Units

When using a sensor that transforms a physical quantity into a voltage signal, it is sometimes useful to display this signal in terms of the input signal to the sensor. This means that instead of displaying a voltage measurement on the front screen, you might want to display a temperature or even a current on the front screen.

Many scopes, Waveforms included, allow you to change the displayed unit on the front screen and the scaling of the input measured voltage into a rescaled value displayed directly on the front screen. This option can be selected from the gear icon on each vertical scale adjust box for each channel. **Figure 7.16** shows where this menu item can be selected.

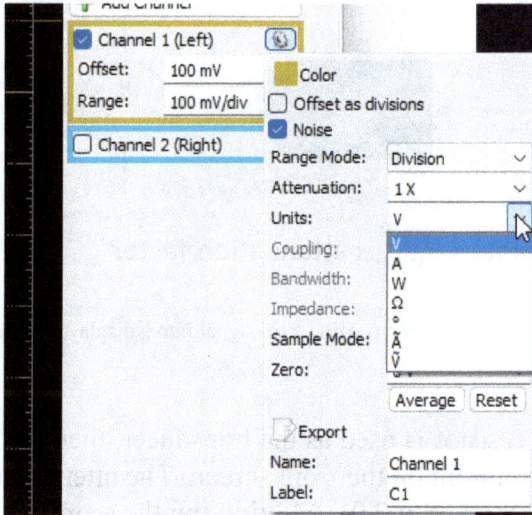

Figure 7.16 Select the units and scaling for each channel from the gear icon.

We usually refer to the signal at the Device Under Test (DUT), where the fixture to the scope connects, as the "tip voltage," as in the voltage at the tip of the fixture. The tip voltage may not be the same as the voltage measured by the scope. For example, the fixture can be a 10x scope probe that decreases or attenuates the input tip voltage by 10x before it is measured by the scope.

When a 10x probe is used as the fixture, the signal as measured by the scope is reduced by 10x from the tip voltage:

$$V_{scope} = \frac{V_{tip}}{10x} \quad \text{or} \quad V_{tip} = V_{scope} \times \text{attenuation factor}$$

This is how to use the attenuation factor when rescaling the input signal to the displayed signal. Generally, it is important to display the tip voltage, the voltage at the DUT, on the screen. This is the measurement we want to know about. This setup is illustrated in **Figure 7.17**.

Tip Signal V_{scope}

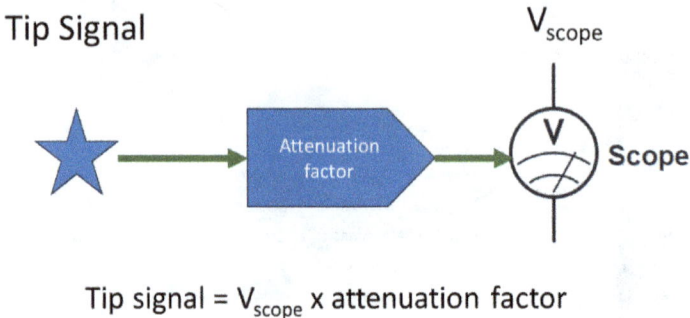

Tip signal = V_{scope} x attenuation factor

Figure 7.17 General setup for converting a tip signal into a displayed scale on the front screen.

If a 100-ohm resistor is used as the transducer, then we would like to display the current on the front screen. The attenuation factor would be 1/R = 1/100 = 0.01. Multiplying the scope measured voltage by this factor would convert the scope measurement into the displayed current on the front screen. What was 200 mV/div would now be rescaled to 2 mA per division.

7.6   Horizontal Scale Adjust

In all scopes, the default horizontal axis is time. This can be changed to an X-Y scale so that the measurements from one channel can be plotted against another channel. In Waveforms, since there is only one channel measured, the horizontal axis will always be time.

7.6.1   Horizontal Scale Adjust and Sample Rate

The horizontal controls in Waveforms are on the right-hand side of the screen, just above the horizontal controls. These allow adjusting just the scale and offset.

In addition to the display of the time base, the horizontal controls can also influence the sample rate of the ADC and the total number of samples. These three settings are important to be aware of in any scope.

- ✓ The full-scale time interval displayed
- ✓ The sample rate of the ADC
- ✓ The number of points in the buffer

In waveforms, the sample rate of the ADC is fixed at 96 kSample/sec and cannot be increased. Likewise, the number of points in the buffer is fixed at 32,768 samples and cannot be increased. These values are displayed in the top, central region of the display screen in waveforms, as shown in **Figure 7.18**.

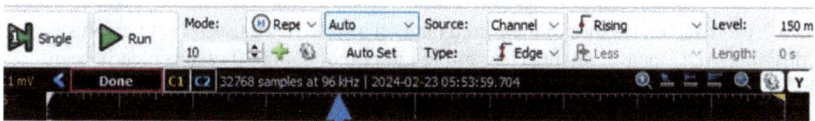

Buffer size and sample rate

Figure 7.18 The current sample rate and buffer size are displayed on Waveforms at the top of the display screen.

The sample rate of 96 kSps is an interval of $1/96\text{k} = 10.4$ usec/sample. Regardless of the timebase settings, measurements are recorded by the scope at this rate. This can be seen by expanding the time base to 10 usec/div. The individual measurement points displayed on the screen at about 10 usec intervals can be seen as + symbols, as displayed in **Figure 7.19**.

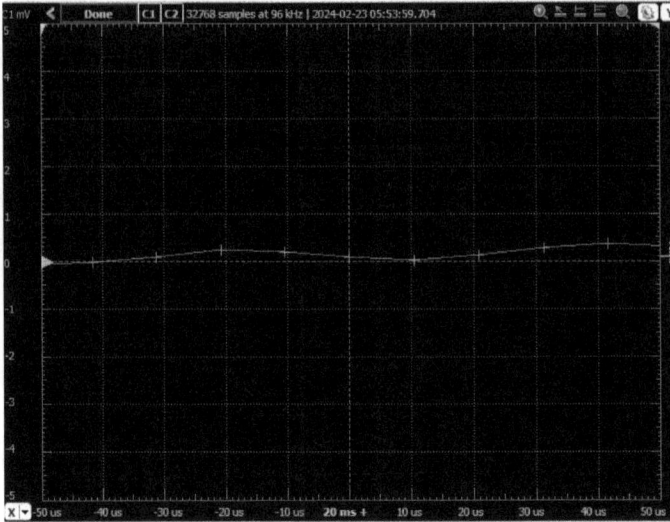

Figure 7.19 A screen capture of measured data on a scale of 10 usec/div. Each plus sign is a different ADC measurement. The sample interval of about 10 usec is clearly seen.

7.6.2   Sample Rate and Buffer Size

In higher-end scopes, the time scale does influence the ADC sample rate. There are three important settings that are all related to the time scale:

- ✓ The displayed time scale of sec/div
- ✓ The number of points acquired in the acquisition buffer
- ✓ The sample rate

Changing any of these values in the controls changes the way the data is acquired, not just how the data is displayed. This principle can be explored using the MAUI Studio scope emulator.

The acquisition buffer size is literally how many measurement points are acquired and saved each time a trigger event occurs. Consecutive measurements at the sample rate are streamed into memory in real time. These measurements are then read out, processed, and displayed on the front screen. They are also available to be stored internally or exported in a csv file, for example.

The number of samples acquired in one acquisition buffer is limited to the maximum memory size of the scope. This is one of the important differentiators of mid- and high-end scopes compared to low-end scopes.

Generally, mid-range scopes start at 100 kSamples in memory and go up to as high as 5 Gsamples in some scopes. In the emulated HDO4000 scope, the maximum number of samples in an acquisition can be adjusted from 500 samples to 25 Msamples. Clicking the box labeled Timebase allows this value to be changed.

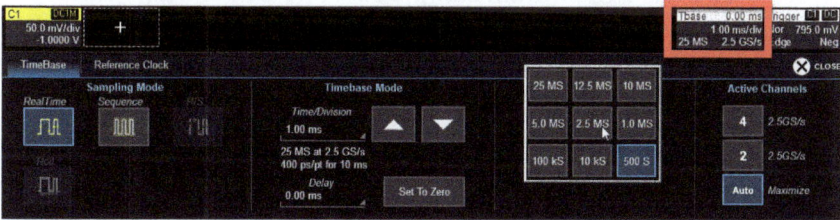

Figure 7.20 Select the Timebase icon, inside the red rectangle on the upper right of the figure, and this dialog box opens up showing the different number of points that can be selected in one acquisition.

The second term to adjust is the sample rate. Every scope has a maximum sample rate based on the fastest sampling of its ADC. This scope emulator has a limit of 2.5 GS/s. This means the time interval between sample readings, at 12-bit vertical resolution, is $1/(2.5 \text{ GS/s}) = 0.4$ nsec/sample. At the highest sample rate, the scope takes a 12-bit measurement of the input voltage every 0.4 nsec.

The third setting is the time scale, in seconds/div. The full-scale time base is 10x the seconds/div scale. There is a connection between the number of points in the acquisition buffer, the sample rate, and the total time scale. These terms are not independent. The relationship is

$$\text{Sample Rate} = \frac{\text{acquisition buffer size}}{\text{total time}}$$

On most scopes, two of the terms can be adjusted and the third is automatically set. As the acquisition buffer size increases, the sample rate will increase for the same total acquisition time base, until a limit is reached.

Many scopes allow two modes of operation: a fixed buffer size or a fixed sample rate. In either mode, the time base is adjusted by the user, and the other terms are automatically set by the scope.

There are two different values for each term to be aware of: the maximum values for which the scope is capable, and the instantaneous values for a specific acquisition of measurements.

Every scope has a maximum sample rate and a maximum buffer size. It is sometimes difficult to find these values, but you should know what they are for the scope you are using.

The strategy is to first decide the time resolution needed. This influences the slowest sample rate acceptable. If measuring a time interval feature of a signal, like its rise time or its period, is important, use a sample rate that is as high as available.

The second step is to determine over how long a time interval you want to measure and display the signal. Generally, this means as much time as will contain the useful information in the signal. The combination of the sample rate and the total buffer time interval determines the number of points saved in each acquisition.

The more data points you record, the more time is required to process the data. Depending on the speed at which the scope can record, process, and display the measurements to be ready to trigger another event, somewhere between 100k and 1M samples in the buffer is a good compromise between processing time and sample rate.

When the acquisition buffer size reaches the maximum value you set, a longer full-scale time base will decrease the sample rate. Always be aware of the sample rate being used for a measurement.

This sets the fundamental time resolution of any measurement. For whichever scope you are using, locate where these three scope settings are displayed and always be aware of their values.

In the MAUI Studio scope emulator, the three important terms are always displayed in the Tbase icon on the lower right below the display screen. This is shown in **Figure 7.21**, showing a sample rate of 2.5 GS/s, 50 k samples in the acquisition buffer, and 2 usec/div time scale: 50 k samples = 2 usec/div x 10 x 2.5 GSps.

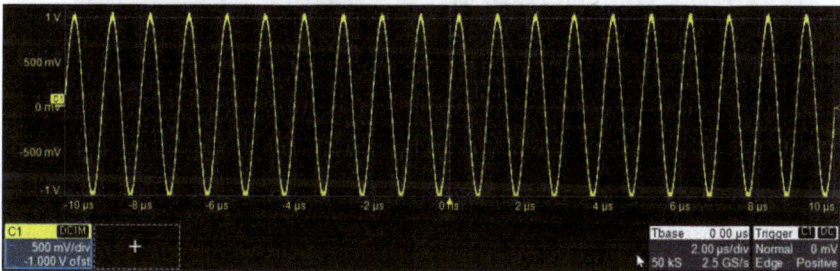

Figure 7.21 All the important time-based information is displayed in the Tbase icon in the lower right corner of the screen.

In the MAUI Studio scope emulator, while the number of points in the buffer can be adjusted, this setting, like the trigger, has no impact on the emulated scope's response. The sample rate is also fixed.

You can always calculate it based on the sample rate and the total time full scale. The Keysight DSOX3024 scope display is shown in **Figure 7.22** with the sample rate displayed in the upper right of the screen.

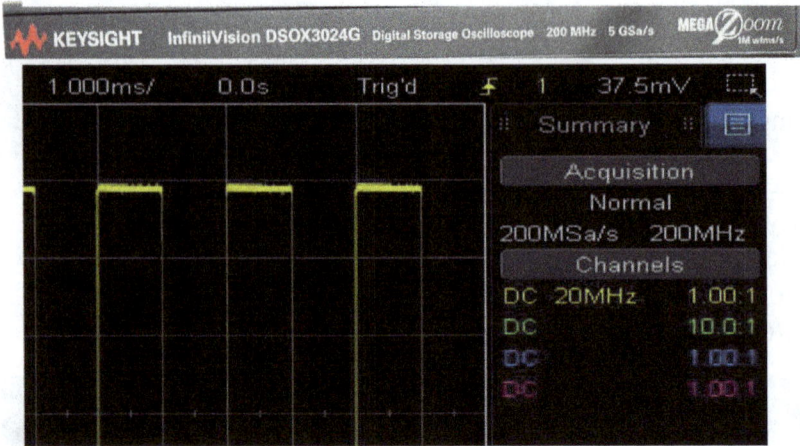

Figure 7.22 Time and sample information displayed on the Keysight 3024 scope.

In this example, the maximum sample rate of the scope is 5 GSps on one channel. On two channels, this drops to a maximum sample rate of 2.5 GSps in each channel. The maximum number of data points in an acquisition buffer is 4 M samples. However, the default value is 2 M samples per channel.

In this specific display, the sample rate for channel 1, with only one channel turned on, is listed in the upper right as 200 MSps. The time base, displayed in the upper left corner is 1 msec/div. This is 10 msec full scale. The number of acquired data points or samples in the acquisition buffer is 200 MSps x 10 msec = 2 MSamples.

This means that the time resolution is only 1/200 MSps = 5 nsec. This is the time interval between successive measurements of the signal. Some interpolation can be used to get higher time resolution, but interpolation is just a technical term for making up a measurement.

If you need the highest resolution in a measurement, you want to make sure to use the highest sample rate. This means adjusting the time base on a short enough full-scale value and positioning the signal on the screen to show the feature of the signal you care about. The best resolution you can achieve in the Keysight 3024 scope, for example, is 1/5 GSps, or an 8-bit measurement of the signal every 200 psec.

7.7 Aliasing of Signal Bandwidth and Sample Rate

The ADC of the scope will sample the measured signal after the preamplifier front end at some sample rate at or below the maximum value for the scope.

In principle, the highest frequency component in a signal the scope can measure is related to the sample frequency. The Nyquist theorem states that the highest frequency that can be measured in the sampled data is no higher than ½ the sample frequency. This highest frequency visible in the sampled measurements is called the Nyquist frequency and is ½ x sample rate of frequency.

If the sampling frequency is 5 GSps, the Nyquist frequency, the highest sine wave frequency component that can be measured in the buffer of data, is ½ x 5 GSps or 2.5 GHz, in principle.

In practice, there will be potential artifacts if the signal being measured is periodic with a frequency at the Nyquist frequency and the signal source and the scope's clock that triggers ADC samples are precision, stable frequencies.

The most extreme case is when the signal is a sine wave with a frequency at its bandwidth. In this case, there would be exactly 2 samples measured per cycle of the incoming sine wave. If you were so unlucky as to sample right at the two zero crossings per cycle, you would measure a zero-level signal for each of the two points per cycle. If the samples were taken at the maximum and at the minimum, the maximum amplitude would be measured. This is a measurement artifact that would depend on the trigger level of the scope that synchronized the sample time and the signal phase.

To avoid this unlikely problem, the sample rate of the ADC should be at least 2.5x the highest frequency component you want to measure. This means there should be at least 2.5 points sampled per period. The Nyquist frequency should be at least 1.25x the signal bandwidth. This is recommended to accurately resolve the amplitude of the frequency component of the signal at the signal bandwidth.

When the ADC samples the measured waveform at a sample rate less than 2.5 x the signal bandwidth, we refer to this as *under-sampling.* In this case, artifacts can be introduced to the measured waveform. We label these sorts of artifacts that can arise when the signal is under-sampled as *aliasing* artifacts.

Aliasing occurs when a signal has frequency components at a higher frequency than the Nyquist frequency.

If the sample rate is 2.5x the signal bandwidth (the sine wave frequency), the sampled points would shift to different phase locations in the measured sine wave each cycle, and over some time, the entire waveform would be sampled.

To recreate the shape of the signal and not just the amplitude at its bandwidth, the incoming signal should be sampled at a sample rate, at least 5 samples per period of the signal. This is a better criterion for the minimum sample frequency. This means a sampling rate > 5 x signal bandwidth.

To illustrate the value of this recommendation, a 1 MHz sine wave was measured by a scope with a sample rate of 10 MSps, 5 MSps, 2 MSps, 1 MSps, and 0.5 MSps. The measured waveforms, shown in **Figure 7.23**, look like a sine wave at 10 samples per period and 5 samples per period, but not very much like a sine wave with fewer samples per period. A signal was still measured, but it does not look much like the sine wave signal present.

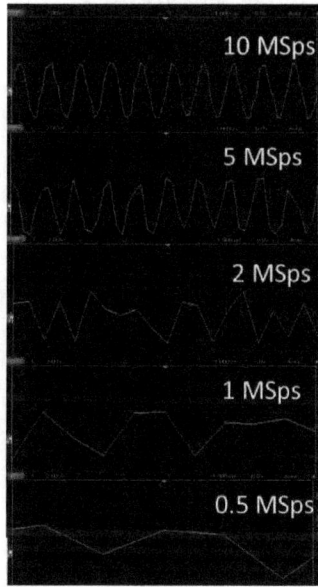

Figure 7.23 The same 1 MHz sine wave was measured with different sampling rates from 10x the signal's frequency to half the signal's frequency. When possible, try to use a sample rate at least 5x the signal's frequency components or a Nyquist frequency = 2.5 x signal BW.

If there are frequency components in the signal above the Nyquist frequency, it does not mean they will not be measured. It means there is the potential for aliasing artifacts that may appear as noise or a modulated signal or even as a clean signal, just very different than expected. Signal frequency components above the Nyquist frequency will introduce aliasing artifacts.

For example, **Figure 7.24** shows two cases of a 1 MHz sine wave measured at a sample rate of 50 kSps or at 20 usec intervals. The sample rate is well below the frequency of the sine wave. The period of the sine wave was 1 usec, while the time between samples was about 20 usec. At most, one sample is taken every 20 cycles. This is an extreme case of under-sampling the waveform.

The sine wave frequency was adjusted slightly from 1 MHz in the two cases. As the frequency of the sine wave changed as a multiple of the sample frequency, the under-sampled pattern changed drastically. Neither of these under-sampled, measured waveforms show the actual input signal waveform. This is an example of an aliasing artifact.

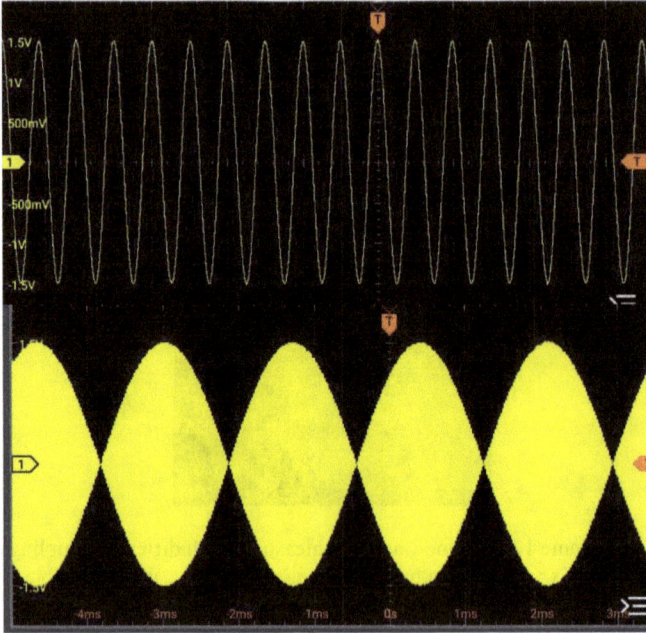

Figure 7.24 The same sample rate of 50 kSps measures a much higher-frequency sine wave at two slightly different frequencies. This is an example of an aliased signal.

The way to avoid an aliasing problem is to make sure the bandwidth of the signal being measured is below the Nyquist frequency based on the sample rate. A better condition is that there are at least 5 samples taken per period or that the signal bandwidth < 1/5 = 0.2x the sample rate.

This is why it is so important always to be aware of the scope's sample rate. If the signal's bandwidth is > 0.2x the sample rate, be very suspicious of possible aliasing artifacts.

When there is a signal or noise present with a bandwidth > Nyquist frequency (= ½x sample rate), there will be an aliasing artifact. Aliasing artifacts can be reduced by adding a low-pass filter between the signal and the scope with a pole frequency at 0.5x the scope's sample rate. This will reduce any frequency components above the Nyquist frequency and reduce the aliasing artifacts. This filter is called an antialiasing filter.

For example, if the sample rate is 2.5 GSps, the signal should be limited to sine waves no higher than 2.5/2 = 1.25 GHz.

Most scopes have a measurement bandwidth that is less than the Nyquist frequency at the maximum sample rate. For example, the Keysight DSOX3024 scope has a measurement bandwidth of 200 MHz and a maximum sample rate of 2.5 GSps on two channels.

The scope's input amplifier bandwidth of 200 MHz acts as a natural antialiasing filter when the scope is sampling at its highest frequency. When the scope is adjusted to sample at 200 MSps, for example, the signal may show aliasing artifacts if it has frequency components above 100 MHz. An antialiasing filter should be added to the scope input.

If the sample rate is reduced from the maximum value, in order to increase the number of sampled points with a longer time span, make sure the signal bandwidth is limited by an antialiasing filter either on the DUT or a hardware filter in the scope. The pole frequency should be no higher than the Nyquist frequency.

If you suspect aliasing to be present, increase the sample rate to as high a rate as possible to push the Nyquist frequency above the signal bandwidth.

7.8   A Strategy for Setting the Scales

There is a lot of flexibility in setting the offset and scale expansion for both the horizontal (time) axis and the vertical (voltage) axis. What approach should be taken? Setting reasonable scales for both the vertical and horizontal axes is a surprisingly important task.

While there are many right ways of selecting the best scale settings, the strategy to follow when adjusting the scale settings ultimately is about the intended purpose of the measurement.

There are always two different goals in any measurement:

1. **View**: Read specific features or figures of merit of the signal directly from the front screen. This helps to calibrate your engineering judgment. It should always be the first approach.

2. **Analyze**: Use the scope as an analysis instrument to extract accurate numerical values from the signal, either using built-in measurement features or exporting the signal and using analysis tools such as SPICE, MATLAB, Python, or even just a spreadsheet.

These measurement goals drive different approaches to setting up the scales. Regardless of the strategy, though, there are three important principles to always keep in mind when setting up the scales:

1. Scopes (and computers) work for us; we do not work for them.

2. The scope is not smarter than you are.

3. If you can't read the figures of merit you care about off the front screen, how do you expect the scope to do it?

This analysis leads to the most important strategy for setting up the scales on a scope:

The starting place for any measurement should always be setting up the scales to make it easy to view the signal and extract useful figures of merit quickly and easily directly from the front screen using your mark 1 eyeball.

Anyone, at a glance at the displayed screen, should be able to read the important figures of merit directly off the front screen without invoking any cursors or built-in measurements. Get in the habit of always setting up the scales to do this. You will use the values you read from the front screen to establish an expectation when using Rule #9, especially when using a built-in analysis tool. Each time you see the pattern made by a signal you care about or measure a figure of merit, you calibrate your engineering judgment. This ultimately feeds your creativity.

Even if your goal is only to analyze a measurement with a built-in function if you can't easily read the figure of merit with your mark 1 eyeball, how do you expect the scope to do this? The same settings that make it easier for you to read the figures of merit off the front screen will make it easier (i.e., less chance of an error) for the built-in scope functions to extract an accurate value from the acquisition buffer.

The less work you have to do, the more productive you will be. If the scales are easy to read, you can get to the answer you are looking for faster and are more likely to get in the habit of using your mark 1 eyeball first.

To read a value directly off the front screen, first use a scale that expands the signal to a large fraction of the screen. Then, use a scale setting with easy-to-read and interpret round numbers for the value of each division.

For historical reasons, the standard scope scales are in multiples of 1, 2, or 5. Use one of these with an offset that is an integer number of divisions. This will make it much easier to read the signal's features directly off the front screen. **Figure 7.25** shows examples of these best practices in adjusting the scales and not-so-good practices.

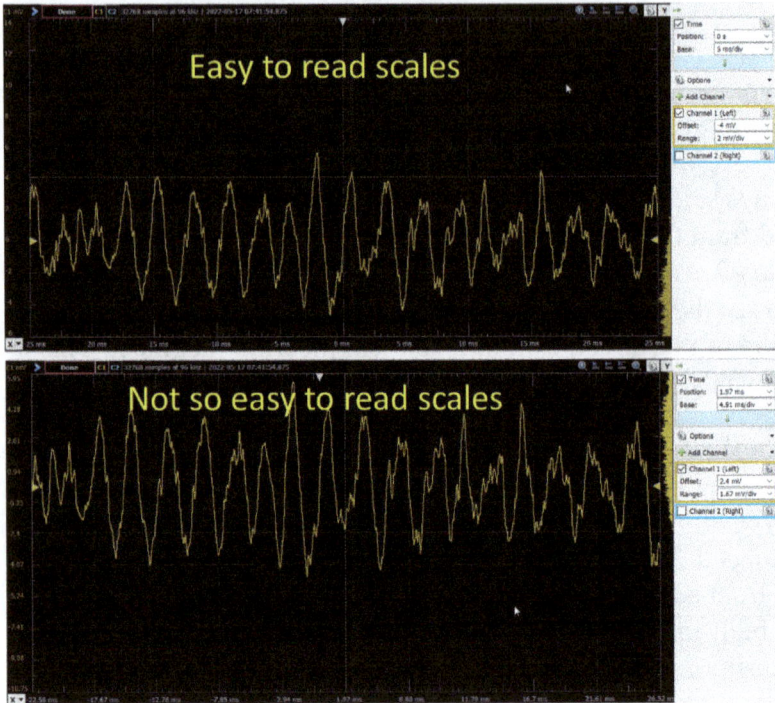

Figure 7.25 Examples of best practices in adjusting the scope scales (top) and not very good scale selections (bottom).

When the scale is whole numbers, with 0 V at the center, it is much easier to read the average voltage of the signal than if the units are 2.7 V/div and centered at -0.21 V, for example. If the scale and offset settings are some random fractions, you have to work at reading the average or the peak or the period off the front screen.

It is sometimes tempting to adjust the scales and offsets to show the signal's pattern with no regard for the resulting values of the scale. It is far better to sacrifice a little bit of screen real estate in order to use scales that are easy to read.

It is also important to make sure the scale is adjusted so that the signal fits within the vertical limits of the displayed window. When it extends to the top or bottom edge of the screen, it may have saturated the input amplifier and the voltage values displayed are an artifact of the scale and not a true measure of the signal.

It is tempting to think that if you use cursors or built-in measurement functions, why bother reading a value from the front screen with your mark 1 eyeball? There are many ways for this extracted number to be wrong. Maybe it is applied to the wrong channel. Maybe the peak-to-peak value is really measuring the peak noise and not the signal you care about. Maybe the rise time is being measured based on long tails and not the region of the waveform you care about.

For example, **Figure 7.26** shows an example of a sine wave measured on a scale that is not sensitive enough to see the signal very well. The scope's measurement function returns a value of the frequency of about 472 Hz. This value has nothing to do with the frequency of the signal but is dominated by noise.

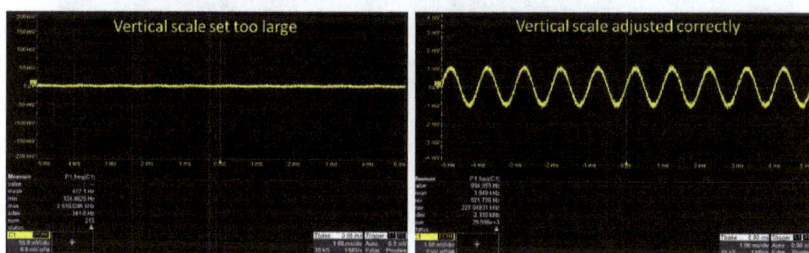

Figure 7.26 The measurement functions will always return a number, but it is not always correct. Left: frequency extracted as 472 Hz but measured on the right as 1 kHz on a more sensitive scale.

A measurement function will apply its algorithm to the measurements in the acquisition buffer. Almost guaranteed it will return a numerical value. If you can't read the figure of merit from the front screen, how do you expect the scope's algorithm to return a value correct to within the measurement uncertainty?

Or the algorithm may measure a feature that is not what you were expecting but meets the algorithm condition. For example, it is easy to write a simple algorithm to extract the figure of merit for the 10-90 rise time of a signal. The algorithm might be to first find the 0% and 100% values, then the time instants of the transition above 10% and 90%. The 10-90 rise time is calculated as this time difference.

Using this algorithm, the scope will almost always display a 10-90 rise time value. But is it close to reality for your signal? **Figure 7.27** shows an example of applying this built-in calculation algorithm to the rising edge of a signal measured with a scope.

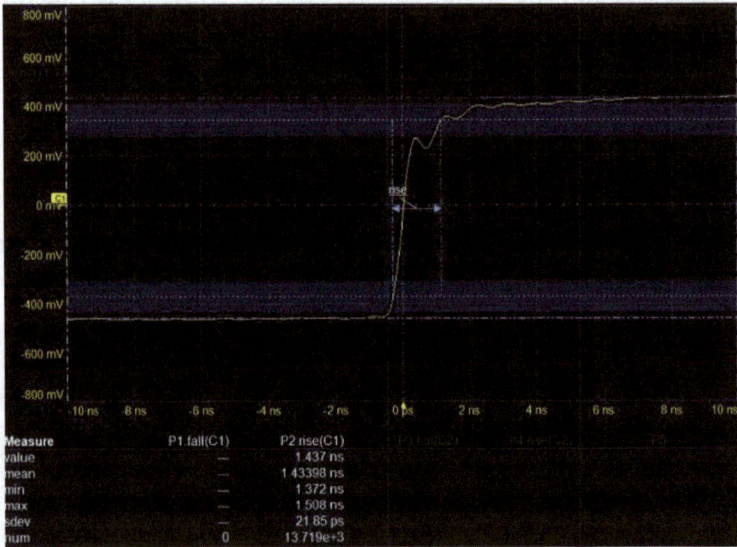

Figure 7.27 Measured 10-90 rise time using a built-in measurement feature.

If you had not looked at the waveform but blindly accepted the figure of merit returned by the scope, you would have believed this rise time to be 1.4 sec. In fact, the rising edge has some structure, and the blind algorithm includes some of this structure in the extracted figure of merit.

If we had looked at the rise time with our Mark 1 eyeball, we would have found that maybe a 10-90 rise time is not a good figure of merit to describe this rising edge. If we cared about the fast-rising edge part of the waveform, maybe the 20-80 rise time was a better figure of merit, or we might have concluded it was closer to 0.5 nsec.

Only after we use our Mark 1 eyeball to estimate the figure of merit should we use cursors or built-in measurement functions on the screen to get a little more accurate value of the figures of merit. The extracted values should match your initial Mark 1 eyeball measurements. This habit of reading a measurement from the front screen directly is an important risk reduction habit.

Good axes will give immediate insight into a signal's behavior, while a poor selection of scales will often keep valuable information off the table and not obvious. Be proactive in using the scales that tell the story of your measurements as clearly and quickly as possible to you and to others.

Here is a summary of the recommended best practices for setting the scales:

1. Use a scale that makes it easy to see the pattern. This means using an expanded scale so that the signal takes up a large fraction of the screen.

2. Start with the 0 V level in the middle of the screen and a scale to show the signal on the screen. Once the signal is identified, further adjust the scales to show off the features you care about.

3. Don't use a scale setting in which the signal is expanded to less than 1 division of the screen. You have the whole screen to work with.

4. Decrease the scale or change the offset if the signal is flat at the top or bottom of the screen. This means the signal is clipped or saturated.

5. Make the vertical scale easy to read. This generally means using a scale and an offset that results in whole numbers on the vertical scale. When you use a fraction or an odd decimal value, you have to work to calculate the amplitude or the peak value of a waveform off the front screen.

6. Adjust the time base to show the time-domain features of your signal. This may require two different measurements on two different time frames.

7. Do not use cursors or other measurement features unless you have first estimated the figure of merit for the waveform. If you can't get a rough measure directly off the front screen with your Mark 1 eyeball, how do you expect the instrument to do it?

8. Adjust the buffer size appropriate to the time scale and sample rate you need and the time it will take to process all of the measurements in the acquisition buffer.

9. Regardless of the settings, always be aware of the sample rate of your ADC and the time resolution you need for your measurement.

10. Be aware of the sample rate and the Nyquist frequency of the scope ADC. If your signal bandwidth is higher than the Nyquist frequency, watch out for aliasing artifacts.

7.9 Comparing Multiple Traces

In a scope with multiple channels, the real time measurements from each channel can be displayed on the same display window or on different windows. In addition, all scopes have built-in memory buffers to store the numerical values of previous acquisition buffers and treat them just like real time measured traces.

The measurements stored in either the real time acquisition buffer or the memory register are called *waveforms* or *traces*. When comparisons between channels are important, these real time or memory traces should be displayed on the same window and on the same scales.

In the MAUI Studio scope emulator, saving an acquisition buffer into a memory location is accessed from the File/Save waveform. This menu selection is shown in **Figure 7.28**.

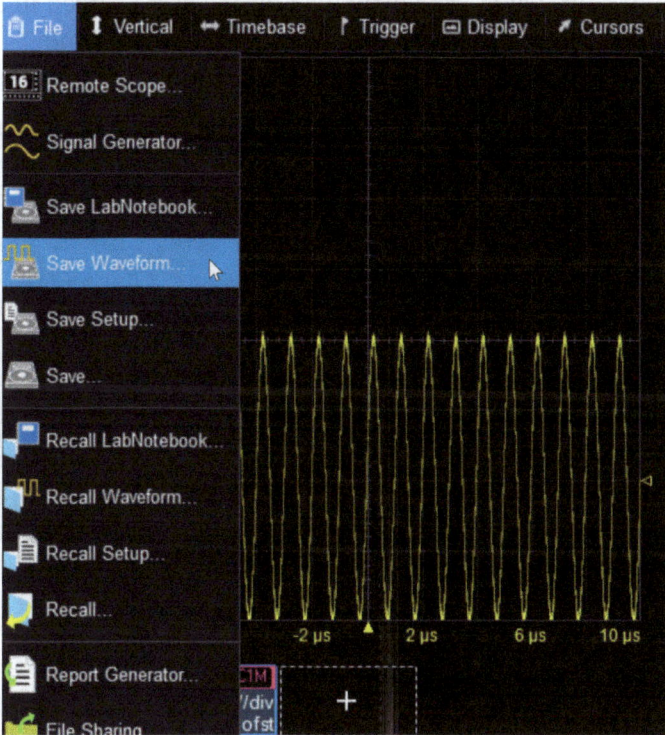

Figure 7.28 Save any measured trace or waveform into a memory location using the File/Save Waveform command.

Once saved in a memory location, the trace can be recalled and displayed on the screen using the current scale setting.

It is possible to show four channels of real time waveforms and as many as 4 to 20 saved waveforms at the same time. Many scopes allow for multiple display windows arranged on the front screen. Every scope has a different way of moving a waveform to a different window.

In the MAUI Studio scope emulator, the waveform is tied to the colored icon on the window with the channel number or memory location labeled at the top. The icon can literally be dragged and dropped to any display window to move the waveform. This makes direct comparison between waveforms easy. **Figure 7.29** is an example of four waveforms displayed on the same screen in a dual grid display. Three of the waveforms are from a real time measurement and one is from a memory location.

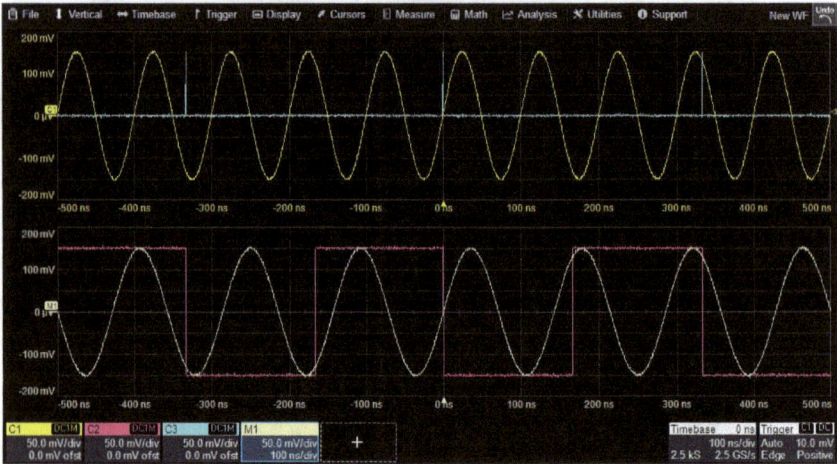

Figure 7.29 An example of multiple display windows with multiple waveforms directly compared.

When comparing two different signals, while their numerical values can always be read off the front screen, the relative comparison on the same scale will often provide engineering insight. It is much easier to see the differences between two waveforms at a glance and calibrate our engineering judgment if they are on the same scale.

In this special case, adjust the scales so the larger or slower signal is well-oriented and then use this same scale for the other waveforms.

Always keep in mind that just because the extracted value of the figure of merit is close to what you observed, there is still no guarantee that the value is correct, but it adds to your confidence.

7.10   Saving and Using Measurements

After you have observed the signal's patterns directly from the front screen, you may need to save the waveform to an external file. All scopes allow saving the acquisition buffers to a file in a variety of formats. There is no universally accepted format for exporting the measured data, but all scopes can export as comma-separated values (CSV). The format you should use depends on the final application.

In most cases, the time and voltage CSV data can literally be imported into an Excel spreadsheet for plotting and some analysis. The CSV file can be written to a thumb drive plugged into the USB socket found on almost every scope.

7.10.1 Saving Waveforms to Text Files

One way of processing a waveform after viewing it on the front screen is to export the time value and voltage value of each point stored in the acquisition buffer. All scopes enable the measured voltage data to be output in a CSV file format, among other formats.

For example, in the MAUI Studio scope emulator, the synthesized sine wave, with all of the scope settings, can be exported to a CSV file by selecting the File/Save Waveform command and then selecting the CSV option. This menu selection and the resulting screen are shown in **Figure 7.30**.

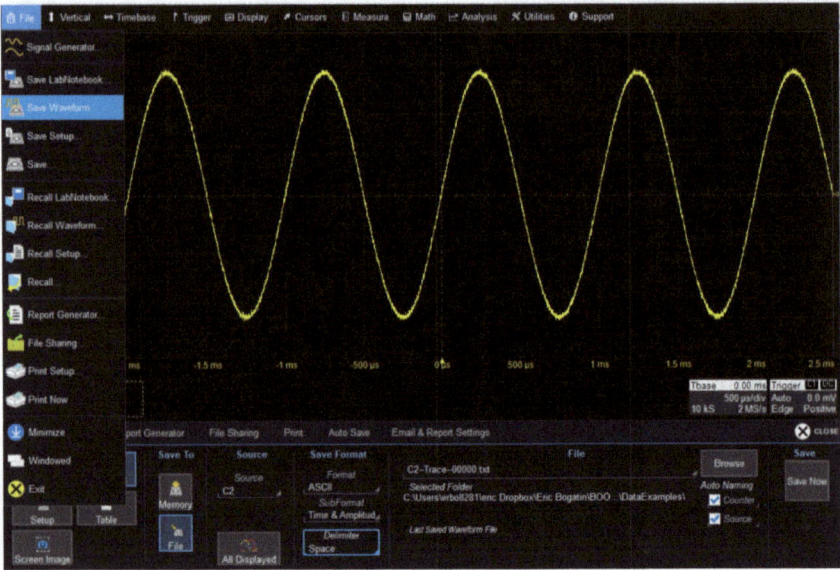

Figure 7.30 Example of the Files/Save Waveform option to save a text-formatted file with the time and voltage data.

When imported into Excel, the same sine wave as on the scope screen can be viewed in the Excel plot. The time column and the voltage column can literally be highlighted and plotted by selecting the insert plot option. The result of plotting the CSV file directly opened in Excel is shown in **Figure 7.31**.

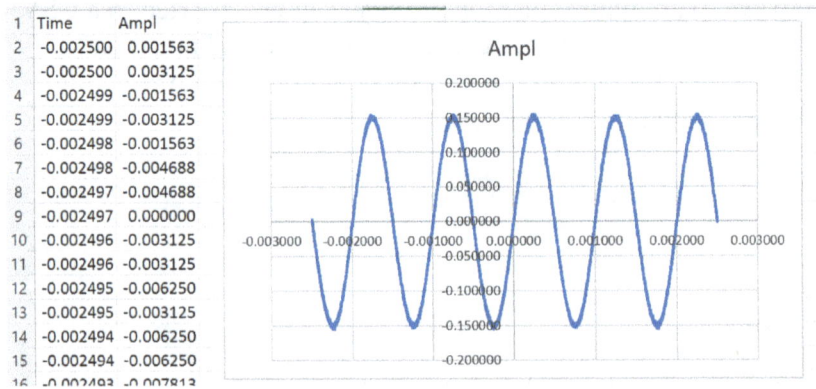

Figure 7.31 The exported CSV files were imported into Excel and plotted in a simple x-y plot with no other cleanup.

Exporting measured scope traces is a feature of all scopes but located in slightly different menu options.

This is one way of enabling further analysis of the measured data. It is effectively using the scope as a high-performance ADC. Once the measurements are in excel, the power of the spreadsheet can be used to analyze the measured waveforms.

7.10.2 Importing Scope Waveforms into a Simulation Tool

The ability to import the measured waveforms into any simulation tool means that the full power of simulation and data analysis can be applied to any measurement. This opens up unlimited possibilities for the analysis of any measurement. This is essential for measurement-simulation correlation.

Each simulation tool has its own requirements for the format needed to display external voltage and time data. Usually, the circuit element used to do this is the piece-wise linear (PWL) element.

In some low-end tools, such as the online SPICE simulator, CircuitLab, the PWL element is called a CSV element. The time and voltage data are literally pasted into this element. It takes two mouse clicks to bring a CSV file into Circuit Lab. The only two important conditions are the time must be greater than 0 and there must be a comma separating the time, voltage. The units for time are in sec and the unit for the voltage data is volts. An example of this scope waveform plotted in CircuitLab is shown in **Figure 7.32**.

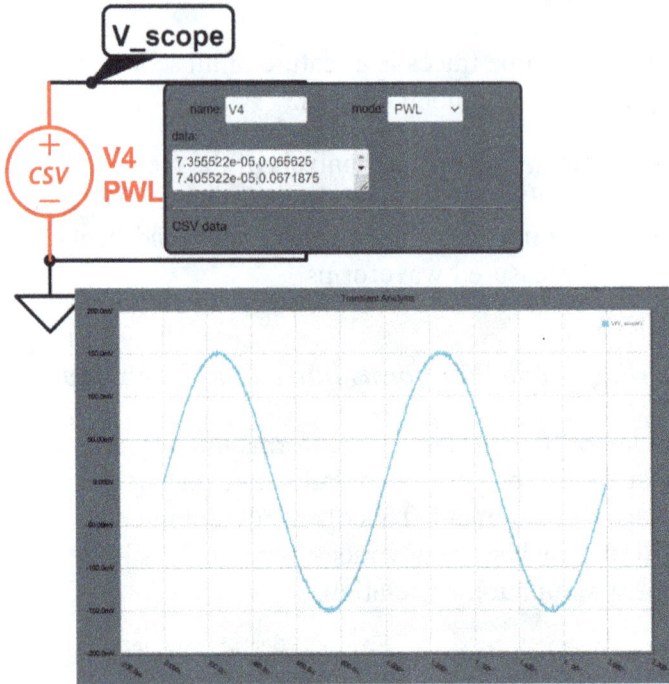

Figure 7.32 An example of the CSV ideal voltage source element and the resulting transient simulation in CircuitLab, showing the measured data exported from the Maui Studio scope emulator.

Another popular and free SPICE simulator is LTSPICE. The circuit element to display measured data in LTSPICE is the voltage source but with the advanced options selected. In this menu, the PWL circuit elements based on a file can be selected as the source of the time and voltage data. This menu option is shown in **Figure 7.33**.

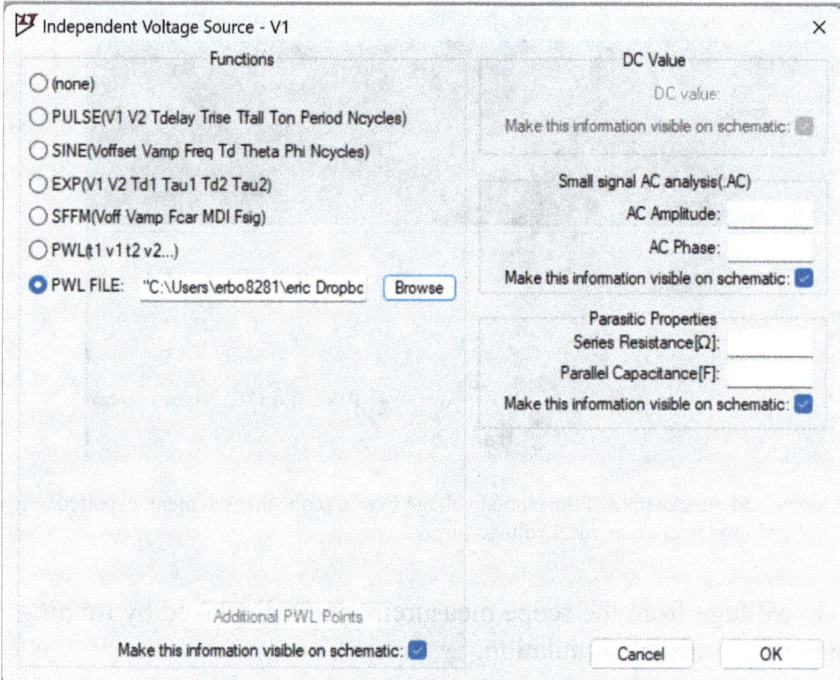

Figure 7.33 Example of the advanced option of a voltage source in LTSPICE showing the option to use a file as the source of the time, voltage data to display.

The one condition in the formatting for the text file in LTSPICE is that there should be some space as the delimiter between the time and voltage data. A space delimiter can be selected when exporting the CSV file in the MAUI Studio scope emulator and in most scopes. In LTSPICE, there is no restriction on having negative time data in the file; only the positive values of time data will be plotted.

Once the text file is selected as the source of the measurements, this circuit element can be used to drive the voltage into a resistor element, for example, to replot the data. Since the data file is what drives the ideal voltage source, it can be used as a source in any circuit simulation. **Figure 7.34** shows the result of simulating this voltage source using the exported scope file as the PWL source file.

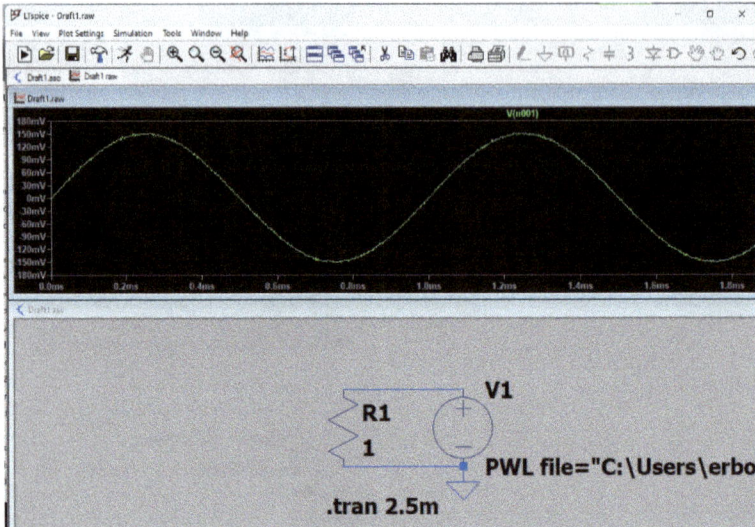

Figure 7.34 An example of the plotted voltage from a scope measurement, exported as a CSV and used to drive an ideal voltage source.

The voltage from the scope measurements is displayed by running a simple transient simulation.

This same exported CSV file can also be imported into a high-end tool such as Keysight's Advanced Design System (ADS). It requires just a few modifications:

Step 1: Change the header in the text file

Step 2: Convert the txt file into a dataset (ds) file

Step 3: Use the dataset as a dataset voltage source

Step 4: Simulate the voltage source in a transient simulation

The first step in using a data file in ADS is to change the header information in the text file. Any header information at the top of the file which contains the time and voltage data, should be replaced with the following:

begin blk

% time(1) v(1)

It is ok to have negative time data in the file, but only positive time measurements will be plotted.

This new text file needs to be converted into an ADS proprietary dataset (ds) file format. This is done using the Tools/Data File Utility/Data File Tool, found on any schematic page, as shown in **Figure 7.35**.

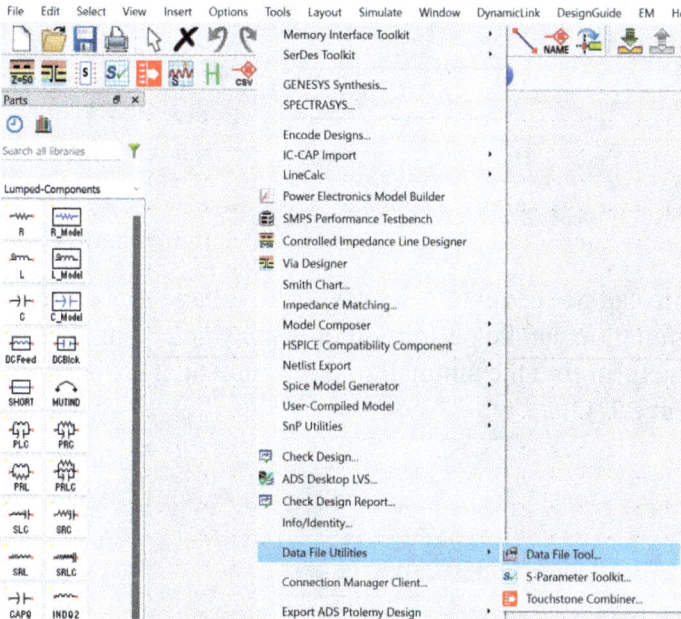

Figure 7.35 Convert the txt file into a ds file format using the Data File Tool.

The data file tool reads the text file as a MDIF file format. When using this tool, the MDIF and then the Generic MDIF format are selected. These options are shown in **Figure 7.36**.

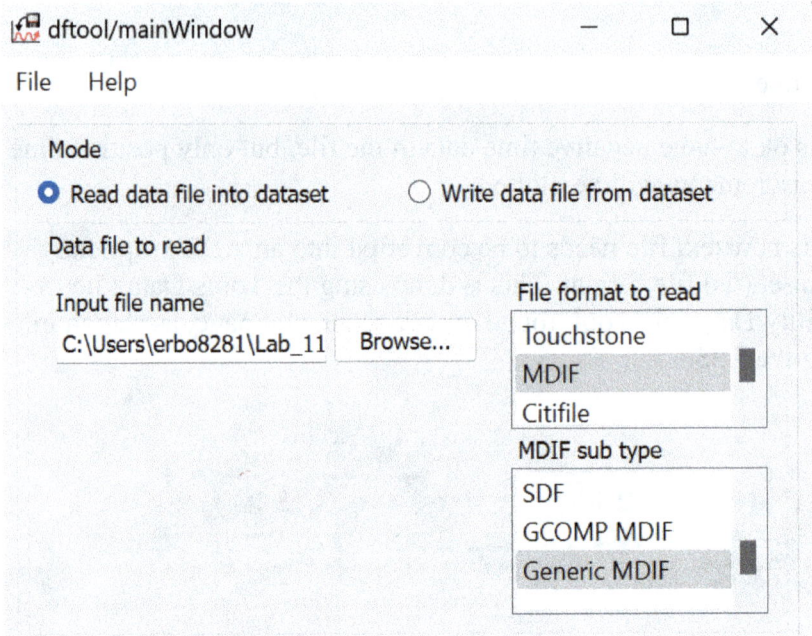

Figure 7.36 The text file is considered an MDIF file format in ADS.

The dataset file can be read into a dataset ideal voltage source, and a transient simulation can be performed to display and bring the measured waveform into the simulation environment. This is shown in **Figure 7.37**.

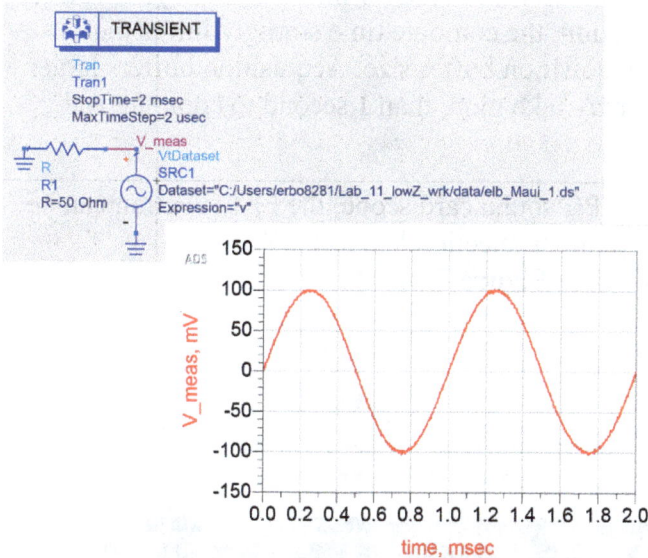

Figure 7.37 The ADS circuit to bring measured data into the simulation environment and an example of the measured sine wave from the MAUI Studio scope emulator.

7.11   Quick Start with the FFT Function

A real time scope fundamentally measures and displays the measurements in the time-domain as the V(t) waveform. Valuable insight can often be gained by looking at the same measurement in the frequency domain. This is enabled by performing a Fourier transform on the measurements in the acquisition buffer and displaying the spectrum of the signal.

When implemented as a fast Fourier transform (FFT), this operation can be done in real time so both the time-domain and the frequency domain view of the same waveform can be displayed for each acquisition.

Every DSO available today has the ability to calculate and plot the FFT of the signal in real time after an acquisition. The only difference between the scopes is the calculation speed and the user interface, which allows the scales to be changed and the spectrum to be displayed in a useful format. An important differentiator between scopes is the user interface for the FFT display. If you are using the FFT function, the compute time strongly limits the practical, useful acquisition buffer size. Acquisition buffers larger than 1 Msamples may take more than 1 second to compute and display the FFT.

On the Waveforms PC sound card scope, the FFT function can be accessed in the top menu selection between +Zoom and Spectrogram, shown in **Figure 7.38**.

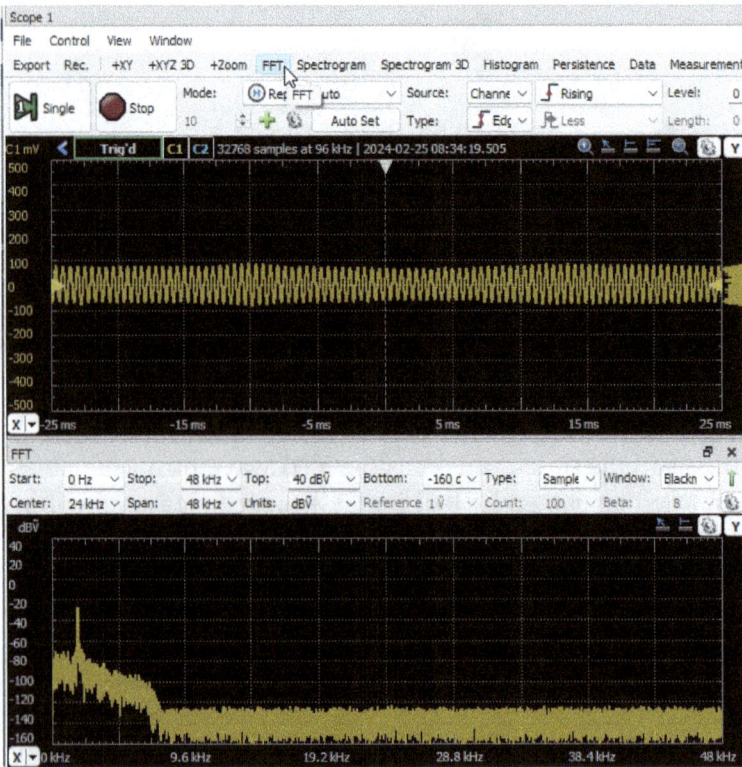

Figure 7.38 Select the FFT function from the top menu items. The amplitude of sine wave frequency components is displayed on a dB scale.

In this example, I recorded the real time waveform and its FFT while I whistled. The peak in the spectrum is at 1.6 kHz.

The spectrum has two important features, which are dependent on how the time-domain scales are set up.

The highest frequency that can be calculated by any Fourier transform is the Nyquist frequency, which is ½ the sample rate. In this example, the sample rate is fixed at 96 kSps. The Nyquist frequency is 48 kHz. This is the highest frequency displayed in the FFT.

The second feature is the lowest frequency that can be measured and the frequency resolution. This is set by the full-scale time base. In this example, it is 50 msec full scale. The lowest frequency component calculated is 1/time base = 1/50 msec = 20 Hz. This is also the resolution of the frequency components.

If you wish to see a lower frequency or a higher resolution, you have to increase the total acquisition time.

Once calculated, the spectrum can be plotted on a linear frequency scale or a log frequency scale. When looking at the harmonics of a waveform's repeat frequency, it is useful to plot the spectrum on a linear scale. Harmonics are evenly spaced. When looking over a wide frequency dynamic range or looking at the rate at which the amplitude of the harmonics drop off with frequency, it is most useful to use a log frequency scale.

The vertical scale is the amplitude of each frequency component. It can be plotted in a dozen different formats, such as a voltage amplitude, a peak value, an RMS, or in a dB scale. Generally, for viewing the spectrum on the highest dynamic range, using values as dBV amplitude and a log frequency scale lets you see the entire spectrum and all the important features all at once.

The spectrum of a signal in the time-domain is a fingerprint of the frequency components that stand out to characterize the waveform. A periodic component in the signal, such as from an SMPS or from interference from an RF signal, stands out as a peak in the spectrum. The more constant the frequency, the sharper the peaks.

This is another method of taking a complex waveform and characterizing it with a few figures of merit.

An example of the time and frequency domain display of the same signal is shown in **Figure 7.39**. On the log-log scale, this shows the highest frequency as 48 kHz, but the amplifier has a sharp cutoff at 9 kHz. The lowest frequency displayed is 20 Hz.

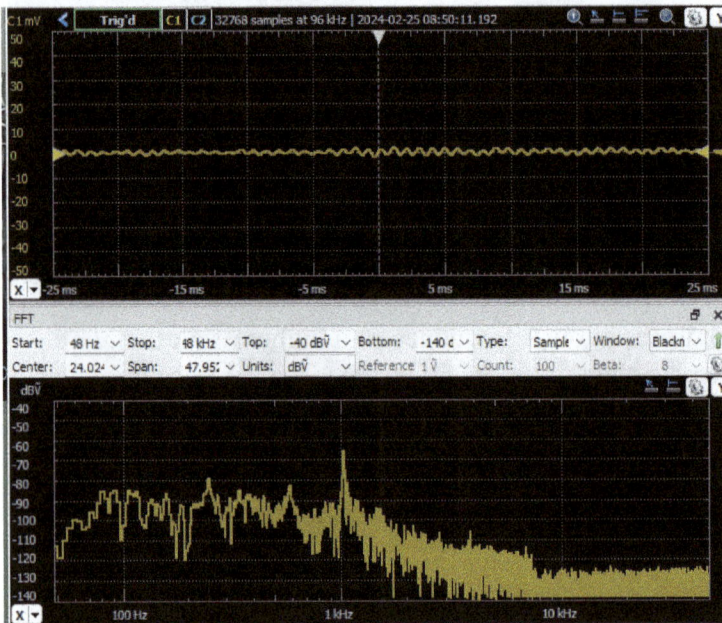

Figure 7.39 The real time signal in the time and frequency domains on a log-log scale.

There is a sharp peak at 1 kHz corresponding to a frequency component in the music being recorded during this time interval. There is also a broad range for the frequency components from roughly 20 Hz to about 3 kHz. There is little sound recorded at frequencies above 3 kHz and the amplifier does not pass frequencies above 8 kHz.

The Nyquist frequency of the ADC, 48 kHz, is 6x higher than the amplifier bandwidth of 8 kHz. This prevents the possibility of aliasing of the signal frequency components above the Nyquist.

The details of mastering spectral analysis are covered in a later chapter.

7.12   Cursors and Measurement Functions

An important value of a scope is to measure a waveform and display it on the front screen to show the signature of the waveform or the pattern it creates. This by itself is important information. Is it increasing, decreasing, or repetitive, or does it look like a sine wave? A square wave? A data pattern? Or does it have a dip, a peak, or is flat, or with lots of noise?

Based on matching the pattern you see to an ideal, well-defined pattern, you can extract a few figures of merit when this pattern is mapped into the ideal waveform. Once you have estimated a figure of merit from the front screen with your Mark 1 eyeball, a more precise value of a figure of merit might be extracted using built-in features of the scope.

A cursor or measurement function can dramatically increase the precision of the figure of merit extracted from the measurement buffer.

7.12.1 Cursors

The cursor functions and features are different on every scope, but they generally fall into two classes. Horizontal cursors define voltage values across the entire screen. They are reference values placed on the front screen. They can be manually moved to read the voltage value of a feature, such as a peak, directly from the front screen. **Figure 7.40** shows two horizontal cursors manually positioned in the MAUI Studio scope emulator.

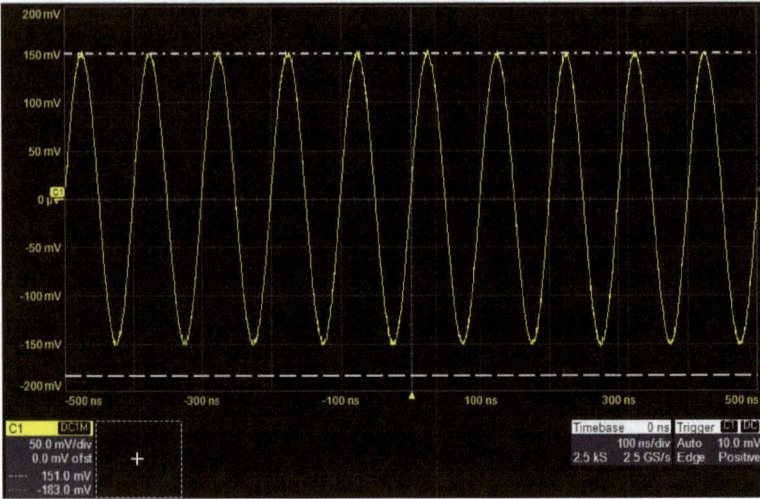

Figure 7.40 Horizontal cursors define a reference voltage on the front screen.

One of the important applications of a horizontal cursor is to set its value at a minimum or maximum value of a signal and use this as a reference to compare all subsequent measurement values. You will instantly see if an incoming waveform exceeds the range established by your cursor.

The second type of cursor is a vertical cursor. As it is manually moved across the front screen, it will read and display the voltage and time value of each point at that position in the acquisition buffer. The vertical cursor is effectively displaying the measured buffer point by point. **Figure 7.41** shows an example of the vertical cursor in the Waveforms PC sound card scope.

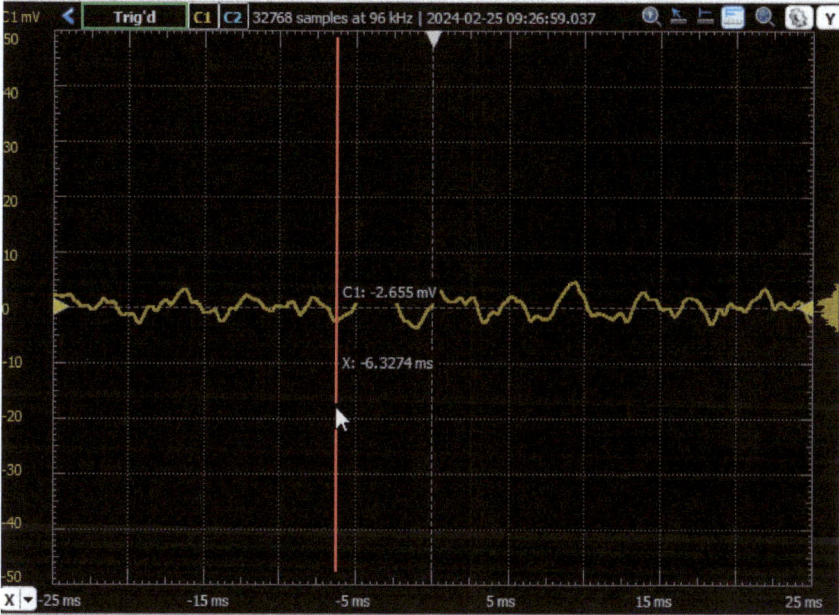

Figure 7.41 The vertical cursor is reading out the voltage and time value of the measurements in the acquisition buffer.

Many scopes allow multiple cursors to be displayed at the same time, and differences in both the voltages and time positions are displayed.

Before you use either cursor, you should have an idea of roughly what value will be measured just by reading the numbers off the front screen with your eye. You can estimate the value expected to within ½ a division. If the value the cursor reads is far off from your eyeball measurement, there is something wrong in your measurement. Double-check it. This is always an important consistency check.

7.12.2 Measurement Functions

When you extract a few figures of merit from measured data, you always have an algorithm in mind. For example, if you are extracting the peak-to-peak voltage values of the measured data, you look for the maximum value and the minimum value in the displayed data. The peak-to-peak is the difference between them.

If you are measuring the period of a repetitive waveform, you first identify the repeating pattern and then one cycle. You then find the start time of one cycle and the end time of that cycle. The period is the time difference between these instants.

All scopes have a selection of automatic measurement algorithms built in. Generally, they all use the same algorithm, but not always.

For example, the mean, true RMS and AC RMS values are calculated over the acquisition buffer time interval using

$$\langle V \rangle = \frac{1}{n} \sum_{\text{time period}} V(t)$$

$$\text{True RMS} = \sqrt{\frac{1}{n} \sum_{\text{time period}} V^2(t)}$$

$$\text{AC RMS} = \sqrt{\frac{1}{n} \sum_{\text{time period}} \left(V(t) - \langle V(t) \rangle \right)^2}$$

where

<V> = the average value of the voltage

time period = the time over which the average is recorded

n = the number of points recorded over the time period

V(t) = the individual voltage values recorded at each sampling time

Of course, in the case of the Waveforms PC sound card, since the input to the preamp is AC coupled, the mean value is always 0 V and the true RMS equals the ACRMS value.

These calculations are done in real time from the raw measured data and displayed at the end of each updated time interval.

There are always hidden assumptions used in the calculation. You should first have confidence the assumptions apply to the measurements you are performing. Otherwise, you will absolutely get a number, but it may not be related to the quantity you really care about. This is how artifacts are introduced to your measurements.

In the Waveforms software, the measurement functions are accessed from the row of functions above the display window, as shown in **Figure 7.42**.

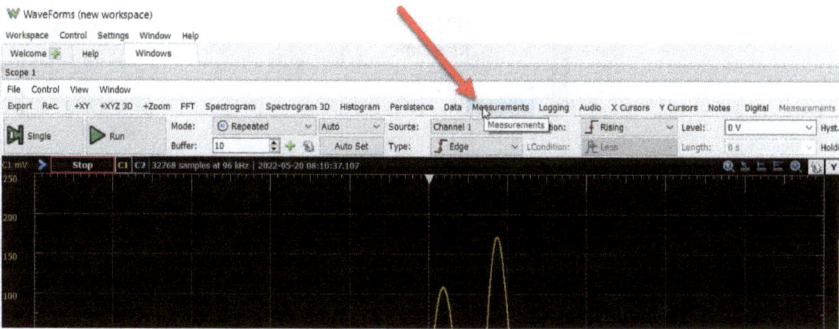

Figure 7.42 Access the measurement functions from the tab of functions under the file pull-down menus, shown with the red arrow.

Generally, a measurement is either going to extract a voltage or vertical measurement or a time or horizontal measurement. The list of built-in functions available in Waveforms, for example, is shown in **Figure 7.43**.

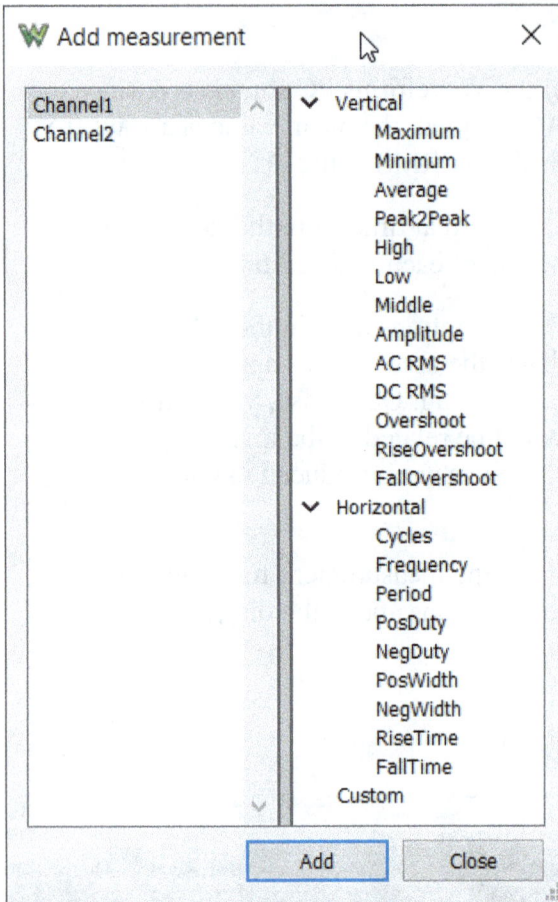

Figure 7.43 List of the built-in functions available in Waveforms. Each one has a specific algorithm.

A selection of just some of the measurement functions available in the MAUI Studio scope emulator tool are shown in **Figure 7.44**.

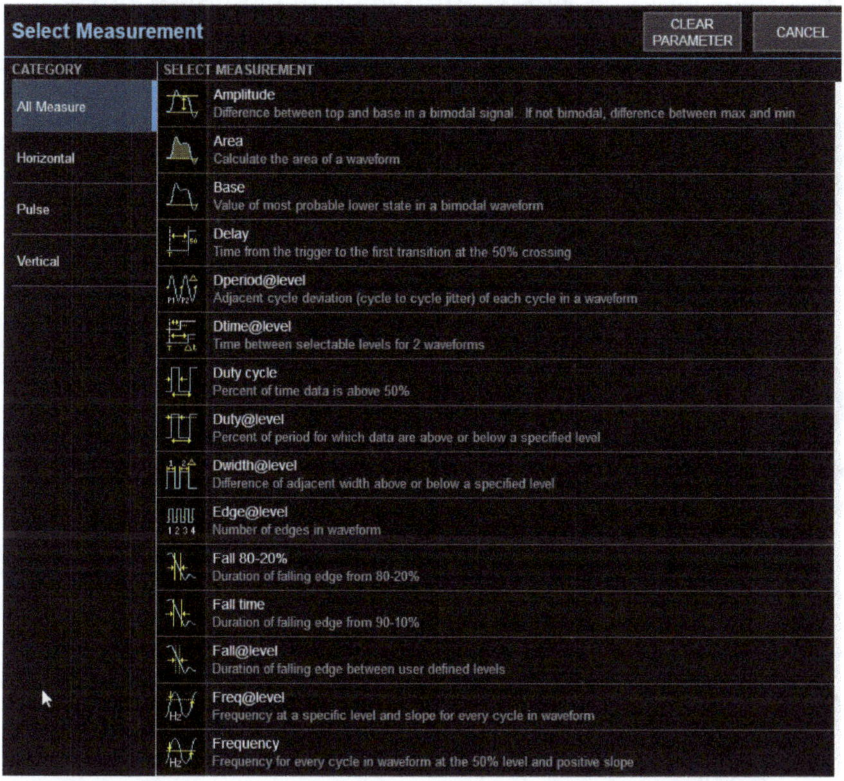

Figure 7.44 Some of the measurement functions are available in the MUI Studio scope emulator. Note that they can be organized vertically, returning a voltage value, or horizontally, returning a time value.

The measurement calculated from the acquisition buffer can be displayed in four different ways.

The *value* of the figure of merit after each acquisition can be displayed as a single value.

As more acquisitions are taken, the *statistics* of each measurement can be displayed. This gives a feel for the average value and the standard deviation of the value over many acquisition buffers. Some tools allow the distribution of values from consecutive acquisition buffers to be plotted as a histogram.

The trend in these values can be plotted over time or just over the acquisition buffer, as a track. An example of this display of measurements in the MAUI Studio scope emulator is shown in **Figure 7.45**.

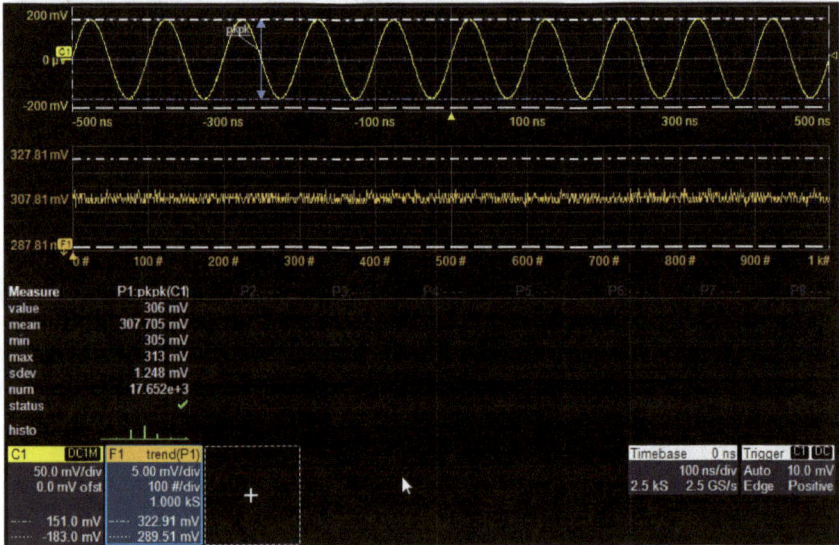

Figure 7.45 Example of the four different ways of displaying a measurement extracted from an acquisition buffer of data, displayed in real time. The top trace is the measured voltage. The middle trace is the trend in the peak-to-peak value over consecutive acquisition buffers. The bottom table shows the statistics in the peak-to-peak value with a small histogram below it.

This type of analysis and display of the results can be applied to any of the built-in measurement functions available in the scope.

7.12.3 Zoom Function

The typical screen resolution at which individual voltage measurements can be displayed is about 1000 pixels across the full scale. When an acquisition buffer holds as many as 1 M or more samples, it is not possible to see all the detail in the measurements displayed for the whole buffer.

All high memory depth scopes have a feature to zoom in on the measured buffer. The zoom function selects a time window in the acquisition buffer and displays the measurements within this window on the front screen. After each acquisition, the same time interval of the acquisition buffer is expanded and displayed. The vertical scale can be adjusted for this zoomed waveform as well.

Since the time stamp for each measurement is referenced to the trigger event, if the waveform is repetitive, the zoomed region will have the same offset from the trigger event in each acquisition. **Figure 7.46** is an example of 25 M raw measurements over a 10 msec time period and the zoomed region spanning only 1 usec full scale. This is using the MAUI Studio scope emulator.

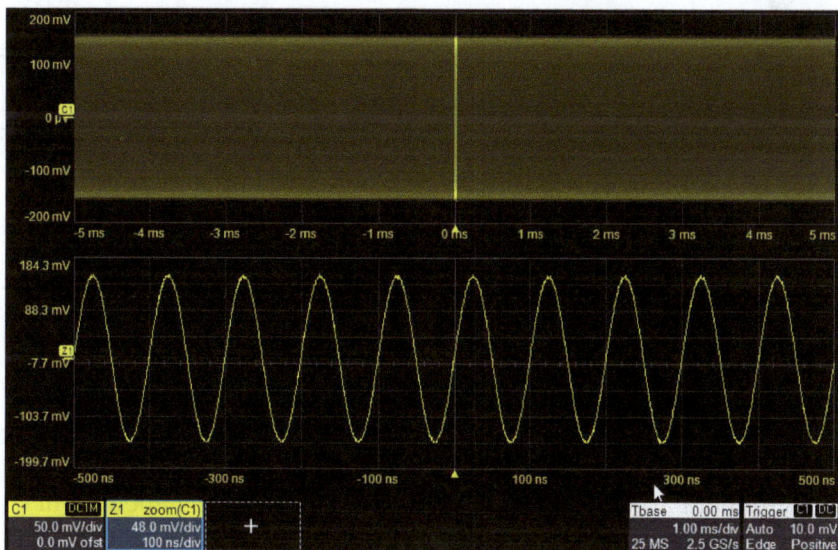

Figure 7.46 Taking a zoomed-in view of part of the measurement buffer using the zoom function. The narrow, highlighted region in the upper trace is expanded in the bottom trace. This is the zoomed-in trace.

7.13 A Checklist of Do's and Don'ts

Now that you have access to two versions of a free scope, you can practice some of the most important best measurement practices that apply to all scopes.

Most scopes have a panic button. This is called *auto setup*. It is sometimes referred to as a "guilty pleasure" because it is so easy to use. *Never* use this button. It assumes the scope is smarter than you are.

If you don't know how to set up a scope and are drawn to use the auto-setup button, learn how to use the scope instead. Some features of a scope transform it into a "smart" scope. This is a misnomer. Never assume the scope is smarter than you are. Remember the important principle:

If you cannot set up the scope yourself or measure a value with your Mark 1 eyeball, how do you expect the scope to do it for you and get it correct?

Before performing a measurement, it is essential to have a clear understanding of what to expect. This is Rule #9. This will give you a feel for the scale settings you would need on the scope to see the signal. Apply Rule #9 to all your measurements. Sometimes, you will not have a clue what to expect. The more you learn about your DUT, the better you will be able to use Rule #9.

When you see what you anticipate, you will have higher confidence that you really understand your DUT. Rule #9 is a confidence builder. When you don't see what you expect, there is always a reason for it. Make sure you have eliminated measurement artifacts. Maybe you have more to learn about your DUT.

The following are the most important best practices for routine scope measurements.

1. When you come to your scope the first day, always press the *default* setup button.

2. Never press the auto setup button. The scope is not smarter than you are.

Many scopes will store the setup conditions after they are turned off. If your scope is also used by others, you may not know how it is set up when you turn it on. In order to get it into a known state so you don't have to go through every combination of settings, press the DEFAULT set up button. This places the scope in the factory reset condition, a known state every time.

3. Always try to apply rule #9 and anticipate what you expect to measure. This will help you set up the scope and understand your DUT.

4. Always start with cursors and measurements off so you can estimate any values you need from the front screen.

5. Do not use the cursor unless you have first measured the value with your Mark 1 eyeball.

More often than not, the measurement or cursor functions are a distraction. *Always* keep them off initially unless you have already used your Mark 1 eyeball to measure what you need from the front screen so you have an estimate. In some situations, this initial estimate is good enough to use without turning on cursors or measurements.

One of the common artifacts using a cursor is that the numerical value you see is for a different channel than the one you think you are on.

6. Do not use a measurement function unless you know exactly how it is calculating what you think it does.

Just because a measurement function says rise time does not mean it is measuring the rise time feature you think it does. If your signal edge has some structure to it, where is the rise time measured? The measurement function will always give you a number. It is just not clear what this number means. If you can't measure the figure of merit with your eye from the screen, how do you expect the scope to do it?

7. Center the voltage and time scales on the center of the screen.

8. Adjust the vertical and horizontal scales to expand the displayed signal on displayed scales with nice, whole numbers, so that figures of merit can be easily extracted from the front screen.

9. Adjust the scale values to use most of the scope screen to display your signal of interest.

In order to make it easy to measure an important figure of merit of a signal off the front screen, like peak-to-peak value or average value, adjust the scales in order to read the signal directly. Press the vertical or horizontal adjust button to auto-center the zero positions.

Anything you can do to make it easier to directly measure a figure of merit from the front screen will encourage you to use your mark 1 eyeball first before you reach for the cursor or measurement button. Only *after* you have estimated the figure of merit so you know what to expect should you use an automated measurement feature. This is an important consistency test.

Use the coarse scale adjust setting for the time and voltage and change their scales so the signal is expanded to most of the screen and the time base shows the features you care about. The scales should be whole numbers like 1 V/div or 0.2 V/div or 1 msec/div or 50 usec/div. This will make it easy to read trace features directly off the front screen.

You have the entire screen height available. If you want to measure some features of your signal, adjust the vertical scale so it uses most of the screen rather than keeping the trace within a fraction of a division.

10. Always start with the trigger on auto mode. Only use normal mode in special cases.

11. Select the correct source channel, the slope, and then the DC threshold level.

12. Know the difference between the auto and the normal trigger mode settings.

13. Do not use the stop feature for repetitive signals. Adjust the trigger instead.

When displaying a repetitive signal, you should get it to be stationary on the screen. This is done with the trigger controls. If you are viewing what you think is a repetitive signal and it is not stationary but smeared out, it may be a trigger problem. Consider using the trigger HoldOff setting to wait for a trigger event at the beginning of a pattern. Only use the run/stop button after trying other trigger methods. The stop mode will give you a snapshot in time with no indication of how the overall pattern is changing.

Avoid these common mistakes:

14. A saturated signal. If the signal is displayed with any of its features near the top or bottom of the screen, there is a good chance it is saturated, and you are not recording a real value. Keep the signal's peak values within +/- 95% of the vertical full scale. Decrease the scale value and then the offset to position the signal so it does not flatline on the screen.

15. Poor resolution. There is a fundamental limit to the voltage resolution of a DSO scope set by the 8-bit or 12-bit ADC. If the scale is set too coarse, the features of the signal may be limited by the ADC resolution. You will get a number from an analysis function, but it could be measuring just digitizing noise. Always expand the scale to view the signal with the highest resolution suitable for your application.

16. Do not adjust the scales on saved data.

After a buffer of measurements is taken and is displayed, changing the scales will just change the display of the measurements. It will still be limited by the digitizing noise of the ADC in both the voltage and time scales. Always adjust the scales *before* you save the measurement so that the scope dynamically changes the amplifier and acquisition settings to give the highest resolution.

17. Getting lost in the scale settings. Sometimes, as you arbitrarily adjust the vertical and horizontal scales and offset, the displayed measurements will end up with very awkward scale values. They may be so far off that it is difficult to adjust them to reasonable values. An example is shown in **Figure 7.47**.

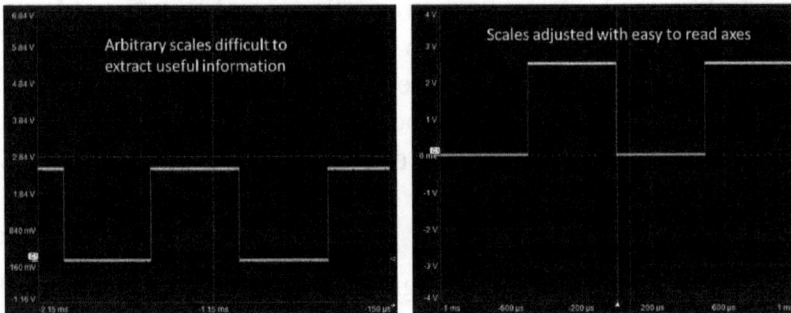

Figure 7.47 An example of the same comp signal on scales that make it very difficult to read its figures of merit directly off the front screen and then adjusted in the two-step operation: centering the 0 V and t = 0 and adjusting the scale settings in multiples of 1x, 2x, and 5x.

To bring the scales into reasonable values, press the offset knobs to place V = 0 and t = 0 in the center. Then, adjust the coarse voltage scale and time scale to display the signal on whole-number scales and read the values directly off the front screen. Any additional offsets can be applied to shift the displayed signal's focus, trying to keep the scale values as easy-to-read whole numbers or decimals.

18. Watch out for aliasing of measured data.

Be aware of the bandwidth of your measured signal, the bandwidth of the scope and the sample rate of your ADC for each measurement. If the sample rate is at least 2.5x higher than the bandwidth of the scope, you should not have aliasing artifacts.

If the sample rate of the scope's ADC is reduced from its maximum value, make sure the bandwidth of the signal measured is lower than half the sample rate. This can sometimes be accomplished by selecting a bandwidth-limiting filter that is either added to the DUT or selected in the scope's input.

7.14 The Bottom Line

1. Practice using a scope with the free sound card scope.

2. The vertical controls adjust the voltage scale factor and the offset.

3. The horizontal controls adjust the time per division where the t = 0 position is.

4. The t = 0 position is when the trigger event occurred.

5. Always adjust the vertical scale so most of the screen is used to display your signal.

6. Always adjust the scales to make the signal value easy to interpret immediately from the front screen. This means whole numbers for the scales.

7. In normal trigger mode, the scope will only trigger the display when it receives a valid trigger event, such as a rising signal passing through a threshold level.

8. In Auto mode of triggering, the scope will trigger on a trigger event normally. But, if it does not see a valid trigger event in about 100 msec, the scope will trigger anyway. This way you are guaranteed to see the signal into the scope.

9. Always start a new measurement on auto trigger.

10. Never use auto setup. Why would you think the scope is smarter than you are?

11. A scope emulator is a powerful tool for practicing using a pro-level scope interface.

12. Most large-scope manufacturers, such as Keysight, Tektronix, LeCroy, and Rohde and Schwarz, offer a software emulator of their user interface. Each vendor offers different features in their tool.

13. The MAUI Studio interface from Teledyne LeCroy can emulate a live computer screen.

14. The emulator can be used to practice setting the vertical controls, both the vertical scale and offset.

15. The emulator can be used to practice setting the horizontal controls, both the center position and the time per division.

16. Multiple traces can be saved into memory and then displayed to compare with live scope traces.

17. All scopes enable exporting a CSV version of a measured trace.

18. A measured scope trace can be imported into any spreadsheet and plotted or analyzed using the power of the spreadsheet.

19. Many free SPICE simulators can be used to import scope measurements and use them to drive an ideal voltage source.

20. Many emulator tools also allow importing measured scope traces to view in the emulator.

7.15 Some Experiments to Try

1. Find the three control regions of the scope.

2. Change the trigger level to trigger on a large signal, then adjust to normal and see the difference.

3. Trigger the scope only on the snap of a finger or a clap.

4. How long does a clap last? A snap of the finger?

5. Increase the time base until the sample rate decreases.

6. How many points are in the acquisition buffer compared to how many are displayed?

7. Output a sine, square, or triangular signal on the speaker. What is the difference in how it looks on the scope.

8. Change the sine wave frequency and look at the scope waveform.

9. What is the DC component of every signal?

10. Display the FFT response of a measured waveform.

11. Drive your speaker with a sine wave and look at the FFT spectral response.

12. Change the waveform to be a swept sine wave with a 20 second sweep time and watch the peak of the spectral response move across the spectral range.

Chapter 8
The 10x Probe

A probe interface between the DUT and the scope has five features that characterize it:

- ✓ The geometry at the tip
- ✓ The load impedance the DUT sees looking into the probe
- ✓ The DC signal attenuation
- ✓ The bandwidth
- ✓ No susceptibility to RF pickup or cross-talk

The ideal probe connection between the DUT and the scope has the following requirements:

- ✓ Transforms the geometry of the DUT to the coax geometry of the scope
- ✓ Has infinite impedance to the DUT
- ✓ Has an attenuation of 1x at DC
- ✓ Has infinite bandwidth
- ✓ Has no susceptibility to EM pickup

Of course, no real probe meets these conditions, but we try to engineer the probe interface between the DUT and the scope to come as close as practical to these requirements at an affordable price.

Of the various low-cost options, the 10x probe is a good approximation to an ideal probe for signal bandwidths less than 500

MHz and DUT output impedances below 200 ohms. In addition, it is suitable for input voltage levels as high as 400 V.

Unless you have a compelling reason otherwise, the 10x probe should be your go-to probe for general scope measurements. But there are a few limitations to be aware of. By understanding how a 10x probe works, you will get the most out of using a 10x probe and avoid some common artifacts.

For bandwidths > 500 MHz, a transmission line probe or an active probe is an alternative to consider. These are described in later chapters.

8.1   Why It Is Called a 10x Probe

The 10x probe has three parts:

- ✓ The tip
- ✓ The special coax cable
- ✓ The base with the BNC connection to the scope

Each part of the probe plays a special role.

An example of a 10x probe with these three parts is shown in **Figure 8.1**.

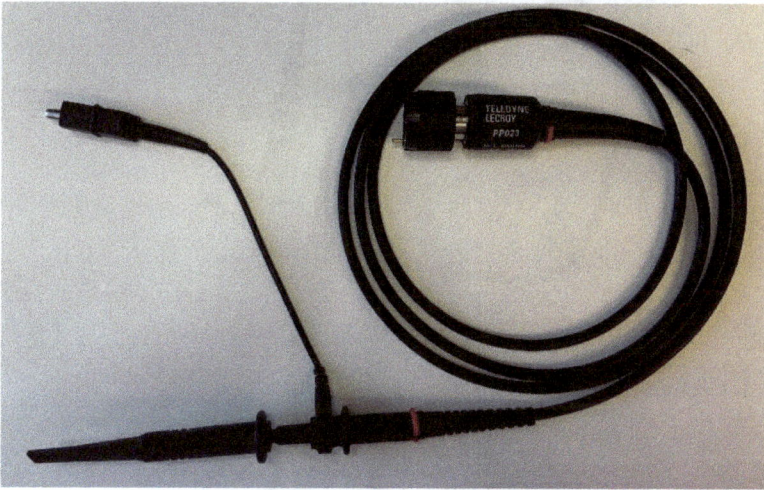

Figure 8.1 An example of a typical 10x probe showing its three parts: the tip, the cable, and the base that plugs into the scope.

There are three important differences among 10x probes:

- ✓ Their bandwidth
- ✓ The tip diameter
- ✓ The option for 10x or 1x in the same probe

Three probes with these differences are shown in **Figure 8.2**.

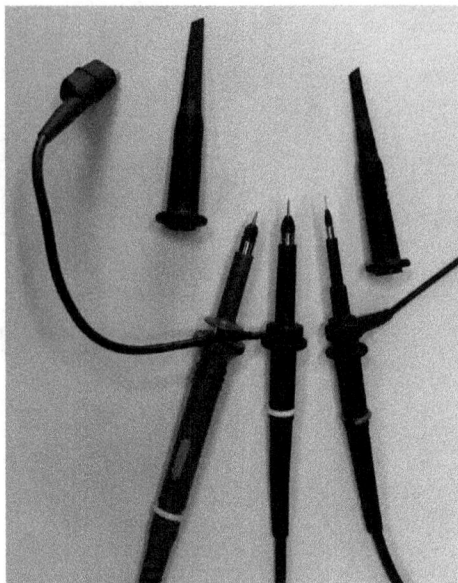

Figure 8.2 Three different 10x probes. The left two have a 5 mm diameter tip. The right is 2.5 mm. The far left has a 1x option. The caps or "hats" have been removed to show the tip diameter.

Generally, the wide tips with 5 mm diameter are more robust and are used predominantly in Europe and Asia. The narrow diameter tips with a 2.5 mm diameter are used in North America. The highest bandwidth probes have tip diameters of 2.5 mm.

A probe with the option for 1x or 10x generally has a maximum bandwidth of 200 MHz, while higher bandwidth probes do not have this option.

The name of the 10x probe is confusing. It does not provide a gain of 10x but an attenuation of 10x of the tip voltage as measured by the scope. This is due to a 9 megaohm resistor built into the tip of the probe. The operation of this probe assumes the scope will be set on 1 meg input coupling. If it is not, the interpretation of the 10x probe measurement will be way off, and its performance will be terrible.

A first-order model of the equivalent circuit of the probe tip and the scope input is shown in **Figure 8.3**. A DC voltage appearing at the tip will be measured at the scope as one-tenth the tip voltage due to the voltage divider circuit.

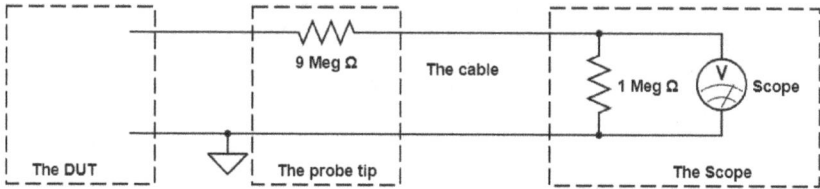

Figure 8.3 A simplified, first-order model of the 10x probe illustrates why it is a voltage divider.

This first-order model also illustrates that the input impedance looking into the probe at DC, will be 10 megaohms. This means that the DUT will have minimal DC loading when the 10x probe is connected to it.

If this were the only feature of a 10x probe, there would be a significant problem measuring higher bandwidth signals. The coax cable connecting the tip to the scope has some capacitance in it, on the order of 50-100 pF. In addition, there is also some input capacitance of the scope amplifier channel. This is on the order of 17 pF in most scopes.

If the input signal to the 10x probe tip is a step edge, the step response of the probe would be a 1-pole response with an RC = 1 Meg x 100 pF = 0.01 msec. This is a low pass pole frequency of

$$f_{pole} = \frac{1}{2\pi \times RC} = \frac{1}{2\pi \times 1 \times 10^6 \times 10^{-10}} = 1.6\ kHz$$

This estimate is very close to the simulated pole frequency of the circuit, where the transfer function drops by -3 dB from the passband at about 1 kHz. **Figure 8.4** shows an example of the equivalent circuit and simulated transfer function of the circuit of a 10x probe if there were just a 9 Meg input resistor along with the cable capacitance.

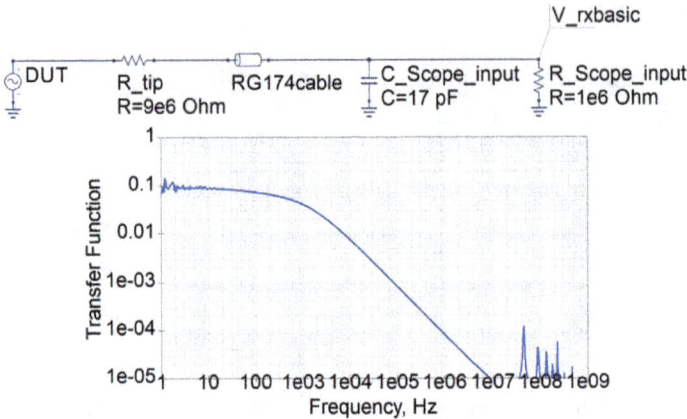

Figure 8.4 An example of the circuit and simulated transfer function of a 10x probe if there were only a 9 meg input resistor at the tip.

We see the passband response is a 10:1 voltage divider. The low pass frequency response of the probe begins to drop off at about 1 kHz. This is a terrible performance and completely unacceptable as a scope probe. If this were how the 10x probe behaved, it would be worthless as a probe. It would limit measurement bandwidths to less than 1 kHz.

There are two important features built into the 10x probe, which enhance its high-frequency properties. The first is a high-pass filter in parallel with the low-pass filter. The second important feature is a very special cable between the tip and the base, which eliminates high-frequency transmission line reflections.

8.2   Second-Order Model of a 10x Probe

The 9 megaohm resistor and input capacitance of the cable and scope create a low-pass filter. The way to fix this very low-frequency roll-off is to add a high-pass filter in parallel. This will "equalize" the probe response across the frequency range. This is implemented with a shunt capacitor across the 9 M resistor.

When this is added, the high-frequency response is now brought up with the high pass filter. In principle, to match the 10:1 divider of the resistors and create a broadband flat response, the goal is to make the shunt capacitor one-ninth the capacitance of the cable and scope input.

A 9 pF capacitor shunted across the 9 megaohm resistor acts as a parallel high-pass filter, which equalizes the channel and gives a higher bandwidth to the probe. This second-order equivalent circuit is shown **Figure 8.5**.

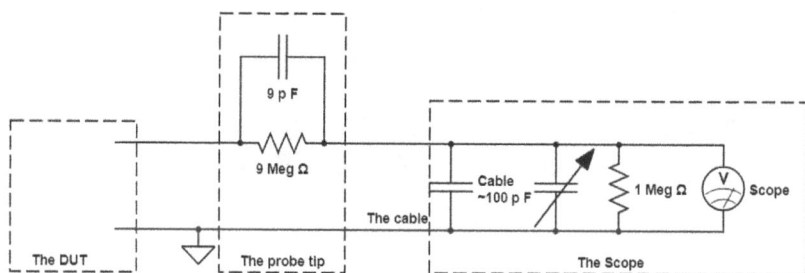

Figure 8.5 The second-order model of the 10x scope probe shows the equalizing high pass capacitor at the tip and the cable and input capacitance in the scope.

For a flat frequency response, the 9 pF capacitor needs to be matched to about one-ninth the capacitance of the rest of the system. In some 10x probes, the tip shunt capacitor is adjustable. In most 10x probes, there is an additional adjustable capacitor in the base part of the probe that plugs into the scope. This capacitor value is tuned to give a flat response of the 10x probe, balancing the low pass filter with the high pass filter. Adjusting this capacitor is called compensating the 10x probe.

8.3   Probe Loading and Probe Input Impedance

At DC, the 10x probe looks like a 10 megaohm resistor. This means it will have minimal loading to any DUT if the DUT output impedance is less than about 100 kOhms.

However, the second-order model points out that the 10x probe does not look like a 10 Megaohm resistor when looking into the probe at frequencies above about 1 kHz.

Above about 1 kHz, the probe's input impedance looks like a 9 pF capacitor. This is important to know when we apply situational awareness and analyze the impact the loading of the probe has on the DUT. **Figure 8.6** shows the simulated input impedance of the 10x probe using a shunt capacitor value of 9.5 pF.

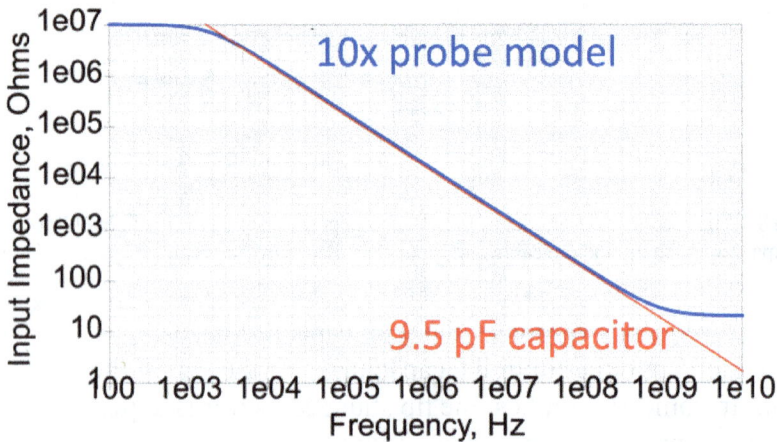

Figure 8.6 The simulated input impedance of the 10x probe compared with a 9.5 pF capacitor.

Knowing the equivalent circuit model of the probes means we can evaluate when the 10x probe will load the DUT and potentially create a measurement artifact. The impact of a roughly 10 pF capacitive load on a DUT will be most apparent as the output impedance of the DUT increases.

For example, when the DUT has an output impedance of 1 kOhm, the probe's 10 pF of capacitance creates an RC with a time constant of about 10 nsec. If the rise time of the DUT signal is less than 10 nsec, the measured rise time will be limited by the RC artifact. The 10x probe's input impedance will load the DUT and distort the measured signal.

One way of evaluating the impact of the probe loading and whether this will affect the signal from the DUT is to measure the signal from the DUT with one probe attached and then add a second probe. If the probe does not load the DUT, the signal measured with one probe or two probes connected will be the same.

But, if the second probe further distorts the signal, this is an indication the probe is loading the DUT and a measurement artifact is introduced. **Figure 8.7** shows an example of a fast signal from a 1 kOhm output impedance DUT measured first with one probe and then the measurement with this same probe when a second probe is attached. The fact that the second probe changes the rise time is an indication the 10x probe is loading the DUT signal and a measurement artifact is introduced.

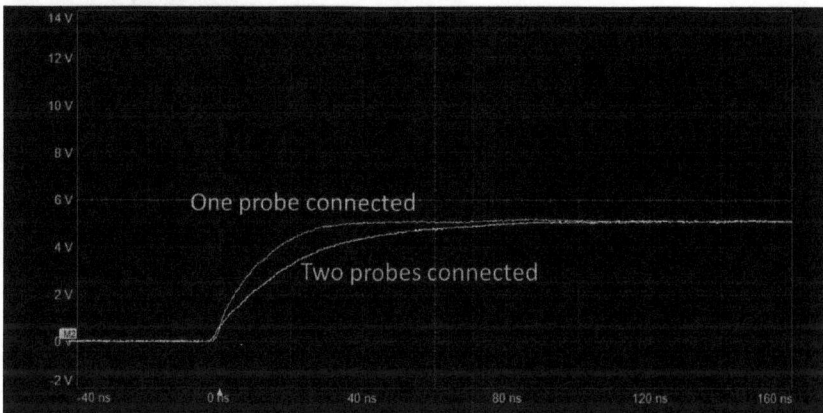

Figure 8.7 Examples of measuring the loading of the DUT signal when a second probe is connected. This indicates a measurement artifact from the probe distorting the signal.

8.4 Compensating a 10x Probe

The compensation capacitor has to be adjusted for a specific probe and scope to balance the high pass pole frequency with the low pass pole frequency. This gives the 10x probe a flat response across frequency.

The process to compensate a 10x probe is to measure a square wave signal and adjust the probe's compensation capacitor until the displayed square wave looks like an ideal square wave. This means the low frequency and high frequency are matched and the overall response is flat.

Every scope is equipped with terminals in the front to which a 10x probe can be connected. The signal on the terminals is a square wave at 1 kHz and about 1-3 V amplitude. While this signal is measured, the internal variable capacitor inside the 10x probe is adjusted or compensated. **Figure 8.8** shows the tip of a 10x probe and the probe base with a special screwdriver adjusting the internal capacitor of the 10x probe.

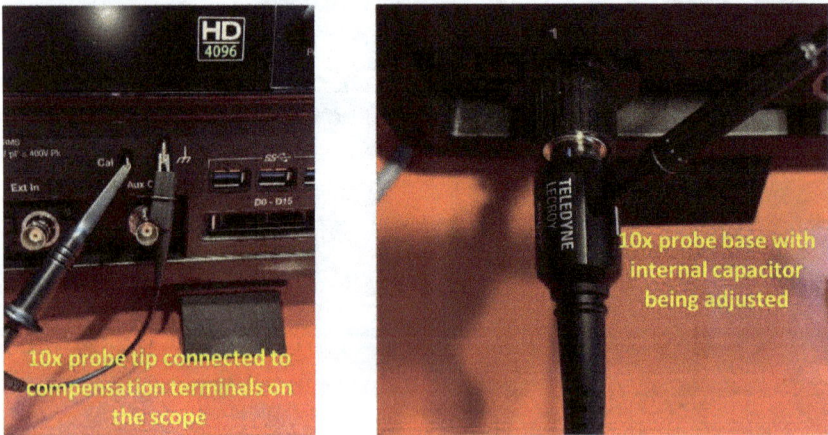

Figure 8.8 A 10x probe connected to the compensation terminals of a scope and the compensation capacitor being tuned with a screwdriver.

The square wave on the screen is measured as the capacitor is adjusted. The probe is adjusted when the measured square wave looks like an ideal square wave. **Figure 8.9** shows what the frequency response of the 10x probe would be and the measured time-domain response of the square wave when the capacitor is adjusted too far off and when it is just right.

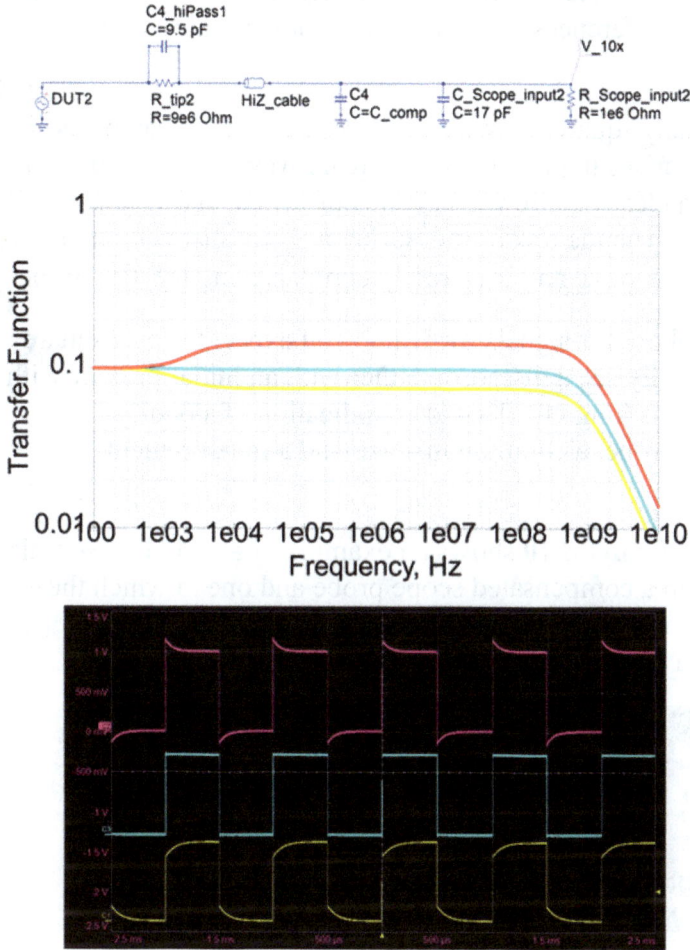

Figure 8.9 Top: the simulated frequency domain response of the probe under the three compensation settings. Bottom: the measured square wave with different compensation settings corresponding to the transfer function color code.

Once the 10x probe's capacitor is adjusted to compensate the response of the probe, it generally does not need any later adjustment. A scope probe only needs compensation once, but when in doubt, especially if others use your 10x probe, always recheck the compensation signal.

As a best practice, only use the compensation reference signal on the front of a scope to compensate the signal. This is because these signals are designed to be very good quality square waves, which can be used as references to optimize the compensation of a 10x probe.

In principle, any square wave signal at roughly 1 kHz can be used. The problem arises in practice if the square wave source you use is not actually a clean square wave. If it is distorted, and you don't know this, you run the risk of measuring a signal that is really a distorted square wave and incorrectly compensating the 10x probe.

Be on the lookout for signals for which you expect to see a square wave but actually see a distortion either with an initial peak or with a fast edge and a long tail. This is an indication of a poorly compensated probe, a common mistake and a measurement artifact.

For example, **Figure 8.10** shows an example of a digital pulse train measured with a compensated scope probe and one in which the probe was poorly compensated. This sort of distortion is a signature of a poorly compensated scope probe.

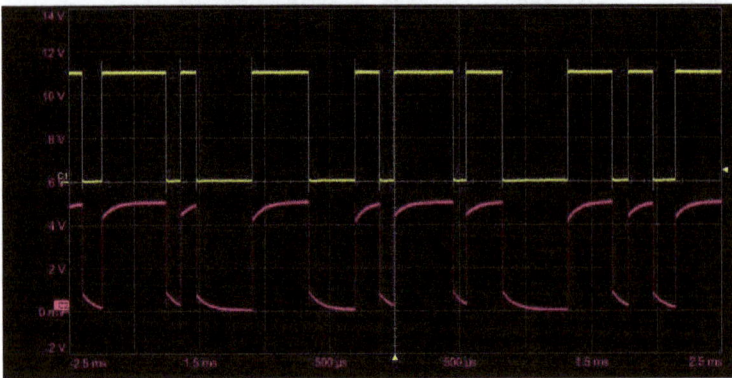

Figure 8.10 Measured digital signals using a scope probe well compensated (top) and a scope probe that was not correctly compensated (bottom). This sort of distorted signal is an indication of a poorly compensated scope probe.

8.5   A Special Coax Cable

A 10x probe has three features to enhance its high bandwidth performance:

1.   Provide low input capacitance at the probe tip so it does not load down the DUT.

2.   Increase the characteristic impedance of the cable to reduce its capacitance and reduce the impedance mismatch with the high impedance at the ends.

3.   Add distributed high series resistance to damp out the reflections from the cable.

The three goals are accomplished with a very special coax cable. It is unlike any other cable. **Figure 8.11** shows a cross section of the 10x probe's cable, which reveals some of its secrets.

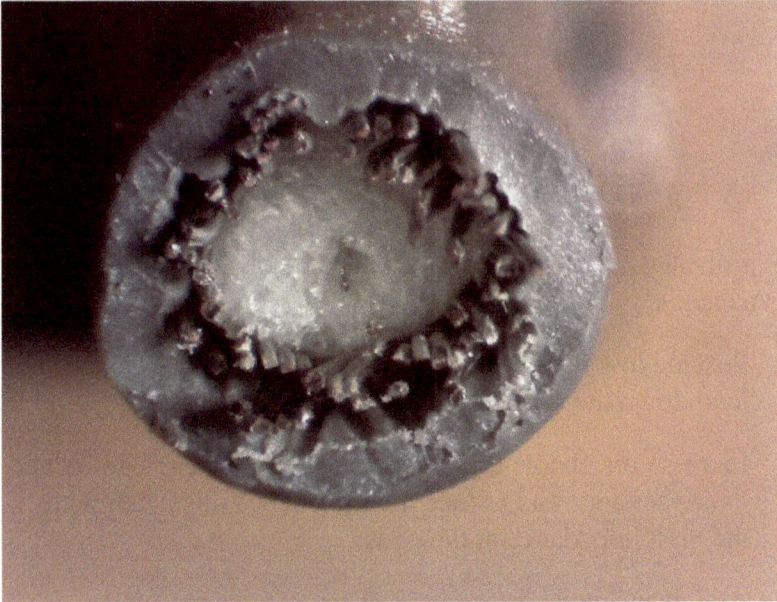

Figure 8.11 A cross section of the special coax cable used in 10x probes.

The dielectric is foamed polyethylene. This means its dielectric constant, Dk, is about 1.3. The ratio of the outer conductor to inner conductor is about 20:1. These two features dramatically decrease the capacitance of the cable and increase its characteristic impedance. The cable's characteristic impedance is on the order of 200 ohms.

The center conductor is not copper, but nichrome, a very resistive metal, with a bulk resistivity almost 40x that of copper. The center conductor's outer diameter is about 4 mils. This combination of narrow conductor and high resistivity means the series DC resistance is about 200 ohms/m. At frequencies above 50 MHz, the series resistance increases even more due to skin depth effects. This effectively damps out the cable reflections with distributed series resistance.

If you sacrifice a 10x probe and cut its ends, you can measure the 200-500 ohm DC series resistance of the center conductor. This distributed series resistance damps out the reflections that would bounce between the ends of the coax cable. This boosts the effective bandwidth of a 10x probe above 200 MHz.

This idea of using a distributed lossy transmission line in a 10x probe to damp out reflections and dramatically increase the useful bandwidth of a 10x probe was patented in 1959 by John Kobbe and William Politis while at Tektronix. Without this innovation, the reflections in the 10x probe cable would limit the bandwidth of the probe to less than 50 MHz.

8.6   Never Use a 10x Probe Other Than from the DUT to the Scope

A 10x scope probe is not the same as a coax cable with convenient connectors on the ends. The internal structure of a 10x probe makes it suitable *only* for use between a DUT and the scope.

If you use it to connect between a function generator, for example, and a DUT, there will be no signal left at the tip. The 9 meg series resistor built inside the 10x probe will dramatically attenuate any DC signal from the source. The 10 pF shunt capacitor built into the 10x probe will act as a high-pass filter and let through the high-frequency, edge part of the signal.

Three cases demonstrate the dangers of introducing significant measurement artifacts if the 10x probe is used as a general-purpose cable from a source to a DUT. In each case, a function generator with a 50 ohm source Thevenin resistance is the signal source, driving a 1 V amplitude, 1 kHz square wave.

Case 1: A best practice: a coax cable connects the function generator to the scope directly. In this configuration, the scope measures the Thevenin voltage of the function generator. **Figure 8.12** shows the equivalent circuit model, the connection to the function generator, and the measured waveform at the scope. The scope measures the 1 V amplitude of the Thevenin source inside the function generator.

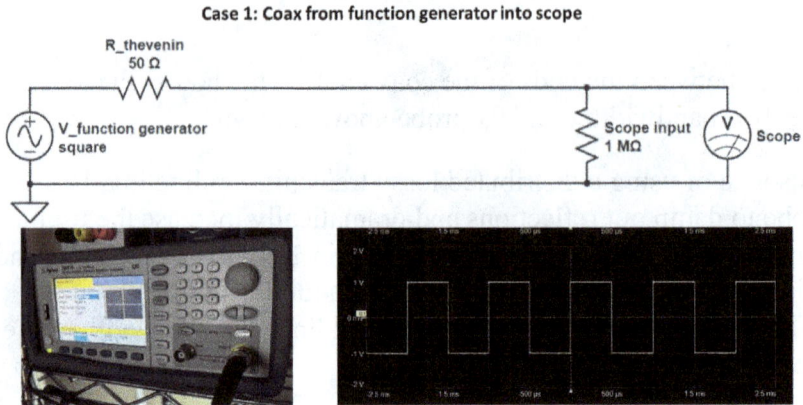

Figure 8.12 Case 1: direct coax connection between the function generator and the scope.

Case 2: A 1k ohm resistor is added as the DUT. A coax cable connects the function generator to the DUT and a 10x probe connects the DUT to the scope. This is the recommended best practice configuration to connect the DUT signal across the 1k ohm resistor to the scope.

In this case, a voltage divider is created by the Thevenin resistance of the function generator and the DUT. The voltage across the DUT, as measured by the scope should be:

$$V_{meas} = 1 \text{ V} \frac{1k}{1k + 0.05k} = 0.95 \text{ V}$$

The DUT will load the function generator's signal slightly resulting in a voltage across the DUT slightly less than the Thevenin voltage inside the function generator. This is precisely what is measured by the scope. This configuration is shown in **Figure 8.13**.

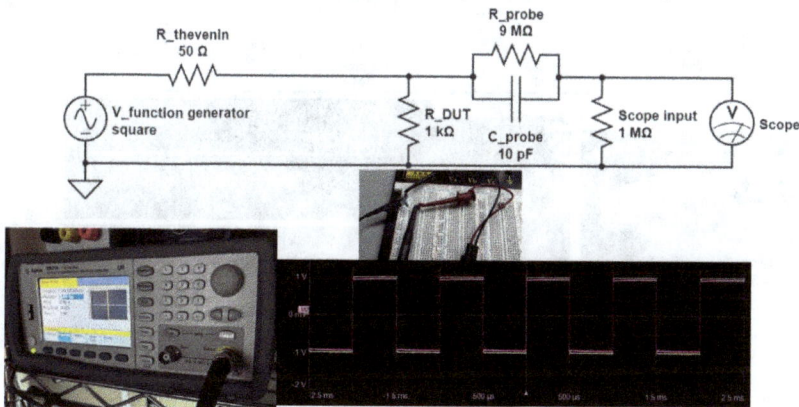

Case 2: Coax from function generator to DUT, 10x probe from DUT to scope

Figure 8.13 Case 2: The function generator connects to the DUT with a coax cable and from the DUT to the scope with a 10x probe. The scope display is the signal directly measured by the scope in case 1 and when measured in case 2.

Case 3: The pathological configuration. A 10x probe is connected between the function generator and the DUT. This means a 9 megaohm resistor is now in series between the function generator and the DUT. At DC, this is a voltage divider that drops the voltage across the DUT to be about 1 kOhm/9 megaohm = 0.01% of the Thevenin voltage from the function generator. This is about 0.1 mV amplitude, below the noise floor of the scope.

In addition, there is a shunt capacitor in series between the function generator and the DUT. This will act as a high pass filter to the DUT. This means any signal with a rise time longer than about 10 nsec will not be measured by the scope. In the square wave output from the function generator, only the edges will pass through to the scope. **Figure 8.14** shows this configuration.

Case 3: 10x probe from function generator to DUT, 10x probe from DUT to scope

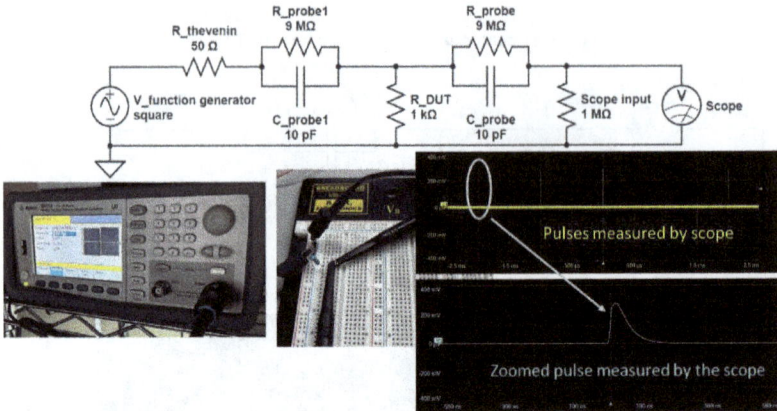

Figure 8.14 Case 3: Not recommended; a 10x probe connects between the function generator and the DUT. A 10x probe connects between the DUT and the scope.

If a 10x probe is used between any signal source and a DUT, the signal applied to the DUT will not be what is expected. It will introduce a significant artifact.

8.7   Rarely Use a 10x Probe on the 1x Setting

The internal structure of a 10x probe also points out why there are severe performance limitations when using a 10x probe on the 1x setting. On this setting, the parallel 9 meg resistance and 9 pF capacitor are shorted with a switch. This connects the tip directly to the 1 meg input of the scope through the very special cable.

To first order, it would seem that if the 9 meg resistor were shunted, the signal from the DUT would just see the coax cable and the 1 meg input to the scope. This would be the case if the coax cable in the 10x probe were a normal RG174 cable. But it is not.

The cable is really a high series resistance driving the input capacitance of the cable. This series resistance is on the order of 200 ohms and the cable and input capacitance to the scope is about 100 pF, the RC time constant of this equivalent circuit model would be 200 ohms x 100 pF = 20 nsec. The 10-90 rise time would be 2.2 x 20 nsec = 44 nsec. The pole frequency is 10 MHz. This means the bandwidth of a 1x probe is only about 10 MHz.

On the 1x setting, the 10x probe has a very low bandwidth. **Figure 8.15** shows the measured frequency response of a 10x probe set as 10x and on the 1x scale.

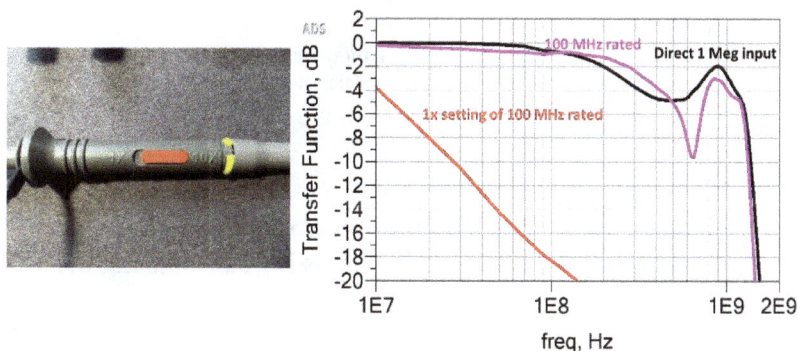

Figure 8.15 The measured frequency response of the scope-probe combination. The bandwidth of the 100 MHz rated 10x probe is actually 250 MHz. When on the 1x probe setting it drops below 10 MHz.

The impact of this lower bandwidth means that when you measure a fast edge signal, its rise time will be slowed down to the intrinsic step response rise time of the cable's RC response. This is expected to be a 10-90 rise time of about 44 nsec. **Figure 8.16** shows the measured response of a fast-rising signal with a measured 10-90 rise time of 9 nsec, measured as 36 nsec with the scope probe set for 1x. This is close to the estimated value of about 44 nsec.

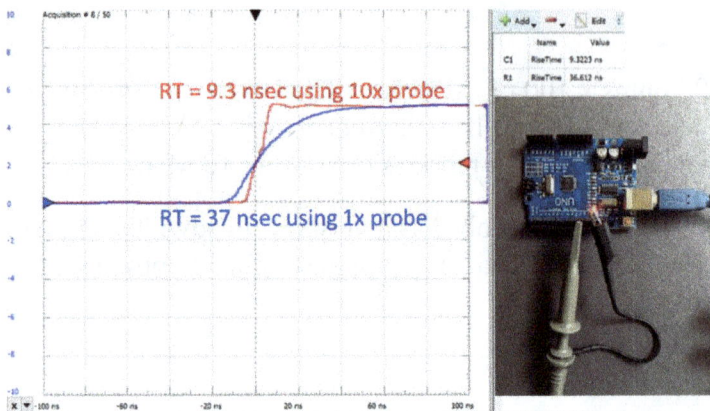

Figure 8.16 Measured rise time of a signal from a microcontroller board (shown in the inset) measured by the 10x probe and the 1x probe. The longer rise time is an artifact of the 1x probe.

If the signal rise times you expect to measure are 100 nsec or longer, then using the 1x probe will not introduce an artifact. However, you must always be aware of this limitation using the 1x setting of the probe.

Unless you have a strong compelling reason, do not use the 1x setting. If you do, be aware of the dramatically lower 10 MHz bandwidth of the 1x probe.

8.8   The Tip Voltage and the Displayed Voltage

The 10x probe attenuates the signal between the tip and the connection into the scope. It is sometimes confusing keeping track of what is actually displayed on the scope versus what is at the tip of the probe on the DUT. As a user, you always want to see the tip voltage, the voltage that appears on the DUT, as the voltage displayed on the scope screen. When the probe attenuates the signal, the scope display has to be scaled up by a factor of 10 to display the tip voltage.

This means that when the scope is adjusted accordingly, the voltage as measured by the scope is internally multiplied by 10x to display the tip voltage on the screen. This scale factor can be entered manually, or some scopes will read the probe and adjust this automatically.

Many scopes are designed to automatically identify when a 10x probe is plugged in. On the scope, there is a ring around the coax connection to the channel. On higher-performance 10x probes, there is a pin sticking out of the probe on the end that plugs into the scope. These features are shown in **Figure 8.17**.

Figure 8.17 Close-ups of the scope connection and the probe connection show the features for the scope to sense that a 10x probe is plugged in.

In the probe base, between the pin and the shield of the coax, is a 10k ohm resistor. The scope measures the resistance between the outer sense ring and the ground shield. Normally, it is an open resistance. When a 10x probe is plugged in, this resistance drops to 10k ohms. The scope sees this and knows a 10x probe is plugged in. It will then automatically scale the measured signal by a factor of 10 to display the tip voltage.

In the channel setup you can confirm that the scope is aware of a 10x probe and has adjusted the scale. With a coax cable plugged in, the attenuation factor is set by default to 1. With a 10x probe plugged in, this indicator will say Probe and identify a divide by 10 factor. This is shown in **Figure 8.18**.

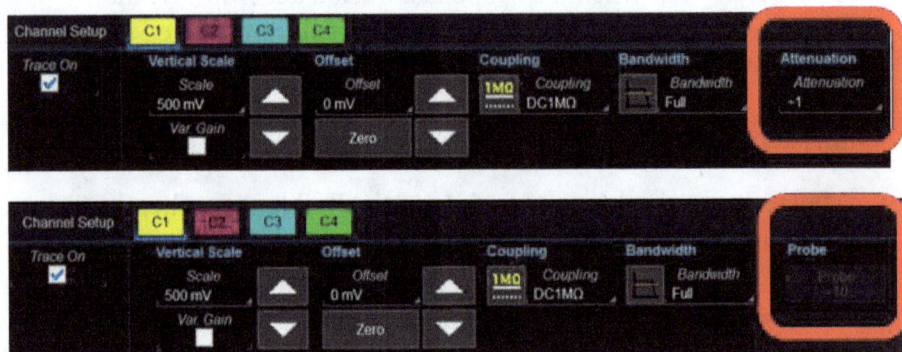

Figure 8.18 In the channel setup look for the identification of the probe and divide by 10 to confirm the scope is scaling the input voltage to display the tip voltage.

This smart feature of a scope-probe combination means that we are always displaying the tip voltage on the scope, which is what we want to see. If your scope does not set this automatically, or there is no pin on your 10x probe, be sure to adjust the probe attenuation setting manually so that the scope always displays the tip voltage. Forgetting to do this is a common problem.

A common mistake in misreading a DUT voltage is to forget to scale the scope voltage to the tip voltage if your scope does not do this automatically. Get in the habit of checking the probe scaling factor when you set up to use a scope. It is usually displayed somewhere on the screen. **Figure 8.19** shows a display on a Keysight DSOX3024 scope with the identification of the probe attenuation setting on the four channels. Only two of the channels are set for a 10x probe.

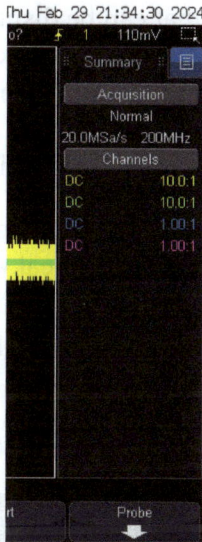

Figure 8.19 Display on this scope showing the 10x probe manually set for channels 1 and 2.

8.9   The 10x Probe and Noise

The 10x attenuation of the probe has two important consequences. Most scopes are capable of measuring an input voltage range of ±40 V. With the 10x probe this is still the scope's input voltage range, but the tip voltage range will be a factor of 10 higher. With a 10x probe, a scope can measure a tip voltage range of ±400 V. This dramatically increases the applications of a scope.

Always be aware that voltages above 40 V can be dangerous and take precautions when performing higher voltage measurements.

The second consequence of the 10x attenuation is at the low voltage end. The highest sensitivity scale of a scope is typically 1 mV/div. At the full bandwidth of the scope, the peak-to-peak amplifier noise of a scope is about 1 mV.

With a 10x probe plugged in, a measured voltage of 1 mV at the scope corresponds to a tip voltage of 10 mV. This means the most sensitive tip voltage scale that can be displayed using a 10x probe is 10 mV/div. On this scale, the scope is actually on the 1 mV/div scale. The internal amplifier random noise of roughly 1 mV peak-to-peak will be scaled to an equivalent tip voltage value of 10 mV peak-to-peak with the 10x probe plugged in.

If the tip voltage is limited by the scope amplifier noise, it will look like a 10x probe, with a 10 mV peak-to-peak noise displayed on the front screen, is noisier than a 1x probe. An example of this is shown in **Figure 8.20**. The 10x probe tip is shorted to the ground lead and displayed on channel 1 as the tip voltage and channel 2 is just left open showing the amplifier noise. They are displayed on the same scale of 10 mV/div.

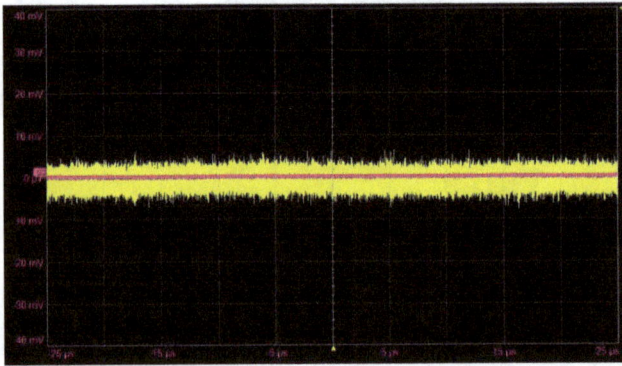

Figure 8.20 Comparing the probe tip voltage and channel noise on the same scales. It looks like the 10x probe is noisier, but this is a display artifact due to the scope scale of 1 mV/div being scaled in the display to the tip voltage of 10 mV.

The 10x probe, displaying a tip voltage scale of 10 mV/div, really has the scope scale set on 1 mV/div. But the measurement in channel 2 with no probe connected is with the scope set for 10 mV/div, showing the real noise in the scope of 1 mV peak-to-peak.

The practical impact of this artifact is that when measuring small voltage levels, the tip voltage will be limited by a noise floor of about 10 mV peak to peak. This means that if your application requires measuring voltage changes of 10 mV or less, your signal-to-noise ratio (SNR) will be less than 0 dB using a 10x probe. A 10x probe is not designed for low-voltage measurements.

When you are measuring voltage changes less than 10 mV, consider using another probing method that does not attenuate the signal. While it is tempting to use the probe setting on the 1x, which will enable you to measure 1 mV level signals, remember that this will reduce the bandwidth of your measurement to less than 10 MHz.

8.10 The 10x Probe Only Measures Single-Ended Voltages

Any voltage measurement is always a voltage difference between two points. All scopes with a coax connection in the front of the scope always measure a voltage between the center conductor and the outer shield of the coax. In almost every situation, the shield of the coax connector in the front of the scope is connected to the chassis of the scope, referred to as chassis ground. The chassis is connected to earth-ground through the third wire in the power plug. This means that in a scope, chassis-ground is connected to earth-ground.

In addition, the chassis-ground is connected to the circuit ground of the scope inside the scope. This means the circuit ground of the amplifier circuit inside the scope is connected to chassis-ground and to earth-ground.

Any measurement by the scope is a measurement of the voltage between the signal at the tip of the probe and the circuit ground connection at the tip of the probe, which is connected to the circuit ground of the scope through the shield of the coax cable. When a voltage measurement is referenced to circuit ground as the second point, we refer to this as a *single-ended voltage* measurement.

A voltage measurement with a 10x probe is inherently a single-ended measurement. Never connect the ground clip end of the 10x probe to any part of a circuit that you do not want connected to earth-ground. The mere act of hooking a 10x probe into a circuit will connect your DUT's circuit ground to earth-ground through the chassis of your scope.

A 10x probe cannot be used for differential measurements. In principle, using a 10x probe tip to measure the voltage between any two points in a circuit may be possible if the circuit being measured is completely floating from earth-ground. However, your 10x probe will connect the circuit under test to earth-ground. Problems can arise when your circuit is not really floating and currents of more than 100 mA flow through the 10x scope probe's cable shield to earth-ground through the scope chassis.

If you need to make a differential measurement, such as when measuring the voltage across a sense resistor, either measure the single-ended voltage across each end of the resistor relative to local circuit ground with two scope probes and subtract them with a math function in the scope, or use a true differential probe. Both of these methods are described in detail in a later chapter.

A common mistake is to not be aware of your circuit's grounding connections and mistakenly connect the scope probe's ground to a node at a higher voltage above earth-ground. At the least, this will introduce a misleading measurement. At worst, it will damage your circuit, damage your probe, or damage your scope. This is why it is so important to always be aware of the grounding connections in your DUT and your instruments. That is why this topic is covered in detail in the next chapter.

8.11   Use Color Bands to Reduce Confusion with Multiple Probes

When one probe is used to connect to your DUT and the scope, there is no ambiguity about which channel is connected to which node in your DUT. As soon as you add two 10x probes, there is the chance of confusion and misinterpreting a measurement. This is a surprisingly common error.

The way to avoid this error is to use color-coded bands on your 10x probe. All scope probes are shipped with a collection of different color bands that slip on the tip end and the base end of the 10x probe. This is shown in **Figure 8.21**.

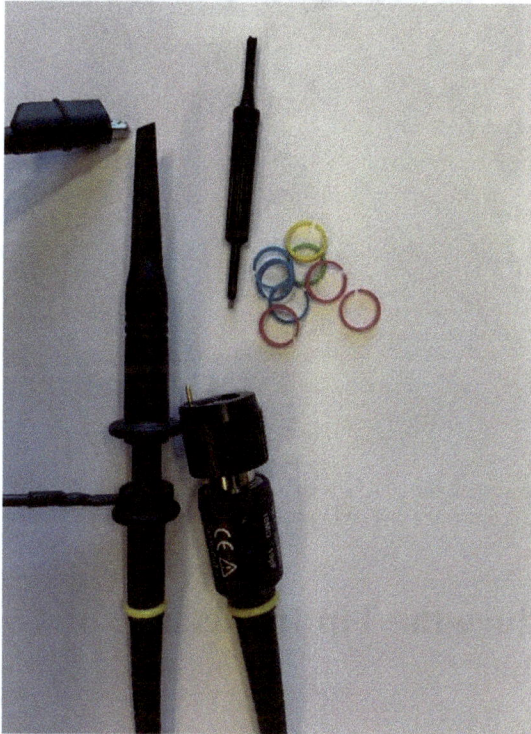

Figure 8.21 Example of the color bands that are shipped with every 10x probe. Use them to reduce confusion in a scope measurement with multiple probes.

It is even more convenient if you connect a scope probe with a yellow band into the channel that has a yellow trace on the screen, or a red band to the 10x probe that is connected to the scope channel with a red trace on the screen. When using four probes, for example, this is essential to avoid confusion. **Figure 8.22** shows an example of four scope probes measuring a DUT with color bands that match the channel trace colors.

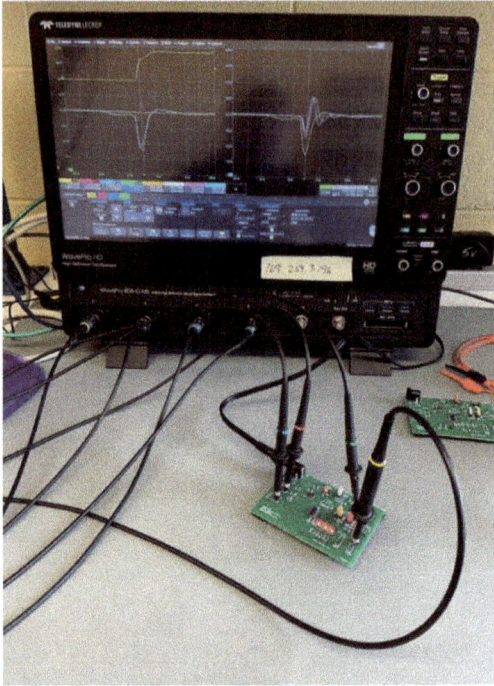

Figure 8.22 A test system for a DUT with four 10x probes each with a unique color band corresponding the scope trace displaying the voltage on that probe.

8.12 Reduce the Tip Inductance of Your 10x Probe

A 10x probe is rated to a specific bandwidth. This varies from 60 MHz to more than 700 MHz for some scope probes. Generally, the bandwidth of a scope probe means the highest frequency at which a sine wave will be transmitted through the probe and measured by the scope and be at least 70% the amplitude of the sine wave at the tip. This assumes a flat frequency response.

This rated bandwidth is only achieved using a coaxial connector adaptor on the tip to the DUT and measured by a scope of higher bandwidth than the probe. This coax adapter is generally shipped with every 10x probe, along with a ground spring tip. **Figure 8.23** shows the coax adaptor that connects the 10x probe tip to a BNC connector. This configuration provides the highest bandwidth connection and the shortest rise-time signal.

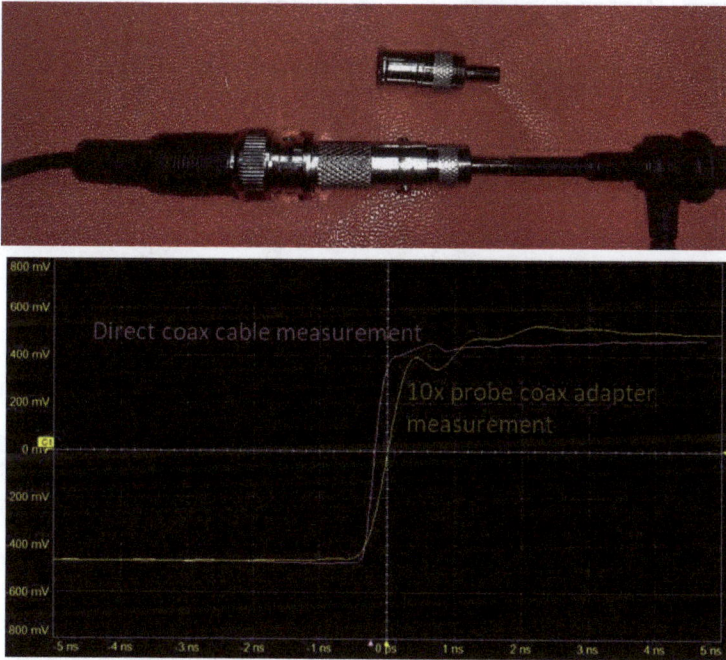

Figure 8.23 The ultimate bandwidth of this 10x probe rated for 500 MHz is achieved with a coax adapter. Even then, the best-case rise time is about 0.7 sec in this example. The 10x probe with a coax adapter is compared with a direct coax cable connection to the fast-rise time source.

Any other connection to the tip of the 10x probe will reduce the bandwidth and introduce some distortion. When the signal and the return path of the tip are pulled apart, loop inductance is added to the tip. This new equivalent circuit is shown in **Figure 8.24**. While the inductor is added to the ground clip, this is a loop inductance that is a property of the entire tip-ground lead path.

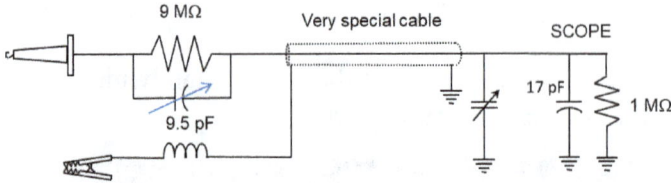

Figure 8.24 Second-order model of a 10x probe showing the shunt 9.5 pF capacitor and the adjustable capacitors at the scope end of the probe with some tip loop inductance.

The loop inductance of the tip has four consequences:

- ✓ It limits the bandwidth due to the source impedance and the series inductance.

- ✓ It adds ringing from the inductance and probe capacitance LC circuit.

- ✓ It contributes to cross-talk from other probes due to the loop mutual inductance with other loops.

- ✓ It picks up RF interference from the environment and elsewhere from the DUT because it is an antenna.

When measuring a DUT with a low source resistance, the combination of the tip loop inductance and the input 9 pF capacitor creates an LC circuit that has a parallel resonance and a higher impedance at the resonant frequency.

The expected resonance frequency for the tip loop inductance and input capacitance of the 10x probe is about 100 MHz,

$$f_{resonance} = \frac{1}{2\pi\sqrt{LC}} = \frac{1}{2\pi\sqrt{200nH \times 10pF}} \sim 100\,\text{MHz}$$

This resonance distorts the transfer function and results in apparent ringing on signal edges. **Figure 8.25** shows the simulated transfer function with a tip loop inductance of 200 nH and the measured step response of a fast edge signal with a peak at about 100 MHz.

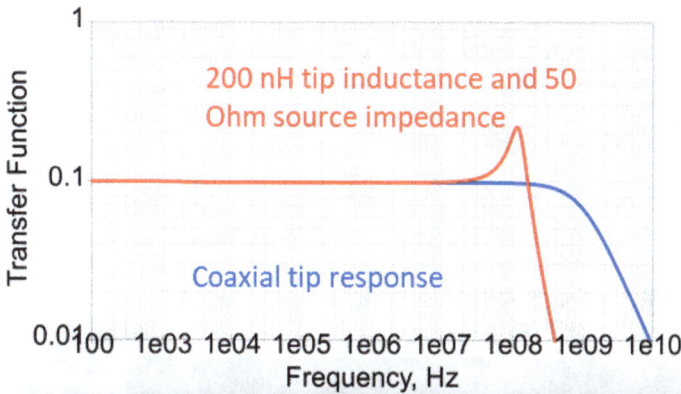

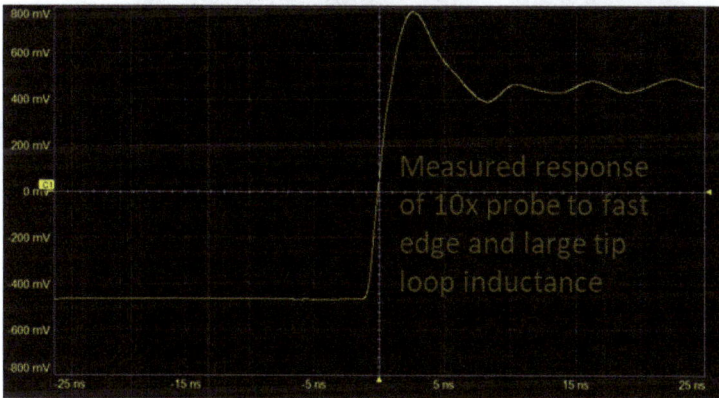

Figure 8.25 The transfer function is distorted when the tip loop inductance is added, which results in ringing of the measured waveform.

The bandwidth of the probe is decreased, and a ringing artifact is added to the measured signal at the DUT.

When the source impedance of the DUT is reduced, the Q of the LC circuit composed of the tip inductance and the 9 pF input capacitance of the 10x probe increases. This increases the peak of the transfer function and contributes to higher ringing. This is illustrated in **Figure 8.26**.

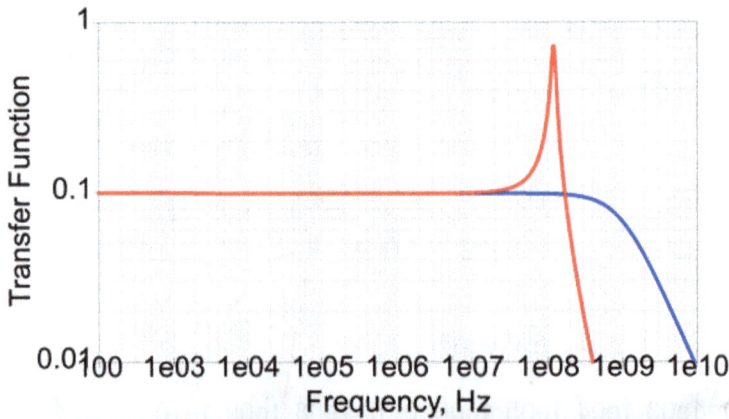

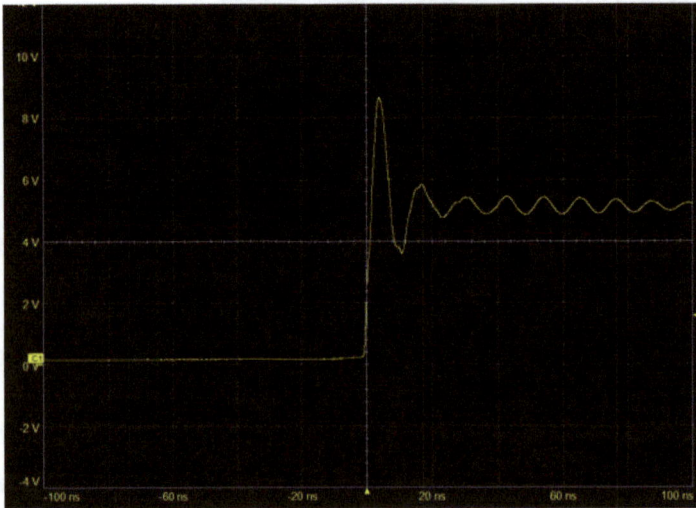

Figure 8.26 Top: Simulated transfer function of the tip with a large loop and low source resistance. Bottom: The measured step response shows the ringing at about 75 MHz from the tip inductance and input capacitance.

This suggests that the best practice to reduce the measurement artifacts from tip loop inductance is always to try to engineer as low a tip loop inductance as possible. **Figure 8.27** shows examples of three different tip connections, with a coax connection, a small spring ground clip, and the long external ground lead.

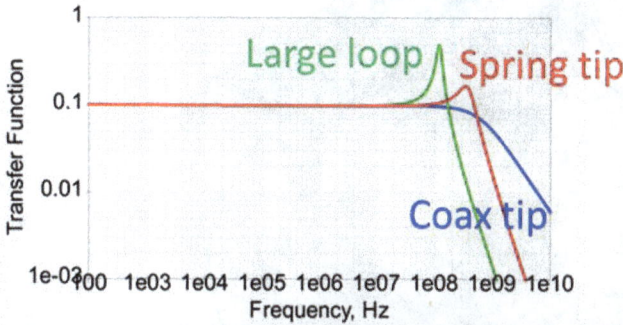

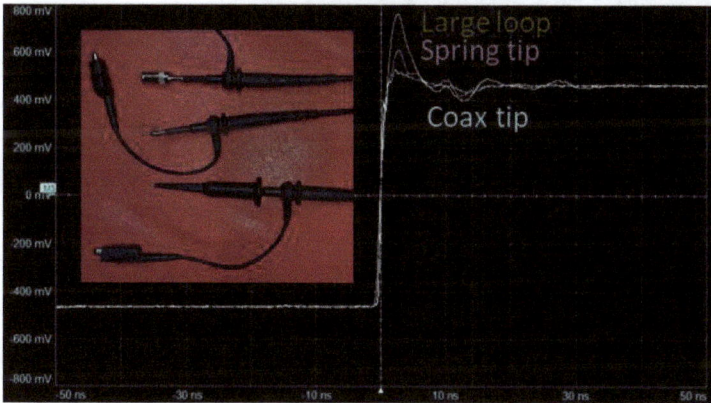

Figure 8.27 An example of three different probing methods with different tip loop inductances and the impact on the signal distortion. Top: Simulated transfer function including the tip loop inductance. Bottom: The measured step response of the same DUT with the three different probe tip geometries.

The worst method of probing a DUT is to use long floppy wires connecting the DUT to the 10x probe. This will dramatically increase the loop inductance of the tip and increase all of the problems mentioned above. An example of what not to do is shown in **Figure 8.28**.

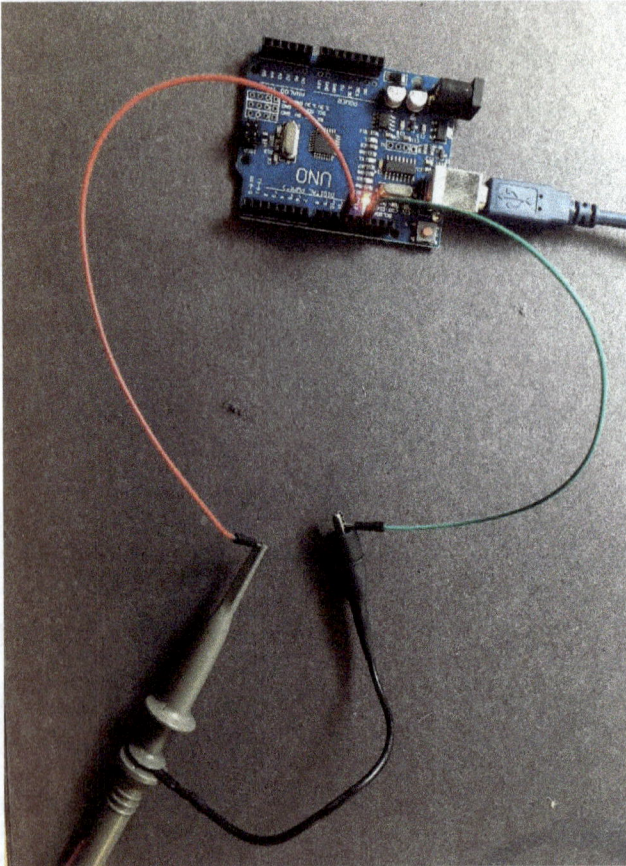

Figure 8.28 An example of the worst way to probe a DUT is using long floppy, high-loop inductance wires between the DUT and the 10x probe.

When it is not practical to reduce the tip loop inductance, and the signal drives ringing noise in the C and L of the tip, a simple addition can reduce the ringing artifact.

The ringing is due to the LC parallel resonance of the tip loop inductance and the input capacitance of the probe. This circuit has a relatively high parallel resistance of 9 meg. This means the LC circuit has a high Q. If a series damping resistor is added to the probe tip connecting to the DUT, the ringing can be reduced. The value of the resistance should be about

$$R_{damping} = 2\sqrt{\frac{L}{C}} = 2\sqrt{\frac{100nH}{0.01nF}} = 200\ \Omega$$

As an example, **Figure 8.29** shows the measured voltage from a low resistance microcontroller I/O pin having a rise time of about 5 nsec, driving the nearly 100 MHz ringing in the 10x scope probe. This is purely an artifact of the large tip loop inductance and not part of the signal from the I/O pin. When a 220 ohm series resistor is added between the I/O pin and the 10x probe, this ringing is completely damped out.

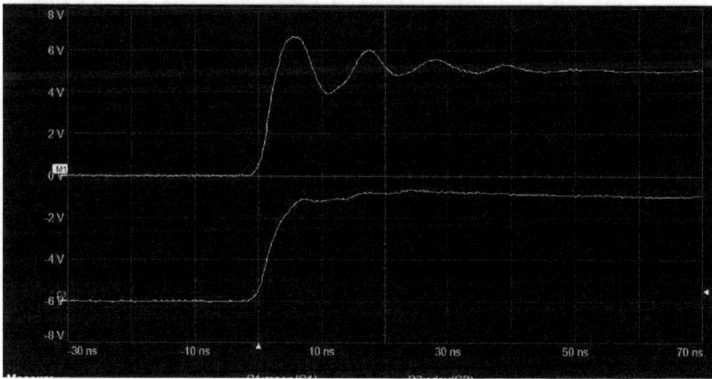

Figure 8.29 Measured signal from a low-impedance digital I/O pin with a large tip loop inductance showing the 100 MHz ringing (top). A 220 ohm damping resistor was added and the ringing is damped out (bottom).

The best practice is to reduce the tip loop inductance to reduce the ringing artifact from the tip. When this can't be done, adding a damping resistor will reduce this artifact.

With a high tip loop inductance, the bandwidth of the 10x probe is reduced to about 10 MHz. The step response of the 10x probe will show a rise time of about 0.35/10 MHz = 35 nsec and a lot of ringing. If the rise time of the signal being measured is much longer than 35 nsec, such as 100 nsec, there may not be high enough frequency components in the signal to drive the ringing of the C and L of the probe tip.

In this case, the large loop inductance of the tip may not introduce an artifact. This is the general boundary between when the probe is transparent and when its electrical properties are important.

For signals with rise times longer than about 100 nsec, it probably doesn't matter how you connect the scope probe to your DUT. Any method that is convenient will work. However, for shorter rise times, the probe connection may introduce artifacts.

This suggests that the best design practice to achieve the highest bandwidth and least distortion is to use as short a connection between the DUT and the tip as practical. The low loop inductance created by the small ground spring that slips on the outside of the 10x probe with the cap removed will not degrade the bandwidth of the scope probe, will not result in ringing, will reduce the cross talk to other probes and will reduce pick up from RF interference from the environment. It should be the tip of choice when signals have bandwidths > 100 MHz and are < 100 mV, when low cross talk is important. A close-up of this spring-tip ground connection is shown in **Figure 8.30**.

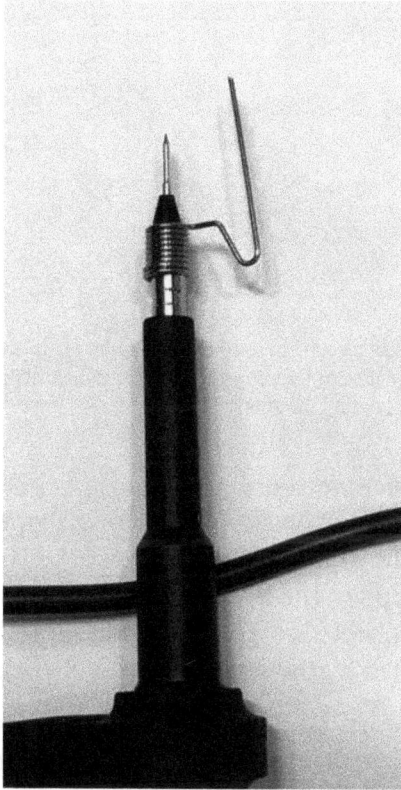

Figure 8.30 The spring tip for the ground connection slips on the probe tip with the hat removed.

8.13 Recognize 60 Hz Pickup

When the tip of a 10x probe is exposed and floating in space, its high impedance at low frequency makes it very sensitive to 60 Hz electric field pickup. This is due to capacitive coupling from the nearby power line wiring and the capacitively coupled currents through the high impedance of the tip. The tip input impedance looks like a 10 meg resistor below about 1 kHz. The equivalent circuit is shown in **Figure 8.31**.

Capacitive Coupling

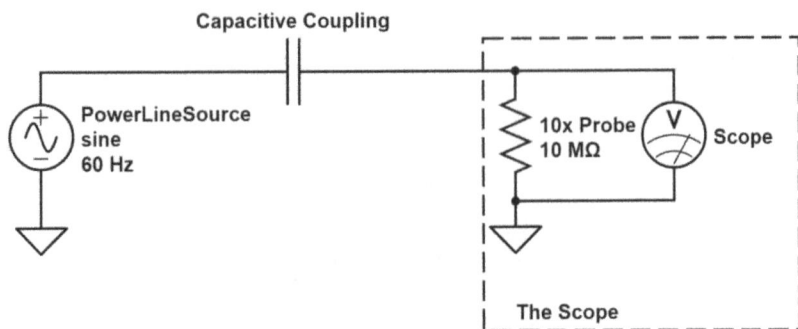

Figure 8.31 The equivalent circuit shows the origin of 60 Hz pickup by capacitive coupling. The high impedance of the scope turns the small capacitively coupled currents into a large voltage measured by the scope.

The larger the capacitive coupling to the power lines, the more the capacitively coupled currents into the scope's input resistance and the larger the 60 Hz voltage induced across the 10 meg input resistance of the scope. The lower the impedance at the tip, as, for example, when connected to a low-impedance DUT, the less the voltage noise created by the same capacitively coupled currents.

The pickup from 60 Hz is very easy to recognize because it has a frequency of 60 Hz. This is a period of 17 msec. When the scope trace is displayed at 10 msec/div, 60 Hz pickup interference has a period of a little under 2 divisions. This is a very distinctive pattern. An example of 60 Hz pickup from a 10x probe tip floating in air is shown in **Figure 8.32**.

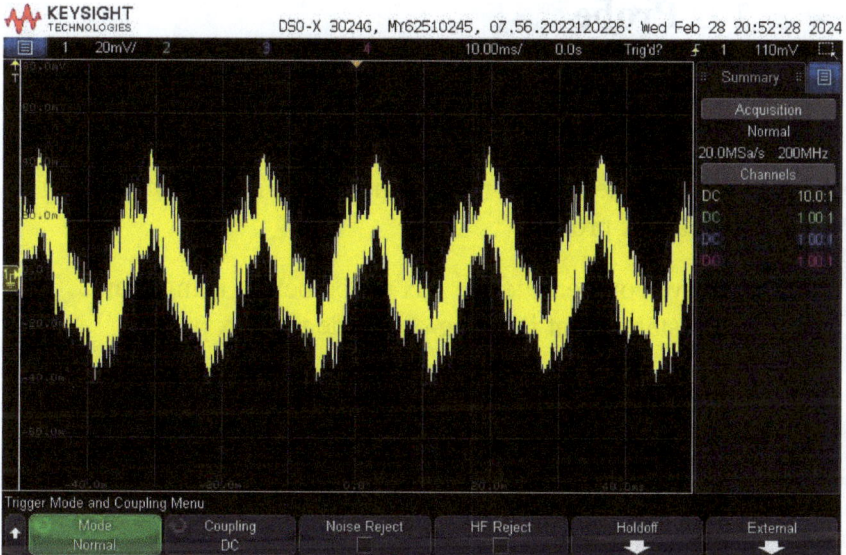

Figure 8.32 An example of 60 Hz pickup from a floating 10x probe tip on a time base of 10 msec/div. Note that the period is a little less than two divisions.

The best way to see 60 Hz pick up is on a time base of 10 msec/div. Get in the habit of always setting the time base to 10 msec/div to check for the presence of 60 Hz pickup. As a corollary, whenever you are observing a signal on a 10 msec/div scale, be aware that periodicity on a scale of about 2 divisions is most likely 60 Hz pickup either in the DUT or in the probe.

This is usually an artifact of the high impedance of the probe tip. Sometimes, this is part of the DUT signal you care about, but most of the time, it is a measurement artifact. The only way to reduce this, as described in a later chapter, is to reduce the tip impedance or add some metal surface connected to the circuit ground nearby to draw the nearly static electric fields from power lines away from the probe tip.

8.14   Summary of the Best Practices Using a 10x Probe

The basic best practices for using a 10x probe for general applications are summarized as:

1. Only use the scope on the 1 meg input impedance setting with a 10x probe.

2. Always compensate the 10x probe before using it and understand what is going on inside the 10x probe.

3. Only use the compensation signal on the front of the scope to compensate a 10x probe.

4. Check to make sure the scope is displaying the signal with a 10:1 attenuation in the probe setting so the displayed voltage is the tip voltage.

5. Never use a 10x probe except between the DUT and the scope. Never use it to connect a signal source to any other instrument.

6. Use the 10x probe on the 1x probe setting only with extreme caution. The measurement bandwidth on the 1x scale is dramatically reduced from the 10x scale.

7. The only situation in which the 1x scale should be used on a 10x probe is when you are measuring a signal below about 50 mV with a rise time longer than 100 nsec. In all other cases, the 10x setting is preferred.

8. Note that the 10x probe always and only measures single-ended signals. never connect the ground lead of the 10x probe to any node in your DUT that might be at other than earth-ground potential.

9. For signals with a rise time shorter than 100 sec, use as small a loop inductance at the tip as practical. This means a small loop.

10. Get in the habit of using the spring ground tip to make connections to your DUT.

11. The 10x probe only measures single-ended signals. Be very careful connecting the ground lead of the 10x probe only to circuit ground nodes in your DUT.

Chapter 9
How Not to Be Confused by Ground

The principles of ground are fundamental to every voltage measurement. Yet, in the real world of measurements, ground is one of the most confusing topics. Keeping track of the various types of ground, what they mean, and what to watch for in any measurement is the best way of not being confused. The terms used to describe the various types of ground and how they apply to signals and measurements are described in this chapter.

There is an entirely different perspective of ground related to how to design and implement a grounding scheme in a system to reduce the noise in the system. This topic is covered in other references such as Bogatin's Practical Guide to Prototype Breadboard and PCB Design.

9.1   Types of Ground

One of the reasons ground is so confusing is that there are really five, very different types of ground. It is not commonly done, but we should get in the habit of adding a prefix to the word ground to distinguish which one we mean when mentioning ground.

In addition to the five terms associated with ground, there is a sixth closely related term, return path. The same conductor in the physical implementation of a system, usually labeled as a ground, is associated with these six terms:

1. Earth-ground

2. Chassis ground

3. EMI ground

4. Circuit ground

5. Local circuit ground

6. Return path

We often use different symbols in a schematic to refer to some of these. These symbols are illustrated in **Figure 9.1**. Return path, EMI ground, and local ground are not distinguished in a schematic.

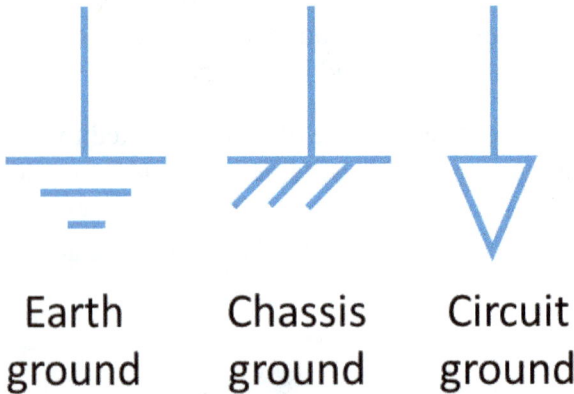

Figure 9.1 The symbols used to refer to earth, chassis, and circuit ground.

Each of these has very different meanings, applications, and best practices associated with their use. They are described in detail in this chapter.

9.2 Earth-Ground

The connection to the ground outside your home or building is called earth-ground. This is literally a connection to the Earth using a copper rod stuck in the ground outside a home or building.

This is a safety issue, with many building codes defining how to make the connection. **Figure 9.2** shows an example of the copper rod outside a home pounded a few feet into the ground, acting as the earth-ground connection.

Figure 9.2 An earth-ground connection using a copper rod inserted deep into the ground. This connection is distributed throughout your home and comes out of the round socket in your 3-hole AC power main sockets.

The powerline wiring inside a home or building routes this earth-ground conductor throughout the house. The round hole in a house AC powerline socket in the United States is connected to this copper rod through the house wiring. The identification of each of the three holes in an AC power plug is shown in **Figure 9.3**.

Figure 9.3 Label of each terminal in an AC socket. The short slot is the highest voltage. The long slot is neutral or return for the hotline. The round hole is the connection to the external earth-ground. The voltage between the neutral and earth-ground should be a low voltage with no intentional current flow.

It is easy to measure the voltages on each pin of the AC plug, with respect to earth-ground, using single-ended 10x probes. Their voltage rating is > 500 V. When plugged into a scope, which itself is rated to +/- 50 V input, the voltage range of the scope and 10x probe is +/- 500 V. This is well above the expected voltage on the AC lines.

In the US, the power lines are rated between 110 – 120 VAC. This is actually the RMS voltage. The voltage amplitude found on US power distribution, assuming sine waves as the waveform, is 1.41 x rms or 155 – 169 V. This is a peak-to-peak of 310 – 338 V.

Since a 10x probe is a single-ended probe, its ground lead must be connected to the earth-ground connection. This means it measures a voltage relative to earth-ground. **Figure 9.4** shows an example of the measured voltage on the "hot" plug and the neutral plug, relative to the earth-ground connection.

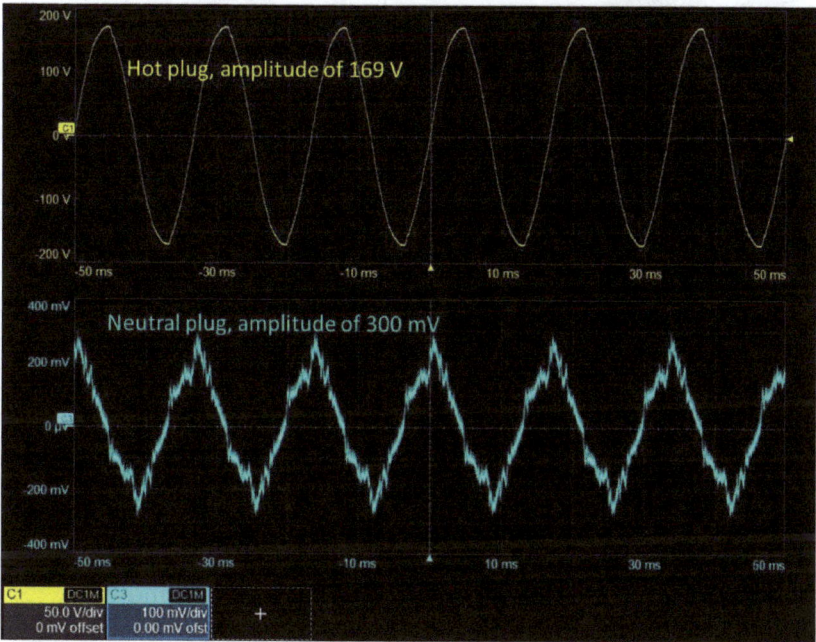

Figure 9.4 Single-ended voltages on the power lines with respect to the earth-ground connection were measured with a 10x probe.

Note that in this case, the time base is 10 msec/div. This means a period of 60 Hz is about 1 ½ divisions. When the scope's time base is on 10 msec/div, 60 Hz appears periodic with a period of about 1 ½ divisions. This is a quick, simple way of identifying 60 Hz pickup.

In addition, the phase of the voltage on the neutral plug is shifted - 90 degrees behind the phase of the voltage on the hot pin. If we assume the voltage is the IR drop from the current passing through the resistance of the neutral wire, this voltage and its phase is a direct measure of the current in the neutral wire.

This means the current lags the voltage by 90 deg, which is exactly as expected if the various loads on the power in the home are mostly inductive from transformers or motors. The equivalent circuit of the measurement with the single-ended 10x probes on the power lines is shown in **Figure 9.5**.

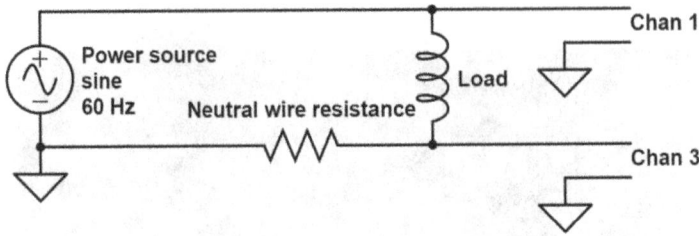

Figure 9.5 Equivalent circuit of the power rails and an inductive load with the neutral wire as a resistor.

The voltage on the neutral wire resistance, V_R, due to the current through the inductor, I_L, is

$$V_R = R \times I_L = R \times \frac{V}{j\omega L} = -j\frac{V}{\omega\frac{L}{R}}$$

This shows the voltage on the resistor -90 degrees behind the voltage on the inductor.

The amplitude of the voltage on the neutral plug is 300 mV. If the current through all the loads is 30 A, assuming a 30 A breaker on this line, then the resistance of the neutral wire is about R = V/I = 0.3 V/30 A = 0.01 ohms or 10 mOhms, a very reasonable value.

Note also that the voltage on the neutral wire is very noisy, an indication of very high-frequency transient loads, such as the brushes on a motor or from switch-mode power supplies connected to the power line. This voltage noise, resulting from the current noise on the neutral wire, routed throughout a house or building like a giant antenna, is a dominant source of RF noise that can insidiously get on every sensitive measurement.

Earth-ground is first a safety issue. The difference in DC voltage between any two earth-ground connections in a home should not be more than a volt or two at most. This means that if all metal surfaces anyone can touch are connected to the same earth-ground connection in your home, you have a much smaller chance of getting electrocuted.

Earth-ground connections are also a way of reducing electrostatic discharge (or damage) (ESD) events. Conductive surfaces connected to earth-ground will not charge up above earth-ground potential due to static charge buildup. If a conductive ESD mat is connected to something else and not earth-ground, while the entire surface may be an equipotential, it is still possible for it to charge up to a voltage above 1 kV, relative to other surfaces.

When components are in contact with the mat and then touched by other conductors charged to a different voltage, the components can be damaged by an electrostatic discharge (also ESD) event. If all surfaces and instruments are connected to the earth-ground, any static buildup can flow through the earth-ground connections to, literally, the Earth, and voltage differences between earth-grounded conductive surfaces have little DC voltage difference between them.

To be effective, an ESD protection surface should always be connected to an earth-ground conductor. If an earth-ground connection is important, verify the instrument to which the mat is connected is using a 3-prong power plug and that the earth-ground connection is actually connected to earth-ground.

To make a connection to the earth-ground round hole, special plugs called earth-ground adaptor plugs, are available to plug into the AC power socket using plastic pins for the power lines, but a socket for a banana plug to the earth-ground connection. An example of an earth-ground adaptor plug is shown in **Figure 9.6**.

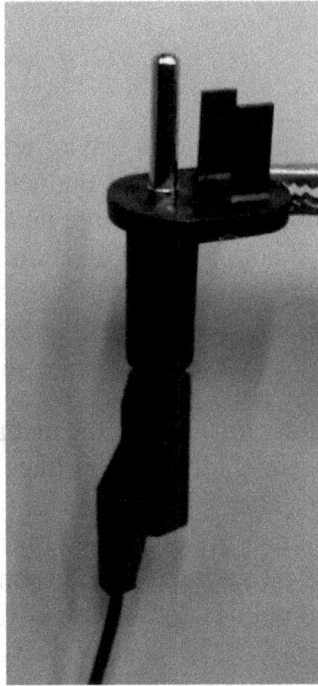

Figure 9.6 An earth-ground adapter plug to make a connection to earth-ground for ESD protection.

One way of telling if a conductor is connected to earth-ground is to use a DMM to literally measure the resistance between the two conductors thought to both be connected to earth-ground. The resistance should be a few ohms. When testing the connections to earth-ground, be sure the instruments are plugged into the wall socket when you measure with the DMM. They do not have to be powered on.

9.3   Floating from Earth-Ground

If a circuit is not connected to earth-ground, like a battery-powered circuit, a car, a plane, or a satellite, we describe the circuit as "floating." This just means there is no connection to earth-ground. If we were to measure the voltage between any point in the circuit and a nearby earth-ground connection with a very high-impedance voltmeter, it could be anything from 0 V to 10 kV. The circuit is floating relative to earth-ground.

A typical DMM has an input resistance between its terminals of 10 megaohms. The act of connecting the DMM between two different floating devices may actually short any voltage difference between them and bring them to the same potential through the 10 M internal resistor. They still float, relative to earth-ground, but the DMM brings the two devices to the same potential through the 10 Meg internal resistor. This is a measurement artifact. As soon as the DMM is disconnected, the devices may charge up to a large, arbitrary voltage difference between them.

If there is 100 pF of capacitance between the two devices, the time constant for which the DMM discharges the capacitance is 10^7 x 10^{-10} = 1 msec. This is very fast.

The difference in voltage between two floating devices can be measured with an electrometer-grade voltmeter, which draws a very low bias current. Many low-cost electrometer grade opAmps are available that draw 2 fA of current. These can measure the voltage between different floating instruments.

The rate of voltage drop between the two devices, with 100 pF of capacitance between them, when using an electrometer to measure the voltage difference between them is

$$\frac{V}{t} = \frac{I}{C} = \frac{2 \times 10^{-15} A}{10^{-10} F} = 20uV/sec$$

This is a very slow rate of discharge and can be used to measure the electrostatic voltage between objects.

A handheld, battery-powered DMM floats. When it is used to measure the voltage of a battery, both the battery and the DMM float. The battery voltage is completely independent of whether the DMM floats or not. This is demonstrated in **Figure 9.7**.

Figure 9.7 An example of a DMM measuring the voltage difference across a D cell battery. Both the DMM and the battery are floating. The measured battery voltage difference is independent of whether or not any lead is connected to earth-ground.

Likewise, a wire connected to earth-ground can connect to any part of the floating circuit. The circuit is then no longer floating, but the voltage difference across the battery terminals in the circuit will stay the same.

Any device that is battery-powered and not connected to earth-ground floats. A car is not connected to earth-ground and floats. An airplane on the ground, or flying, or a satellite in space, all float.

Each time you get in and out of a car, you might be at a different voltage relative to the earth-ground. This is especially the case when you slide over a car seat getting out. This process can generate large static voltages compared to earth-ground. When you now touch a metal surface connected to earth-ground, there is the possibility you can generate a momentary spark. This is why all gas stations recommend you do not get in and out of your car once you are pumping gas.

Large trucks carrying flammable materials will often have chains dragging on the ground behind the wheels to provide some small connection to the local earth-ground and minimize any voltage difference between the truck and the local earth-ground. This reduces the chances of a random spark from a high-voltage ESD arc. An example of these grounding chains is shown in **Figure 9.8**.

Figure 9.8 Chains are dragging behind a truck to keep the truck frame at earth-ground and prevent sparks from ESD.

Usually, if some parts of a circuit float relative to the rest of the circuit, the two circuits are isolated from each other using an isolation transformer. We sometimes refer to the two circuits as transformer-isolated or galvanically isolated. This is an old term that just means there is no direct current path between the two parts of the circuit.

There is usually a limit to how large a voltage can be applied between the two halves of the isolated, floating circuit. This ranges typically from 100 V to 1,000 V. It is usually marked on the outside of the instrument. If the voltage on one circuit, floating compared to another part of the circuit, exceeds this value, there may be a spark, arc, or breakdown of some component between the two circuits, and the circuits can be damaged.

Most DMMs are rated as able to connect to a device that is as much as 600 V floating above earth-ground. This is sometimes referred to as over-voltage protection. The highest over-voltage to which the device is capable of safely withstanding is rated by a CATEGORY or CAT rating.

The CAT ratings of devices are based on their application, intended use, and typical highest voltage they might routinely see. A CAT I device is supposed to be used with low voltage appliances not connected to AC power. A CAT II rating is for devices connected to single-phase 120 V power mains. A CAT III rating is for devices connected to 240 3-phase power mains and a CAT IV device is one rated to be used connected to 3-phase higher voltage distribution systems.

In addition to the CAT rating, there is also a maximum, transient overvoltage rating for the instrument. A DMM rated for CAT II and 1000 V means the DMM can withstand up to a 1000 V overvoltage. This is a measure of the insulation protection the DMM provides the user.

If you were to be connected to earth-ground, such as wearing a grounding strap, for example, you could hold and operate a CAT II 1000 V rated DMM that was connected to a device that was floating at up to 1000 V relative to earth-ground. If the device is at a higher voltage, relative to earth-ground, the DMM might arc over and connect you to the device under test. This could be fatal.

Know the rating of your DMM, and make sure you do not try probing a DUT, which you know to be at a higher voltage rating than the DMM. **Figure 9.9** shows an example of the CAT rating of a DMM. It is always printed on the outside of the DMM. This DMM could be used with devices connected safely to 120 V power mains or even 240 power mains.

Figure 9.9 A DMM usually has printed on the outside the category rating for safe use of the DMM. This is the maximum floating voltage of the DUT for safe operation without the danger of an arc from the DMM to the user who might be connected to earth-ground.

Whether a circuit floats or is tied to earth-ground is first a safety issue and then an issue related to how the circuit will be connected to other circuits.

Always pay attention to whether a circuit floats or not. If it floats, there is the possibility it can be at a high voltage compared to earth-ground. Be very careful: if touched under some circumstances, it could be lethal.

When multiple instruments are connected together, it is important to keep track of which instruments are floating and which are connected to earth-ground. This is both a safety issue and a noise issue. One of the important features of a battery-powered DMM is that it floats. It will always measure the voltage difference between the two terminals on the front of the DMM, no matter if the device to which it is connected is floating or connected to earth-ground.

Generally, all instruments that are battery-powered float. Some instruments are designed with their internal transformer isolated from the earth-ground connection to the instrument and their outputs can still float even if the instrument is plugged into the power line with a 3-prong plug. Most instruments plugged into the wall power with a 3-prong plug are connected to earth-ground and their inputs or outputs do not float. Always check for these connections in any instrument you use.

When an appliance or instrument has a 2-pronged power plug, such as seen in **Figure 9.10**, there is no earth-ground connection. This means conductors on the inside of the enclosure, hidden from prying fingers by the plastic housing, could be at high voltage compared to the voltage of earth-ground elsewhere in the house.

Figure 9.10 Examples of 2-pronged plugs that connect to devices that float above earth-ground. They have no connection to earth-ground.

In this case when the power plug is 2-pronged, the inside of the appliance "floats" from earth-ground. It has no connection to earth-ground and can be at almost any voltage compared to earth-ground.

When a device plugs into the wall with a 2-prong plug, generally, the rest of the circuit inside the device is isolated from the wall power with a transformer. The only connection between one side of the transformer and the other side of it is through magnetic fields from one coil to another in the transformer. We refer to the circuit as being transformer-isolated from the wall power. There is no direct path for current to flow from the wall power to the rest of the circuit connected on the other side of the transformer.

Even if a device plugs into the wall with a 3-pronged plug and has a connection to earth-ground, some parts of the product may float. For example, in some power supplies, some of the output terminals float compared to earth-ground, even though the chassis of the enclosure is connected to earth-ground.

The connections on the front of the instrument to earth-ground are sometimes labeled with a green mark and a small earth-ground symbol. The power supply output terminals may still be transformer-isolated from the rest of the circuit and floating from the earth-ground of the round prong in the cable plugged into the wall.

An example of the front of a power supply with floating outputs is shown in **Figure 9.11**.

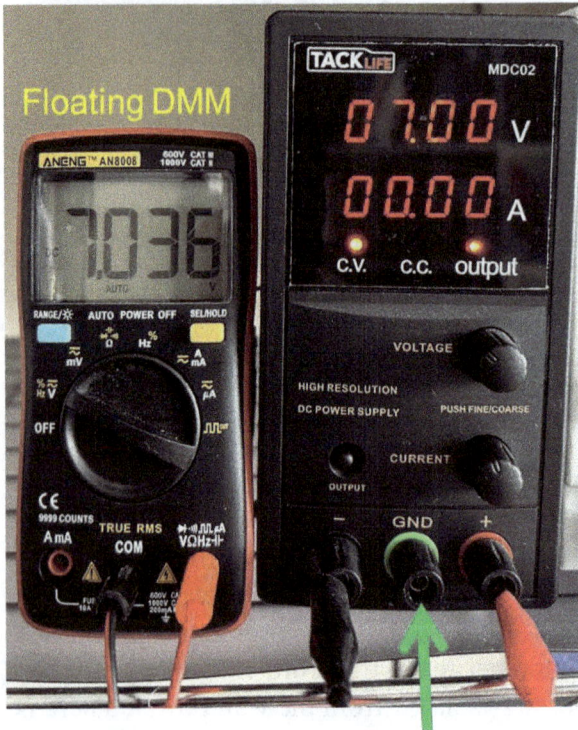

Figure 9.11 An example of an earth-grounded power supply but with terminals that are transformer-isolated from earth-ground and are floating. The green connector labeled GND is actually connected to earth-ground, but the power connections are floating.

In this power supply, the output terminals are floating relative to earth-ground. You could use the black (-) and red (+) output terminals to connect to a circuit that was floating, always providing the set voltage difference between the two terminals. The power supply would maintain the isolation of the device to which it was connected from earth-ground. The DMM is able to measure the voltage difference between the two terminals of the power supply regardless of whether they float or not and without changing the floating status of the output voltage.

This power supply is plugged into the wall with a 3-prong plug. The connection to earth-ground is brought out to the front panel to the center plug labeled as GND with a green band around it. This is an externally available connection to earth-ground.

Unfortunately, many instruments, such as this power supply, do not have their earth-ground connections labeled very well. It would be much less ambiguous and less confusing if this green-banded plug were labeled as earth-ground instead of just GND.

One way of verifying that the GND labeled terminal is connected to earth-ground is to literally measure the resistance between the GND terminal and other conductors known to be connected to earth-ground. The connection to earth-ground will be a continuous conductive path and will not change if the instrument is off or on.

The resistance between the + terminal and the GND terminal will be very high. The resistance between the − terminal and GND will be very high. The resistance between the GND terminal and the round prong of a wall outlet will be less than 10 ohms.

The GND connection is provided on the front of this power supply, and many other instruments, as a safety feature. If there is no compelling reason to use a floating circuit, the circuit should always be connected to earth-ground. This is for safety and consistency. If the circuit is normally floating, this connection on the front panel can provide the connection to the device being powered to earth-ground. This will make it safer to use a single-ended scope probe to measure a voltage node in the circuit.

In a single-supply powered circuit, it is common to connect the (−) low side terminal of the power supply to earth-ground. This will connect any node to which the (−) terminal is connected to earth-ground. In a bipolar powered circuit, it is typical to connect the earth-ground to the central, common connection, labeled in the circuit as circuit-ground.

Two identical single-ended, floating power supplies, could provide +10 V and -10 V power to any external circuit by connecting their outputs in series. The common connection could be connected to earth-ground for safety or just left floating. The local circuit ground connection could be to the common terminal between the (+) and (-) terminals and left floating from earth ground.

This is a simple way of generating a bipolar power supply using two, identical, low-cost AC to DC wall wart power supplies, each floating compared to earth-ground. Simply connect the center conductor of one plug to the outer conductor of the other plug. This is the common connection. The voltage between the remaining outer and center conductors will be twice the voltage of either supply. The center, common connection, could also be connected to earth-ground or just circuit-ground and floating.

9.4   Chassis-Ground or EMI-Ground

We call the metal enclosure surrounding an appliance or instrument the *chassis*, pronounced "chassy." Since this is a large metal surface associated with the entire instrument, we refer to this metal as *chassis-ground*. Every surface inside the enclosure connected to the chassis-ground will be at nearly the same voltage, so prying fingers touching different pieces of metal, all connected to the chassis-ground will see the same voltage, and there is a reduced chance of being electrocuted.

The fact that we label the large metal enclosure as "ground" has nothing to do with any electrical property or behavior of this large piece of metal, other than it is a large piece of metal that spans the dimensions of the instrument.

This large piece of metal can act as the other conductor of an antenna if the product has an external cable attached. For this reason, the chassis is sometimes also called EMI (electromagnetic interference) ground. This just refers to the possibility if there were some other external conductor, these two conductors, with a source of voltage noise between them, acting as a dipole antenna, would radiate and produce EMI.

In principle, the chassis-ground should never be used to carry intentional current. This is more an EMI issue than safety. In practice, many systems are designed by inexperienced designers who use the chassis to carry return currents. This means different parts of the chassis may be at slightly different transient voltages. These voltages are generated by the currents flowing through the impedance of the chassis.

When there are gaps or slots in the chassis, such as at the seams of the enclosure, the poorly engineered intentional currents flowing around these gaps will radiate and are a common root cause for electromagnetic compliance (EMC) certification test failures.

If an external cable is connected to the enclosure, and the chassis has some voltages on it, these voltages will drive currents between the chassis and the external cable. The cable will be an antenna and radiate with the chassis acting as the second conductor. In this sense the chassis acts as an EMI ground for the cable to radiate against. This is also a common reason a product will fail an FCC certification test.

To reduce the chance of radiated emissions between the chassis and some other external cables, always keep a low impedance between the chassis and the shield of external cables. The shield of all cables should be connected to the chassis-ground. This is also why it is bad design practice to use the chassis as a return conductor intentionally. It should be used as an earth-safety ground and as an EMI-ground. Any intentional return currents could cause a voltage difference between some parts of the chassis and any external cable. This would cause unintentional radiated emissions.

Underwriters Lab (UL) safety requirements specify that all consumer appliances with a metal enclosure (labeled as chassis-ground) have their metal enclosures also connected to earth-ground. This means the instrument must use a 3-prong power cable plugged into the house power, and the wire that connects to the round hole in the AC power plug is also connected to the chassis. If the device is UL-approved, the chassis-ground should be connected to the earth-ground.

The alternative requirement is that all electronic elements powered by the AC power line must be separated from the chassis-ground with at least two layers of insulation to prevent the possibility of electrocution by the user.

Figure 9.12 shows an example of the screw connection to the chassis-ground and the chassis-ground connected to the earth-ground of the power cord socket.

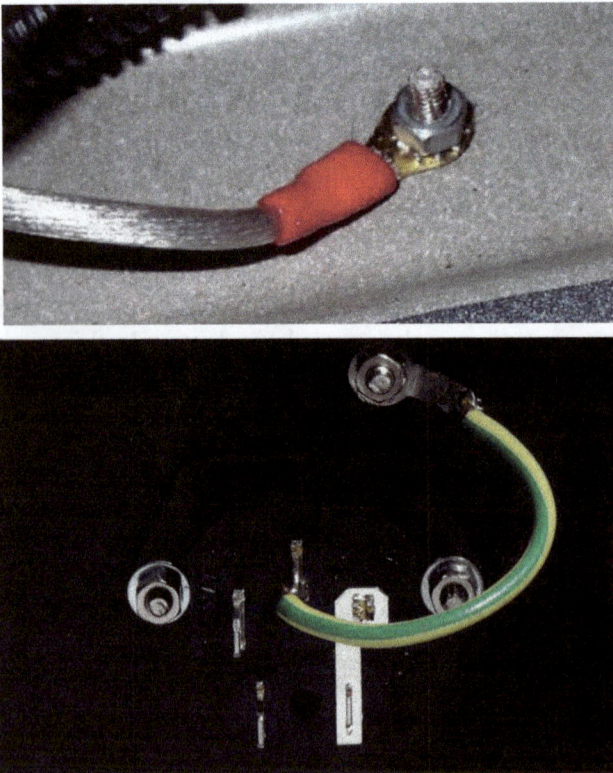

Figure 9.12 A connection to chassis-ground by a screw into the enclosure and how this is connected to the earth-ground of the power cord.

If the product enclosure is plastic, and a consumer would never touch any metal when handling the appliance, there is no chassis, and the appliance does not need to have an earth-ground connection. These appliances can use a 2-pronged plug.

As long as prying fingers are kept away from metal inside the enclosure, it doesn't matter if the device floats or is connected to earth-ground.

It's when the plastic enclosure is open that there is a potential danger. Different pieces of metal inside the plastic enclosure could be at power mains voltage differences from each other and from the house's earth-ground. When one finger touches one conductor and another finger touches another conductor, and they happen to be at a large voltage difference, electrocution is possible.

As a good safety practice, assume all metal inside an appliance is lethal. This is why you should never stick a fork inside a toaster to pull out the toast. If you must, unplug the toaster first. For 99% of the time, you are not touching anything important. It's that 1% of the time that can be lethal. This would be a difficult lesson from which to learn.

9.5   Best Practices for Earth-Ground Connections

When making connections to earth-ground or chassis-ground, there are two basic principles.

First, no intentional current should flow in the chassis-ground conductor. This will prevent two problems: the IR voltage drop between two different reference ground points in the circuit, and potential unintentional radiated emissions from surface currents in the chassis enclosure encountering chassis holes or seams. This is a dominant root cause of EMC compliance test failures.

This means that the chassis conductor should not be used as a return path conductor. It is sometimes common practice to connect a device's circuit ground heat spreader to the chassis for good thermal transfer. This will use the chassis-ground conductor to carry return currents. Do not do this. Electrically insulate the heat spreader from the device that is being cooled.

A second common practice is to connect the chassis-ground to the circuit-ground on a circuit board at multiple locations on the circuit board. This makes the chassis-ground conductor in parallel with the board level circuit-ground. If the chassis is low impedance, there may be substantial return currents flowing in the chassis-ground in parallel with the circuit ground conductor. This may also contribute to unintentional radiated emissions.

When possible, make one connection between chassis-ground and circuit-ground. This will prevent return currents from flowing in the chassis-ground conductor.

Generally, if there are shielded connectors off the board and out of the device to other devices or peripherals, the shield of the cable should connect to the chassis. This is the principle that the cable shield should be an extension of the chassis or enclosure. This provides a path for common currents in the shield to flow back to the chassis and limits their unintentional radiation.

If the cable shield is also the return conductor, a separate connection to the circuit-ground, the return conductor, can be made to the cable shield inside the enclosure. This is in addition to the cable shield connected to the chassis conductor. The signals' return currents will flow on the inside of the cable shield, and the common currents' return will flow on the outside of the cable shield.

As a general principle, when measuring a DUT with a scope, the earth-ground connection to the DUT should not be through the scope probe's cable shield. When the scope probe is not connected to the DUT circuit ground, the DUT may float. This is a potential safety issue.

If the DUT is floating, multiple, low-impedance connections between the DUT's circuit ground and the scope's chassis-ground will reduce any earth-ground-loop noise.

In the rare cases when there is too much noise pickup due to the earth-ground loops, a differential probe can be used to connect between the DUT and the scope. In this case, it is only important to be sure the DUT's common voltage is within the range of the differential probe.

9.6   Circuit-Ground

The third type of ground is circuit-ground. This is the physical conductor in a circuit that the design engineer defines as the point to which the voltage at all other nodes in the circuit will be measured. We sometimes refer to circuit-ground as the voltage reference point for the circuit. This is the ground that is usually thought of when we look at the ground in a circuit schematic.

In principle, every voltage is really a voltage difference between two nodes in a circuit. An amplifier or receiver that is sensitive to a voltage is really sensitive to a voltage difference at its inputs. In some amplifiers both terminals at the input between which the voltage is measured are brought out for access. In this case, the amplifier is usually referred to as a differential amplifier.

When a voltage measured at the input to an amplifier is the difference between a node and circuit-ground, we refer to this type of voltage as a *single-ended* voltage. While we explicitly refer to the voltage at a node, implicitly, we are really saying the voltage is the voltage difference between the node and the circuit-ground, reference conductor.

A circuit-ground node can float or be connected to earth-ground. Connecting circuit-ground to earth-ground is a safety issue, not a performance issue. If the circuit-ground is connected to the chassis and the chassis is connected to earth-ground, then the circuit-ground is connected to earth-ground.

Always try to trace all the connections between the circuit-ground, to chassis ground, to earth-ground from your device to any other device to which it connects, and their connections, eventually into wall power. Be aware of the connections to earth-ground in your circuit and if your device floats or is connected to earth-ground.

In principle, we can select any conductor in a circuit as the circuit-ground reference point from which to measure all other voltages. The circuit designer who owns the circuit gets to decide which node will be the reference point from which to define all other voltages, or the circuit-ground.

There are some nodes that are more convenient than others to use as circuit-ground, and some are selected by convention. In the case of a single-ended power supply with just a 9 V power rail, for example, a circuit-ground is usually selected as the conductor with the lowest voltage between it and all other conductors. This means if you were to measure the voltage at internal points in a circuit, they would mostly appear as small positive voltages, relative to this circuit-ground reference node.

When there are two different voltages powering a circuit, such as a +/- 5 V supply, we usually select the conductor with a midpoint voltage between the +/- 5 V power rails as the circuit-ground reference node.

If the power supply floats, our choice of which node is circuit-ground can be arbitrary. It reduces confusion and makes it easier to understand and work with the circuit if the circuit-ground node is selected as the midpoint voltage of the power rails in bipolar supplies or the lowest voltage node in single ended supplies.

The circuit-ground in a circuit can float or be connected to earth-ground. If there is no compelling reason otherwise, and the opportunity is available, circuit-ground should be connected to earth-ground. This will provide a measure of safety and potential ESD protection. Sometimes, the process of connecting the circuit-ground conductor in a circuit to a scope probe will also connect the circuit-ground to earth-ground through the scope probe.

Unless you have a strong compelling reason otherwise, your circuit-ground connection in a circuit should be connected to earth-ground.

Regardless, always be aware of the earth-ground, chassis-ground, and circuit-ground connections in every application.

9.7 Noisy Ground, Return Path, or Circuit-Ground

In each of the applications of ground, we have assumed that there is no voltage difference between the different locations on the same conductor labeled as one of the various grounds. This is why it does not matter where we connect to the circuit-ground conductor when we measure a voltage using this conductor as a reference. In principle, every location on the same conductor is an equally good reference node.

The only way we will get a voltage difference between two points on the same conductor is if there is current flowing through the conductor. A voltage is generated when the current flows through the impedance of the conductor. The impedance is in the form of series resistance or inductance.

To be used as a voltage reference, it is important to engineer the circuit-ground conductor to either not carry any current or to reduce the impedance of the circuit ground reference conductor so that the voltage drops from I x R or L x dI/dt are acceptably low.

When a conductor is specifically designed to carry a return current for either a signal or the power path, it is more appropriate to refer to this conductor as the *return path*, rather than *circuit-ground*. The term circuit-ground should be reserved for when the conductor is designed to not carry current but only act as the reference conductor to which all other voltages are measured.

In noise sensitive circuits, a separate circuit-ground conductor can be routed to all the amplifiers that measure a single-ended voltage, relative to this node. This circuit-ground conductor would be specifically designed to *not* carry any return current. Other conductors would be used as the return path to carry return currents.

Unfortunately, there is no place in a schematic to label a conductor as a return path. In a schematic, all wires are identical and considered ideal, with zero impedance and as transparent interconnects. Whether a conductor is a return path or a reference ground conductor is a feature that is created during the layout of the board.

When the return conductor is on a wide plane, the return currents can spread out, and the impedance the return currents see is minimal. In this situation, the plane is a good approximation to an equipotential. The plane that is carrying the return currents is also used as the reference node and is also connected to circuit-ground. This conductor is often called the ground plane.

Recognize that the ground plane conductor is playing the two independent roles of carrying the return current and acting as the reference conductor. In extreme cases when the return currents are large, or there are discontinuities in the plane, such as gaps, which increase the impedance of the return conductor, the return currents can generate DC voltage drops from the plane resistance and transient voltage drops from the LdI/dt.

We refer to these sources of transient voltage differences in the ground plane as *ground bounce*. It is a major source of noise in noise-sensitive circuits. The sensitivity of ground bounce noise on measured voltages can be dramatically reduced if differential voltage measurements are used in a noise-sensitive circuit.

The typical DC series resistance between different points in a plane used to carry return currents is on the order of 1-2 squares of sheet resistance. For 1 oz copper plating, this is about ½ mOhm/square. If DC currents on the order of 1 A flow through the plane, there could be as much as 1 mV of DC voltage drop between different points on the plane, used as a reference to local voltages.

When currents in the plane are changing as when the plane is used to carry return currents of digital signals, voltages in the plane are generated by the dI/dt and inductance discontinuities, like gaps, or leads in a package or pins in a connector. When transient currents pass through this impedance, voltage drops from one region to another can be created. Usually, the dominant source of noise in a ground plane is from the switching currents and the inductance in the return conductor. This is the origin of *noisy grounds*. If there is no return current in the circuit-ground conductor, the reference ground will not be noisy.

9.8   Ground Bounce and Local-Ground

In the presence of ground bounce and a noisy ground, where on the circuit ground conductor we connect the reference terminal of the voltmeter or scope influences the voltage difference we measure.

Different locations in the ground conductor may have a different voltage. To describe this location-dependent ground reference, we refer to the specific point on the ground conductor to which we connect the reference terminal as *local circuit-ground* or just *local-ground*. It is the specific location on the circuit ground conductor with respect to which we measure the voltage of any circuit-node in the vicinity. **Figure 9.13** shows an example of few scope probes connected to a circuit board measuring the voltage at different node locations, each with their own local-ground connections.

Figure 9.13 Example of probing a circuit board at different locations and each probe using its local-ground as a reference node.

In principle, if there were no voltages between any two locations on the circuit-ground conductor, then all local-ground points would be the same voltage and equivalent. If there were no currents flowing in the circuit-ground conductor, all points on the ground reference conductor would be at the same voltage and all local-ground locations would be equivalent.

In practice, this is unfortunately not the case. Ground bounce voltages are generated across different locations in the circuit-ground conductor due to DC or transient currents flowing through the series resistance and inductance of the circuit-ground conductor.

Figure 9.14 shows an example of the simulated voltage distribution in a circuit board's circuit-ground conductor at a specific instant in time due to currents flowing through this conductor.

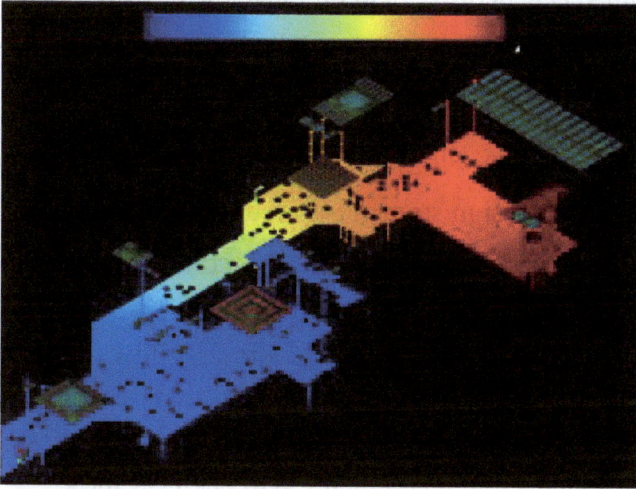

Figure 9.14 Example of the instantaneous voltage distribution in a ground plane when transient current flows. The voltage range from red to blue is 0.2 V simulated with Keysight SIpro.

The voltage at various locations in the circuit-ground conductor will vary as the currents in this conductor change. Even the specific path the currents take, and their resulting voltage drops, may depend on the transient currents' frequency components. This spatial variation in the voltage distribution on the circuit-ground conductor, ground bounce, means that the local-grounds may be at different voltages and the voltage measured by an instrument at one node using a specific local-ground location may see a signal component that has nothing to do with the signal, but is about the ground bounce on the local-ground conductor.

Generally, the transient currents with their large dI/dt, can create much larger ground bounce voltages due to the L x dI/dt than the voltage drops due to currents through a resistance, referred to as IR drop. The IR voltage drop in conductors from the currents through the series resistance of the circuit-ground conductors will appear as voltages that vary as the current varies.

This means ground bounce will exist at DC and low frequency due to the instantaneous current and will get worse with shorter rise time signals. When signals have rise times of microseconds, the ground bounce noise is generally small and just due to IR drops.

When the rise times are on the order of nsec or less, ground bounce noise from the large dI/dt when the signals change can sometimes dominate any signal voltages. Since this occurs at the switching edges of the signals, we also call ground bounce *switching noise*.

We generally apply two design strategies to reduce ground bounce:

- ✓ Reduce the series resistance and inductance of the circuit-ground conductor.
- ✓ Reduce the DC and transient currents that flow in the circuit-ground conductor.

The most effective way to reduce the series resistance and inductance of the circuit-ground conductor is to engineer these conductors as wide, continuous planes. A plane will have the lowest series resistance and lowest self-inductance. This is why so many board designs use "ground" planes.

Unfortunately, we use the term ground plane to refer to a plane used to distribute both the reference voltage node and return current for all power and signal paths. As long as the return currents are small, there is little impact on the local ground voltages in the plane.

The ground plane is usually placed in the stack up of the circuit board in a layer adjacent to the signal conductors for which it is providing a reference node. **Figure 9.15** shows an example of a circuit board with a wide ground plane and a region with just circuit-ground traces. The region of the circuit with no plane, but ground traces show dramatically more ground bounce noise.

Figure 9.15 An example of the same circuit designed with a low impedance circuit-ground plane and another region with just circuit-ground traces. The ground bounce noise is dramatically less with the circuit-ground plane.

If a narrow trace conductor is used as the circuit-ground conductor, such as in a solderless breadboard interconnect system, and there are currents flowing in this ground reference conductor, there is guaranteed to be ground bounce noise. There will be a voltage difference between different local-ground locations.

The second way we reduce ground bounce is by trying to avoid driving currents in the circuit-ground reference conductor. If the same reference voltage conductor needs to be distributed over a board, it can be routed as a signal trace.

As long as it does not carry any current, it will be an equipotential and a low-noise reference. In many low-noise analog applications, the reference voltage conductor, which is designed not to carry any return currents, is routed with the signal traces to any amplifiers or receivers needing a low-noise voltage reference node.

When multiple locations on a board are measured at the same time, each signal voltage may have a different local-ground reference voltage. This makes interpreting the measurements complicated. Special probing and measuring methods, described in a later chapter, can be used to reduce this problem.

9.9   Single-Ended and Bipolar Power Supplies

A single voltage power source has just two terminals between which a voltage is supplied. It can be floating or one of the terminals connected to earth-ground. A battery is a single-supply power source. The two terminals of a single-voltage supply are labeled as the + and − terminals.

If the two terminals are floating with respect to earth-ground, they can connect anywhere in a circuit and provide a voltage difference between the two connection points.

When the negative side of the power supply is connected to circuit-ground, this power supply provides a positive, single-ended, voltage source to the circuit. Alternatively, the positive terminal could be connected to circuit-ground and provide a negative, single-ended voltage power supply to the circuit.

A single-ended supply can also be connected to earth-ground if one of its terminals is connected to an earth-ground conductor. Or the single-ended supply can float.

When the product is battery powered, like a flashlight or a cell phone, the circuit-ground is not connected to an earth-ground and the circuit or device floats.

Two floating, single-ended voltage supplies can be connected in series with their common conductor available for a connection. In this case, the outer (−) terminal is negative relative to the common shared terminal and the (+) terminal is positive relative to the center tap. With the combination of the common, center tap electrode, these three terminals make up a *bipolar power supply*. **Figure 9.16** shows an example of two 9.5 V batteries connected in series, providing a +/- 9.5 V bipolar supply relative to the center, common terminal. In this example, the DMMs are measuring the voltage of each external terminal, relative to the common terminal.

Figure 9.16 An example of a bipolar voltage supply created by combining two batteries in series. They provide a +/- 9.5 V supply relative to the center tap conductor.

When the shared terminal, or center tap, is connected to the circuit-ground of a circuit, the two outer terminals provide a positive and negative voltage supply relative to the circuit-ground.

A bipolar power supply can be connected to earth-ground or it can float. When used to power a circuit, if the circuit floats, the supply can float. If the center tap terminal is connected to earth-ground, connecting the center tap terminal to circuit-ground will also connect the circuit-ground to earth-ground.

A power source can be single-ended or bipolar. It can also be floating or connected to earth-ground.

9.10   Single-Ended, Earth-Ground Referenced Measurements

All voltage measurements measure the voltage difference between the instrument's two input terminals. In this respect, all voltage measurements are differential measurements.

When one of the input terminals of the measurement instrument also connects to its own circuit-ground, we refer to this measurement as a *single-ended voltage measurement*. The voltage measured is relative to or with respect to the circuit-ground of the measurement instrument.

When the instrument does not connect either of its input terminals to the instrument's circuit-ground, we refer to this as a *differential voltage measurement*. This distinction is illustrated in **Figure 9.17**.

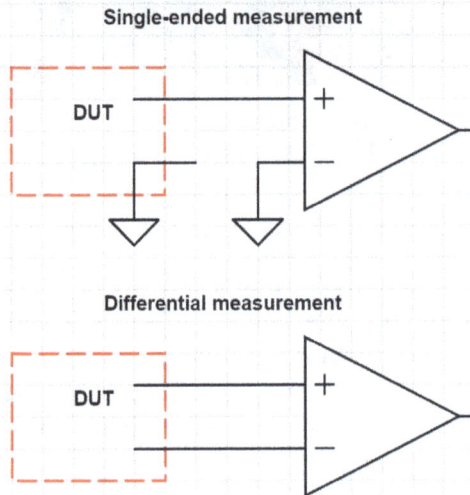

Figure 9.17 Examples of single-ended and differential measurements.

When measuring the voltage difference between two conductors in a DUT, one of the DUT's conductors can be its local ground or not. In a single-ended measurement, the second conductor to which the instrument connects to the DUT is always the circuit-ground of the instrument. The act of measuring connects one conductor of the DUT to the instrument's circuit-ground.

In most benchtop scopes, the power cable that plugs the scope into the wall power connects to earth-ground. This means the scope's chassis is usually connected to earth-ground. Its internal circuit-ground is connected to the chassis which connects to earth-ground.

Most bench-top scopes have coax connectors on their front panel to connect with a cable or probe to the DUT. The outer shield of the coax cable is connected to the chassis, which is connected to earth-ground. **Figure 9.18** shows a closeup of the typical BNC connector on the front of a benchtop scope. This means most scopes, when plugged into wall power with a 3-prong plug, will only measure single-ended signals.

Figure 9.18 An example of a scope front panel. The shield of the BNC bayonet connectors on the front of the scope is connected to the scope's chassis, which is connected to earth-ground. Note there is a small terminal lug on the front panel to the left of the print label, which is identified as another connection to earth-ground.

When a probe is plugged into the coax connector of the scope, the voltage measured is always a difference between the center connector and the outer conductor, which is connected to earth-ground. A single-ended measurement in a scope is always using earth-ground as the reference conductor.

This means never try to use the scope probe as a differential probe in a circuit, especially a circuit that is not floating. Be aware of the invisible connections between the DUT circuit-ground, chassis-ground and earth-ground, and the scope probe's reference conductor's connection to the scope's circuit-ground, chassis-ground, and earth-ground.

All 10x scope probes, such as shown in **Figure 9.19**, use the outer shield of the coax cable as the reference conductor from which to measure the voltage of the tip. These probes only measure single-ended, earth-ground referenced signals.

Figure 9.19 Example of a 10x probe showing the coax shield connected to the long ground lead of the tip. This is connected to the chassis-ground of the scope which is connected to earth-ground through the power cable.

Only use a 10x scope probe with the ground lead of the scope probe in a single-ended measurement with the probe's ground lead connected to the circuit-ground of the DUT.

9.11 Differential Probe Measurements

To measure a differential voltage between two nodes in a DUT that are at some voltage above earth-ground, either use two single-ended probes and subtract their voltages with a math function or use an active differential probe.

In contrast, the two input terminals of a DMM or the Digilent AD2 scope can measure true differential signals. These instruments are shown in **Figure 9.20**.

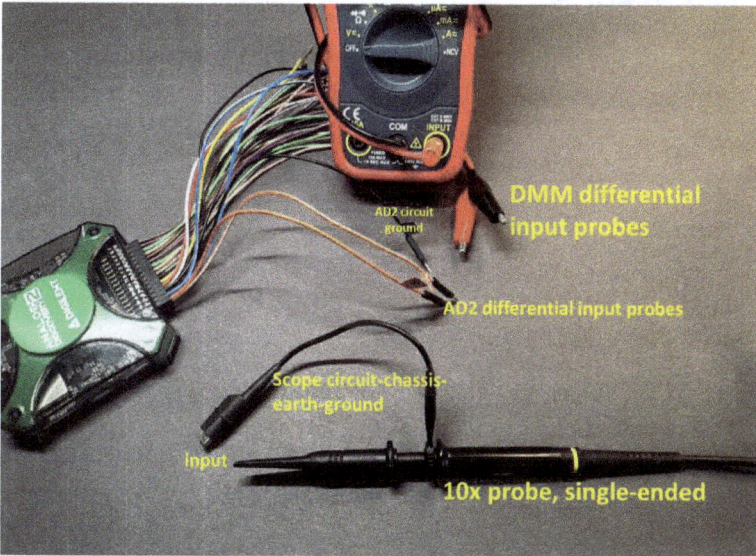

Figure 9.20 Examples of three different probes: floating and differential, floating with circuit-ground reference, and single-ended.

There is always some common (average) voltage between the two nodes of the DUT and its local circuit-ground. When the measuring instrument with differential inputs connects to the DUT, this common voltage will also appear as the common voltage between the instrument's input and its local circuit-ground.

In every instrument, there is a limit to the maximum common voltage that can appear between the average of the inputs and the instrument's local circuit ground. The instrument is capable of measuring true differential inputs, as long as the common voltage is below a limit, usually on the order of 10 V. Otherwise, the common voltage signal may saturate the differential inputs and introduce a measurement artifact. This is why it is always a good practice to connect the circuit ground of the DUT to the circuit ground of the instrument. In this case, be aware of the common voltage on the nodes of the DUT and make sure it is below the limit for the instrument.

When the DUT's circuit ground is floating relative to earth-ground, and the connection to the instrument is with differential probes, and their local circuit grounds are not connected together, there is the potential for an artifact from two sources. This can be understood from the equivalent circuit model for this configuration, in a Digilent AD2 scope, for example, as shown in **Figure 9.21**.

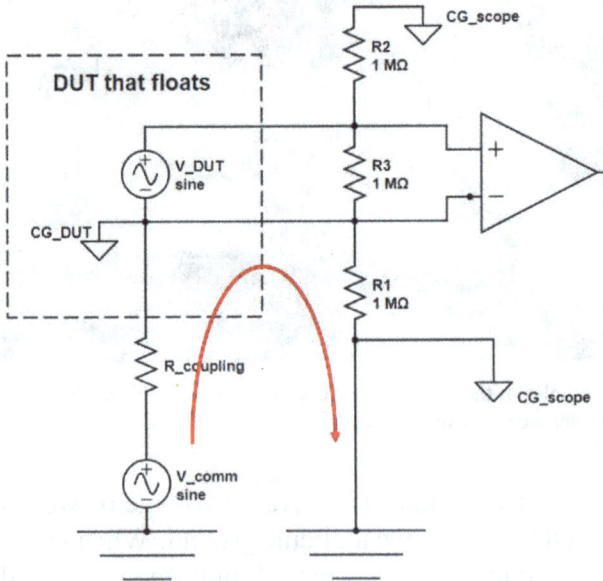

Figure 9.21 Equivalent circuit model of a floating DUT and differential measurements in an AD2 scope. The local-circuit-ground of the DUT (CG_DUT) and local-circuit-ground of the scope (CG_scope) are not connected together.

In this example, there will always be some coupling between the floating DUT and earth-ground. This could be through the hot side of the isolation transformer powering the DUT or just AC pick up from the power lines radiating in the lab coupled to the floating circuit ground of the DUT.

Generally, the source impedance of this coupled noise is high, but the voltage of the source can also be large. The resulting voltage across the 1 meg common resistor will generate some common voltage in the differential amplifier of the scope.

Some of this common voltage will appear as differential voltage, due to the finite common mode rejection ratio (CMRR) of the differential amplifier from imbalances in the resistor networks. If this common voltage exceeds the maximum allowed common voltage, the output of the differential amplifier will be affected.

The way to fix this problem is to make sure the local circuit ground of the DUT is connected to the local circuit ground of the scope. This will short the common voltage generated by the pickup from the floating DUT. By shorting the local grounds together, the DUT no longer floats.

The measurement of the DUT is still a differential signal between the two nodes inside the DUT, but connecting the circuit grounds together ensures that the common voltage on the differential nodes is within the restricted range.

For example, an Arduino board has an on-board 3.3 V low drop out (LDO) linear regulator. This produces a 3.3 V voltage, capable of sourcing as much as 500 mA, relative to the board's circuit-ground.

When the Arduino board is powered by an AC to DC wall wart power supply with a 2-prong plug, the circuit-ground of the Arduino board floats.

An AD2 is used as a differential probe to measure the voltage difference between the 3.3 V rail and the circuit ground of the DUT. When the scope's circuit ground is not connected to the DUT circuit ground, there is enough common voltage to saturate the differential amplifier and distort the output.

When the circuit ground of the AD2 connects to the circuit ground of the DUT, effectively connecting the circuit ground of the DUT to earth-ground through the scope, the AC pickup common voltage is shorted out, and there is no distortion in the differential voltage, measurements.

This example with the differential measurement of the voltage rail of the Arduino board with and without the circuit grounds connected together is shown in **Figure 9.22**.

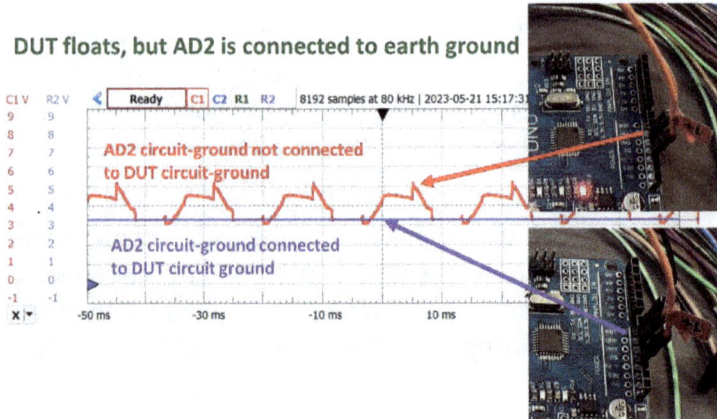

Figure 9.22 Measured artifact of power line noise when the earth-grounded differential amplifier connects to the floating DUT. Eliminate this artifact by connecting the circuit-ground of the scope to the circuit-ground of the DUT.

Do not connect the differential inputs of an earth-grounded amplifier to a floating DUT. This will generate an artifact of considerable noise on the differential amplifier.

A benchtop scope, with single-ended inputs can measure differential voltages using an external differential to single-ended amplifier.

An example of the LeCroy AP033 active differential probe that converts the differential signal at the input into a single-ended signal at its output is shown in **Figure 9.23**. It is also limited in the voltage of either of the two differential inputs of the DUT to be within 10 V of the circuit-ground, chassis-ground and earth-ground of the scope. To avoid the common noise current of a floating DUT, be sure to connect the differential probes' circuit ground to the DUT circuit ground.

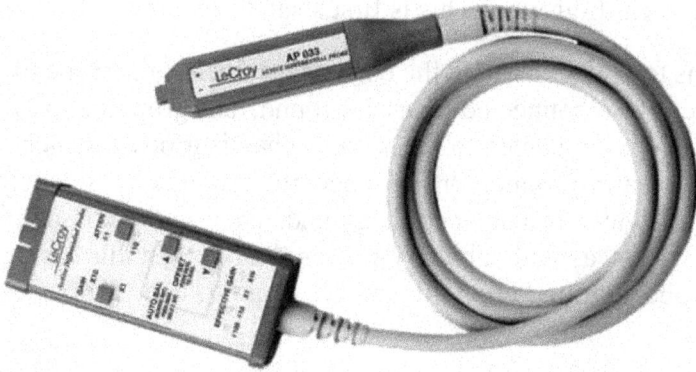

Figure 9.23 An example of a differential to single-ended probe that allows a differential voltage measurement in a DUT. The DUT circuit-ground must be connected to the scope's earth-ground.

9.12   Avoid Common Ground Loop Problems

When we measure a real DUT, we always have to be aware of its grounding conditions and the impact of noise picked up from conducted and radiated emissions due to the power distribution system in the measurement-DUT system. Once this is understood, we can measure the signals from the DUT, recognize measurement artifacts, and engineer a method to avoid them.

The concept of ground introduced in this chapter describes the different types of ground and why it is so important to use the correct prefix for ground. It also discusses how these terms and features apply to the scope, the measurement system, and the DUT.

First and foremost, ground is about safety. Any voltage over 50 V can be lethal. Lower voltage can be lethal as well, under some conditions, so never handle voltages above 20 V unless you understand the safety implications. While it is important to learn from your mistakes, it is difficult to apply any learning after a lethal mistake.

An important safety consideration is to use a scope that has an earth-ground connection. This means it uses a 3-prong plug. Unless otherwise noted, assume the chassis of the scope is connected to earth-ground. This is first a safety feature.

This means to always assume the coax cable shield of a probe to your scope is also connected to earth-ground. The ground clip of the 10x probe is connected to the scope's chassis ground, which is connected to earth-ground. Only connect the ground clip to a point in your DUT which is also at earth-ground. This is not an issue if your DUT floats, as with a battery-powered device. In this case, the 10x probe's ground clip will connect your DUT to earth-ground.

9.12.1 Noise from Ground Connections: Ground Loops

There is a very confusing source of noise that will appear whenever there are multiple earth-ground connections to devices. This is from conducted noise related to noise from the distribution of earth-ground connections, which we lump in the general heading of *ground-loop noise*.

A generic earth-ground connection between two different devices is shown in **Figure 9.24**. This figure is an example of situational awareness: seeing the real physical world through your engineer's mind's eye as an equivalent electrical circuit model. We analyze the behavior of the system and understand why it behaves the way it does based on this equivalent circuit model. It will be the model to understand the root cause and the potential solutions for ground loop noise.

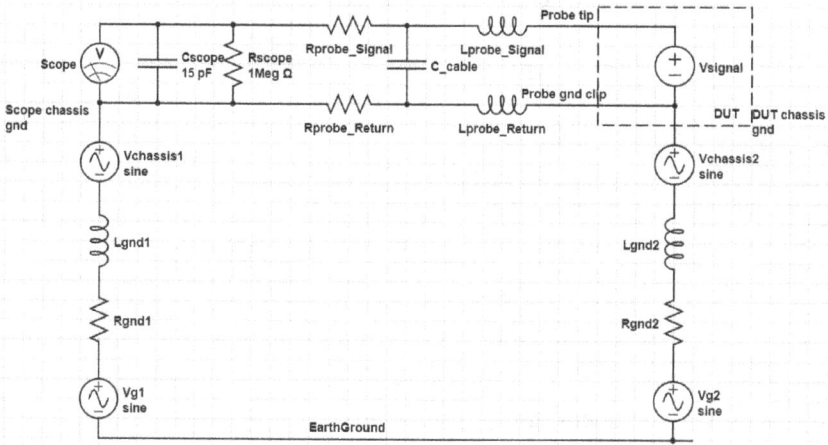

Figure 9.24 The equivalent circuit model of the scope and its connection to earth-ground and another DUT.

In this circuit, the connection to the earth-ground from two instruments is shown at the bottom. Between the external earth-ground location and the scope's chassis, there are some impedances due to the interconnects, as well as two additional noise sources. One source of noise is the noise between the earth's ground and the chassis wiring. This is externally generated noise from the power distribution network powering the instruments. The second source of noise is the noise generated by the instrument itself, which makes its chassis a different voltage than the earth-ground.

The same structure is modeled for the DUT.

The probe is modeled as a simple RLC interconnect. There is some capacitance between the signal and return pins from the cable capacitance and some equivalent series R and L in each path. The coax cable of the 10x probe cable is unbalanced in that the electrical properties of the signal and return path are not the same. This means that a common current generated, for example, by just the probe's ground connecting to a voltage source between the DUT and earth-ground, will be converted into some differential signal and be measured by the scope voltmeter. This is a common signal converted to a differential signal from the unbalanced interconnect.

This is easily demonstrated by connecting the ground lead and signal tip together. Normally, the measured voltage in this case would be the noise floor of the scope. Then, a function generator, with its chassis connected to the earth-ground, producing a 20 V peak-to-peak open circuit voltage and 400 mA peak-to-peak current when short-circuited, is connected to the common connection of the probe tip.

In principle, there should be no voltage measured by the scope. In practice, some fraction of the 400 mA common current flowing in the scope probe is converted into a differential voltage across the 1 meg input resistance of the scope. The actual measured voltage by the scope is about 80 mV peak to peak. This is illustrated in **Figure 9.25**. This is an example of how a common current can be turned into a differential measurement by the unbalanced scope probe. This will happen with any coaxial connection.

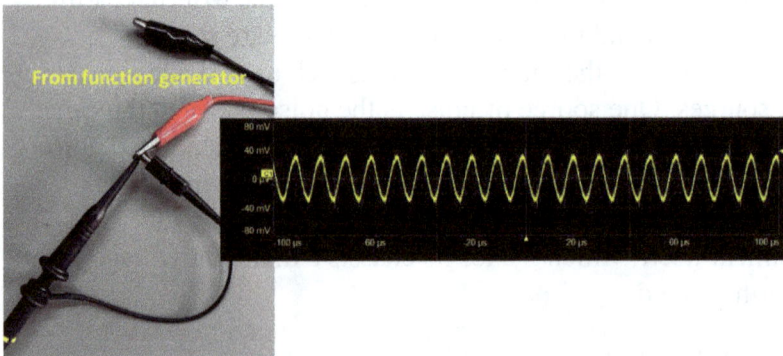

Figure 9.25 An external function generator is used to drive a 100 kHz common signal on the 10x probe, which is measured by the scope as a differential signal across the 1 meg scope resistance.

This means that any common signal on both the signal tip and ground lead tip of the 10x probe, which connects to the DUT, will be turned into an artifact of measured voltage even if the signal at the DUT is zero, but the DUT chassis is driven by the noise on the earth-ground connection. This is a form of earth-ground bounce which is a ground loop problem since it is about the difference in voltage between earth-ground locations.

Generally, the ability of a probe to reject these common voltages is measured by its common mode rejection ratio (CMRR). The higher the CMRR, the less sensitive the measurement system is to common signal interference. Differential probes and amplifiers are designed with a high CMRR. When there are common noise problems, such as from ground loops, using a differential probe and amplifier is a good alternative to using a single-ended probe to measure the voltage on a DUT, even if it is a single-ended signal, referred to its chassis ground.

Unfortunately, the CMRR of a differential measurement usually decreases at higher frequencies. In most amplifiers, the CMRR can be as large as 80 dB below 1 kHz, but as low as 20 dB at 10 MHz. This means high-frequency ground loop noise will often show up as what appears to be high-frequency noise, with a periodicity of 60 Hz.

Using this model for the scope-DUT system, we can identify the source of noise from different system-level grounding problems, which is in the general category of ground loops. From this simple model, we can understand the root cause of the ground loop noise, how to measure it, and how to reduce it.

9.12.2 Ground Loop Noise When the DUT Floats

The best case, the lowest ground loop artifact noise, is when the DUT is floating with a high impedance to any nearby earth-ground. In this case, the chassis noise voltage, Vchassis2 , can be very small, and the impedance at high bandwidth between the DUT chassis and the earth-ground is very high. This is the best case for a voltage measurement on the DUT. Artifact noise will be from RF pickup in the scope probe. **Figure 9.26** shows the noise baseline, as displayed in the spectra on a 2 MHz and 20 MHz full-scale range. There is no difference when the scope's input is grounded or connected to the probe with its signal tip and ground lead shorted together. This is a combination of scope noise and ground loop noise in the system.

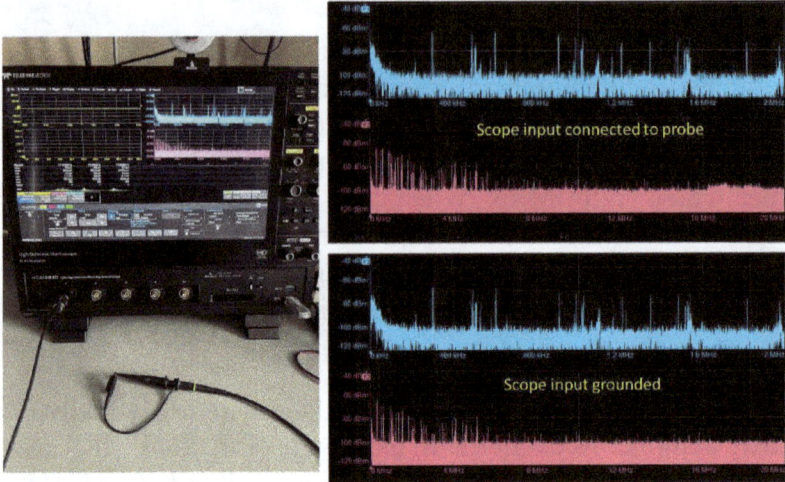

Figure 9.26 Establishing a baseline measurement of the noise of the scope and probe.

Next, the 5 V power rail of an Arduino board was measured when the Arduino was battery-powered. This is the best case of minimal ground loop noise. The only change in the spectrum was the addition of sharp peaks at 8 MHz, 16 MHz, and multiples. The sharp, narrow peaks are the signature of a clock, which is the 16 MHz clock of the Arduino. The subharmonic at 8 GHz is probably due to some operation that happens at two clock cycle intervals.

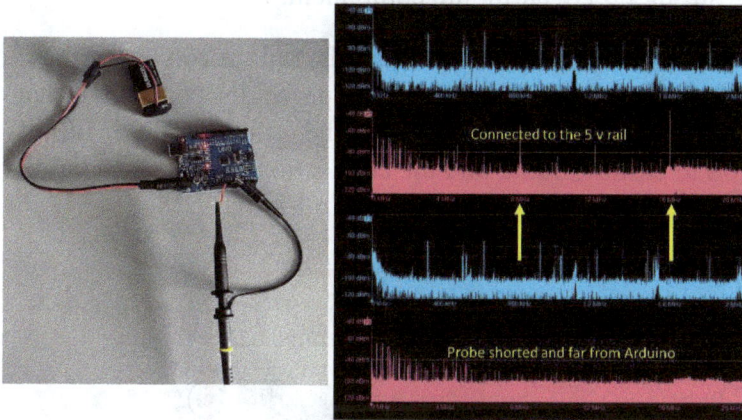

Figure 9.27 Measuring the 5 V rail noise when the Arduino is powered by a 9 V battery. Note that the only difference in the spectrum is a peak at 8 MHz and a very large peak at 16 MHz, which is the clock frequency of the Arduino.

There is no additional noise present in the 5 V rail measurements. This low noise is expected since this is a linear regulator, and there is little load on the supply.

When the Arduino is powered by an AC to DC converter, the noise is dramatically different.

9.12.3 Ground Loop Noise in AC to DC SMPS

All AC to DC converters, either benchtop or small power modules, are "galvanically" isolated from earth-ground. But they are strongly coupled through their isolation transformers to earth-ground and generate their own chassis-to-earth-ground noise. This is a notorious problem in SMPS systems.

In this specific AC to DC converter, its chassis-ground is connected to its circuit-ground on the outside conductor of the power jack. This means it is not possible to directly measure the noise between the circuit ground to its chassis ground. Because of the isolation transformer, the voltage between the chassis-ground and earth-ground of any transformer isolated SMPS is always large and has a strong 60 Hz harmonic component.

When the 10x probe is used to measure the output voltage of the power jack, with the gnd lead of the 10x probe disconnected, what is measured is the ground loop between the output of the SMPS, its connection to earth-ground, and the connection back to the chassis-ground of the scope. This path is shown in **Figure 9.28**.

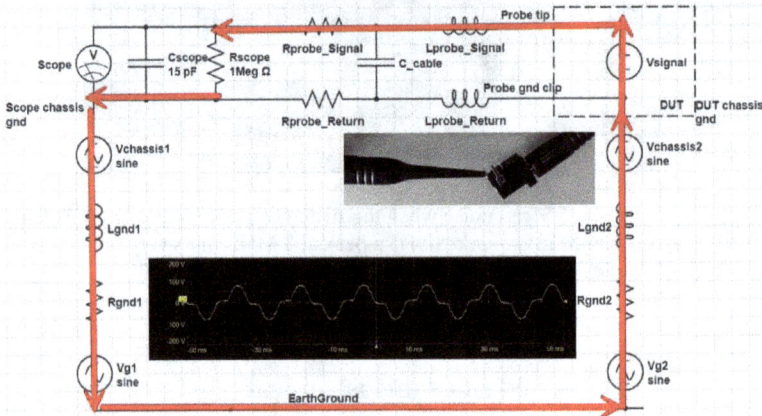

Figure 9.28 Connecting the 10x scope probe tip to the output of the SMPS measures the ground loop voltage shown in the inset.

The measured voltage in the ground loop between the floating SMPS and the chassis-ground of the scope is about 180 V peak to peak. In addition to the strong 60 Hz harmonic component, there is also a strong component associated with the switching noise. This is apparent in the spectra of the ground loop noise. This is shown in **Figure 9.29**.

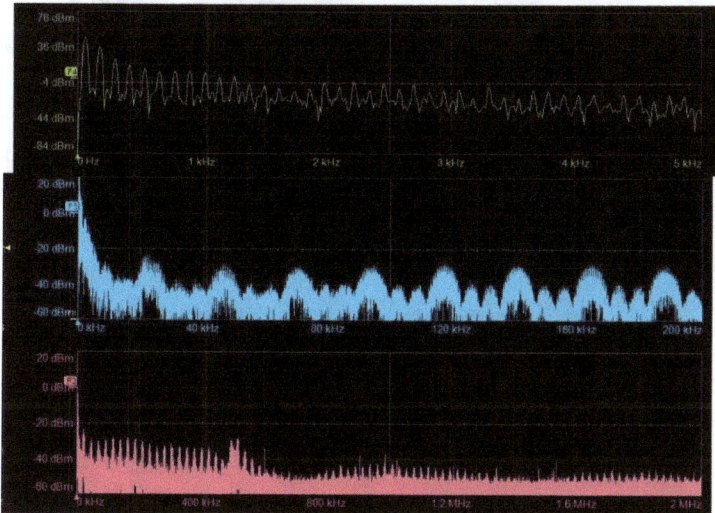

Figure 9.29 Spectra of the ground loop noise on three different frequency scales. This shows the strong 120 Hz harmonics of the power line and the 20 kHz harmonics of the SMPS extending into the 2 MHz range.

When the difference voltage is measured across the 9 V output of the AC to DC converter, the nominally 9 V DC signal has the added high-frequency noise due to the common voltage of the chassis to earth-ground ground loop noise converted to some differential noise by the probe. The magnitude of the spectral components are reduced by 30 dB, due to the CMRR of the probe, but the spectral components are still there. This corresponds to a peak-to-peak voltage noise of about 0.75 V out of 9 V, as seen in the direct measurements in **Figure 9.30**.

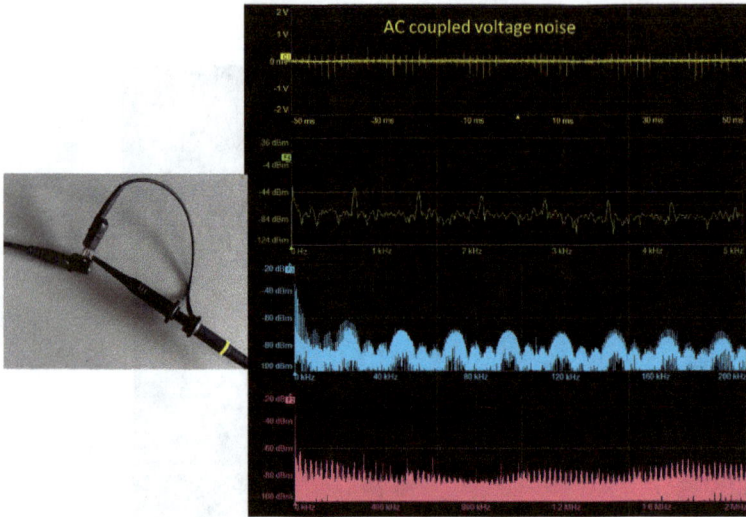

Figure 9.30 Direct measurement of the 9 V output from the SMPS showing the residual ground loop noise on the output, reduced by about 30 dB from the directly measured ground loop noise, due to the CMRR of the scope probe.

When the AC to DC converter is plugged into the Arduino board, the circuit ground and chassis-ground of the AC to DC SMPS will connect to the circuit-ground of the Arduino. This means that the ground loop noise will be present on any measurement of any voltage on the Arduino board, referenced to its local circuit ground, such as the 5 V rail. Even though the 5 V rail is known to be very low noise, relative to its local ground, the scope will measure the ground loop noise from the SMPS as part of its 5 V rail.

An example of the measured voltage noise on the 5 V rail of the Arduino board showing the 16 MHz clock harmonics and the ground loop noise of the SMPS is shown in **Figure 9.31**.

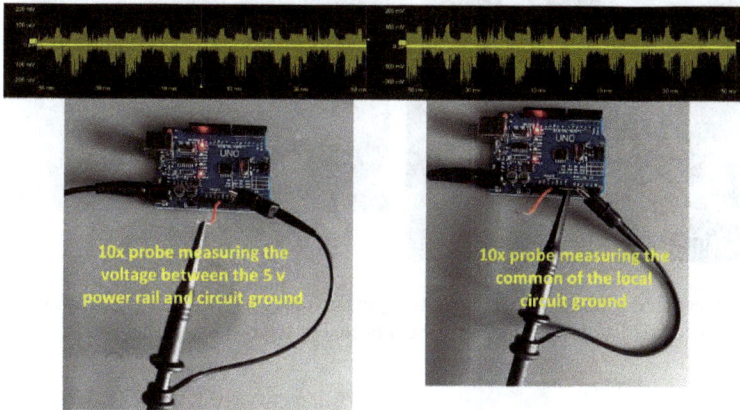

Figure 9.31 The noise measured by the scope is the same when measuring the 5 V rail or when both the signal and ground lead are connected to the circuit ground on the board. The common ground loop noise is measured in both cases.

Ground loop noise is always a problem when using a SMPS. Different supplies have a different amount of ground loop noise. This is sometimes an important selection criterion.

The RDDSPON LW-K3010D power supply is a particularly noisy power supply. When it is powered on, it generates almost 4 V peak-to-peak of ground loop noise. **Figure 9.32** shows the measured chassis-ground noise on the LW-K3010D powered on and the TACK MD02 powered on. The LW-K3010D has more than 20x higher ground loop noise than the TACK MD02.

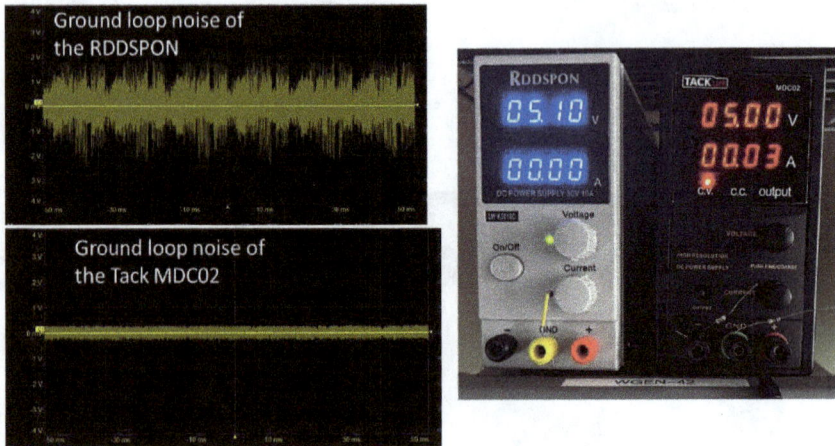

Figure 9.32 Measured ground loop noise of two power supplies when each is turned on, one at a time, measured with the scope probe connected to the chassis of the power supply.

When a device is powered off, its chassis noise with respect to the chassis of the scope is about the noise on the earth-ground connection. All devices connected to the same earth-ground connection will show the same ground loop noise when powered off. When a device generates significant ground loop noise when it is powered on, this noise will often contaminate the earth-ground connection.

For example, the ground loop noise on a function generator is the same when the function generator is powered on or powered off. It does not generate any measurable conducted noise on its chassis relative to earth-ground. However, when the LW-K3010D power supply is powered on, it pollutes the common earth-ground connection, and a higher amount of ground loop noise appears on the function generator's chassis. This is shown in **Figure 9.33**.

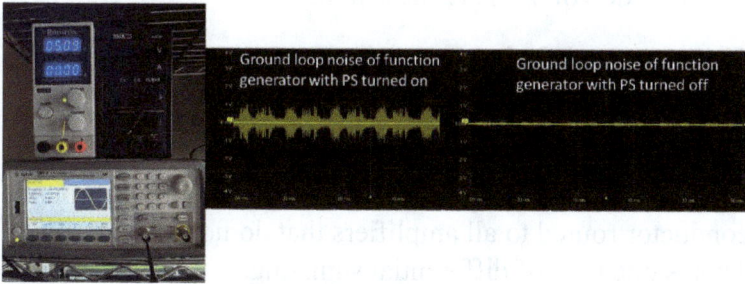

Figure 9.33. Ground loop noise was measured on the function generator with and without the LW-K3010D power supply turned on and off. The scope is measuring the chassis noise relative to earth-ground.

This means that sometimes identifying and turning off the source of ground loop noise on the power line earth-ground can reduce ground loop noise between devices.

9.13 The Bottom Line

1. Get in the habit of using the appropriate prefix to the word ground.

2. There are at least six different types of ground, each with a different meaning: earth-ground, chassis ground, circuit ground, EMI ground, local ground, and return path.

3. Earth-ground is a safety issue. It is literally connected to the Earth.

4. A circuit that is not connected to earth-ground is called floating. Always try to connect floating circuits to the earth-ground for safety reasons.

5. Chassis-ground is the large metal of a product or instrument enclosure. It should be connected to the earth-ground inside the enclosure, but it may not be.

6. Circuit-ground is the reference node to which all other circuit node voltages are measured.

7. The return path is the conductor that carries the signal or power return current. It should not be used as the reference node in the circuit, but it often is.

8. When low noise is important, use a separate circuit-ground conductor routed to all amplifiers that do not carry current. This is one type of differential signaling.

9. When the return path and circuit ground are the same conductor, make them a wide plane with a minimum series resistance and inductance.

10. When large transients are present in the circuit-ground conductor, ground bounce noise will be created, and different points on the same circuit-ground conductor will have a different voltage. The local ground may depend on location.

Chapter 10
Basic Scope Measurements of Common Signals

You can explore and experiment with the most common features of every scope using a free tool that will run on your computer. But ultimately, you will use a real scope. There is no substitute for hands-on experience with a real scope measuring real signals.

10.1 Measure a Signal for Which You Know the Answer

When using a new scope or other instrument or tool for the first time, to apply Rule #9, you must know what to expect to see in your measurement. It is difficult to anticipate what to expect when measuring a brand-new DUT using a brand-new scope or other instrument.

Whenever you are performing a measurement and see signals you do not anticipate and are not sure why you observe them, one source of artifacts may be your measurement system or the methodology you are using to set it up. To distinguish issues in your DUT from your measurement system, you should always go back to measuring a DUT for which you know exactly what to expect.

When you perform a measurement, if the result is not what you expect, always on your list of likely suspects to investigate is the possibility the scope is either not set up correctly or is out of calibration. While this is rare, it is still possible.

The best way to verify you are using the best measurement practices and that your scope is working to its specifications is to measure a signal for which you already know the answer.

10.2 The Seven Most Important Best Measurement Practices

In the following examples of using a scope to measure and analyze some commonly found signals, you can practice applying the seven general, best measurement practices that apply to all instruments:

1. Follow Rule #9.

2. Perform all the consistency tests you can think of.

3. Apply situational awareness to identify possible measurement artifacts by understanding the equivalent circuit of your DUT, the fixture and cables, and your instrument.

4. With a new instrument, start by measuring something for which you know the answer.

5. Always try to estimate a measurement directly from the front screen before you apply built-in, complex measurement functions.

6. Turn the raw measurements into information in the form of a few figures of merit.

7. Use the information to answer the "So What?" question. What decisions will be made based on the new information obtained from the measurement?

These best practices will be applied by measuring two commonly found signals:

✓ *The compensation signal on every scope*

✓ *A built-in or external function generator*

10.3   The Default Setup Button

Sometimes, you are not the only person using your scope. Most scopes save the settings used right before they are powered off. This means that when you turn on your scope, it may have settings someone else used that you are unaware of. It is best practice to always push the Default settings button on your scope to bring it into a known state.

An example of the location of the default setup button on a Teledyne LeCroy HDO6104b scope is shown in **Figure 10.1**.

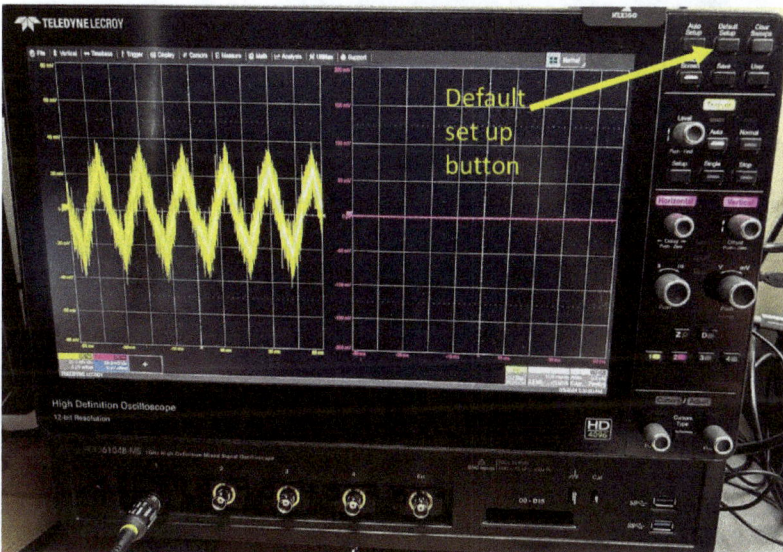

Figure 10.1 The location of the default setup button is usually in the upper right corner of most scopes.

The default settings vary from scope to scope, but they usually have a vertical scale of about 5 V/div, an input coupling of 1 M DC, and a time base of about 0.1 msec/div. Once in the default state, there is less chance of an obscure setting you are unaware of distorting your measurements.

This button is different from the auto-scale button. While the auto-scale button is on all scopes, it should rarely be used. It will perform measurements on each channel to see the incoming signal levels. Based on what its internal algorithm suggests as a signal someone would want to measure, it will adjust the vertical and horizontal scales to show the signal on each channel.

However, how does the scope know what you want to see? The auto-scale button is often referred to by experienced scope users as the "guilty pleasure" button. It is not that the auto-scale button is too easy; even worse, it can be very misleading. It will almost always guarantee you will see a signal displayed on the front screen, but it may not be related to what you are looking for.

If you get in the habit of using it instead of anticipating the signal you expect to see and setting up the scales manually based on your expectations, sometimes, your display will be misleading, and you may miss the important signal features you care about most.

The right way to set up the scales is to understand what you expect to see, apply Rule #9, and manually set up the scales to see an initial signal. This way, you will have confidence that what you see is what you expect.

10.4   Input Coupling to the Scope

Most scopes generally have four options for coupling a signal into any measurement channel. These are:

- ✓ *DC with 50 ohm input resistance*
- ✓ *DC with 1 megaohm input resistance*
- ✓ *AC with 1 megaohm input resistance*

✓ *gnd*

Even though these are often selected from menu items in a scope's interface, they are hardware features of the scope. **Figure 10.2** shows how these are selected in the MAUI Studio scope emulator, as in most scopes.

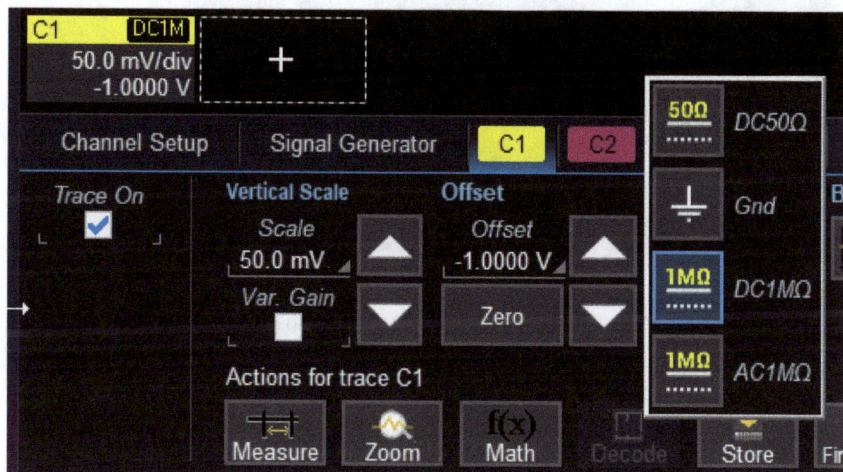

Figure 10.2 The four options for input coupling to a scope are selectable from a menu for each scope channel.

In the DC 1 megaohm coupling, a 1 megaohm resistor is connected between the input signal and ground. The voltage across this 1 megaohm resistor is measured by the scope. It is the coupling to use with a 10x probe and for most measurements with a bandwidth of < 100 MHz. This should become your default coupling to use unless you have a strong, compelling reason otherwise.

You can verify the input impedance of the scope on the 1 megaohm input setting by literally measuring the input resistance of a channel of your scope with a DMM set for ohms. In **Figure 10.3**, the DMM reading for a Keysight DSOX3024 scope is shown as 1.004 megaohm.

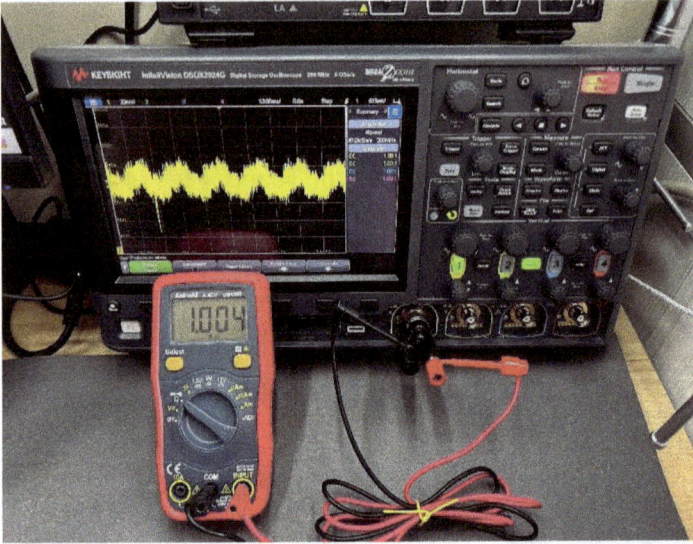

Figure 10.3 A DMM measures the input resistance of a channel set for DC 1 megaohm. The scope scale is offset and expanded to 20 mV/div and 10 msec/div. The 60 Hz frequency is immediately identified on this scale.

In addition, the voltage the DMM applies to a DUT to measure its resistance is measured as 520 mV. The signal was centered in the display, and the scale expanded to reveal the magnitude of the small fluctuations. There is about +/- 20 mV of 60 Hz pick up on top of the 520 mV DC component of the signal. This is not from the DMM itself. It is 60 Hz pickup in the long, unshielded cables from the DMM to the scope. The 60 Hz voltage from the nearby power lines is capacitively coupled to the high impedance of the circuit. This is an example of a measurement artifact.

The AC 1 megaohm setting adds a typically 0.01 uF capacitor in series with the 1 megaohm resistor. This equivalent circuit is shown in **Figure 10.4**.

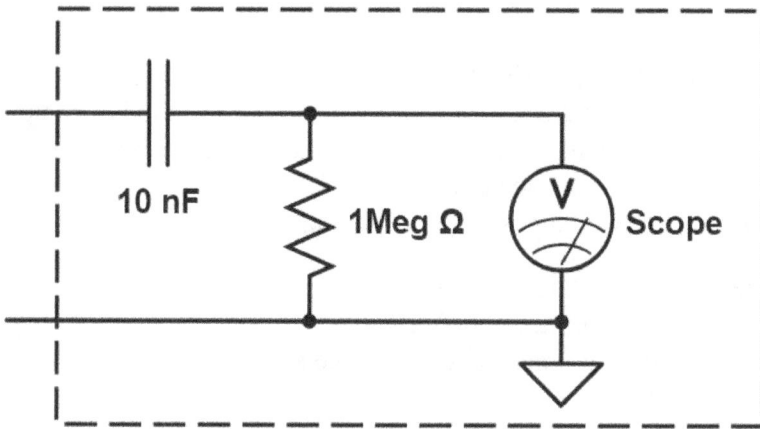

Figure 10.4 Equivalent circuit model of the scope input when set for AC 1 Meg setting.

Picturing this equivalent circuit model of the input to the scope in your engineer's mind's eye is part of situational awareness. This model enables you to evaluate how the scope will respond to input signals.

The input to the scope is a high-pass filter with a pole frequency of 16 Hz. This means that frequencies below about 16 Hz are blocked, and only higher-frequency components of the signal are measured by the scope. A DC value is blocked and only changes on top of the DC value are measured by the scope.

Most scopes have a 16 Hz high-pass filter when AC-coupled. This is why drift is often referred to as slowly varying signals with frequency components lower than 20 Hz. A scope on AC coupling is not sensitive to slow drift in the signal.

In principle, AC coupling should be used whenever you want to measure a small change on top of a large DC component. The downside is you lose information about the DC component, such as drift, and distort the signal displayed in time frames of about 1/16 Hz or 60 msec or longer. This is often important when looking at nominally DC power rails where the low-frequency variation is as important as the higher-frequency noise in the supply voltage.

A much better alternative to AC coupling to see small voltage changes on top of a large DC component is to use the DC coupling setting and the offset control to center the signal on the screen on an expanded voltage scale. In this mode, the signal is DC-coupled, and small, slow changes in the DC voltage can be observed.

Some scopes offer an input setting of ground input coupling, which is useful for checking the offset voltage of the input preamplifier. On this setting, the measured DC value should be within 1 mV of ground, or 0 V.

This setting opens up the connection between the signal from the DUT and the scope's preamp, shunting the input signal through a 1 meg resistor to ground and shorting the input to the preamp directly to the internal local circuit ground of the amplifier board. This setting is rarely needed unless you want to check for slight input offset voltages in the scope preamp. This is also why this setting is not commonly available on all scopes.

The DC 50 ohms option adds a 50-ohm resistor between the signal and ground. The preamp measures the voltage across this resistor. This can also be tested by connecting a DMM as an ohmmeter to the input of the scope channel, as shown in **Figure 10.5**.

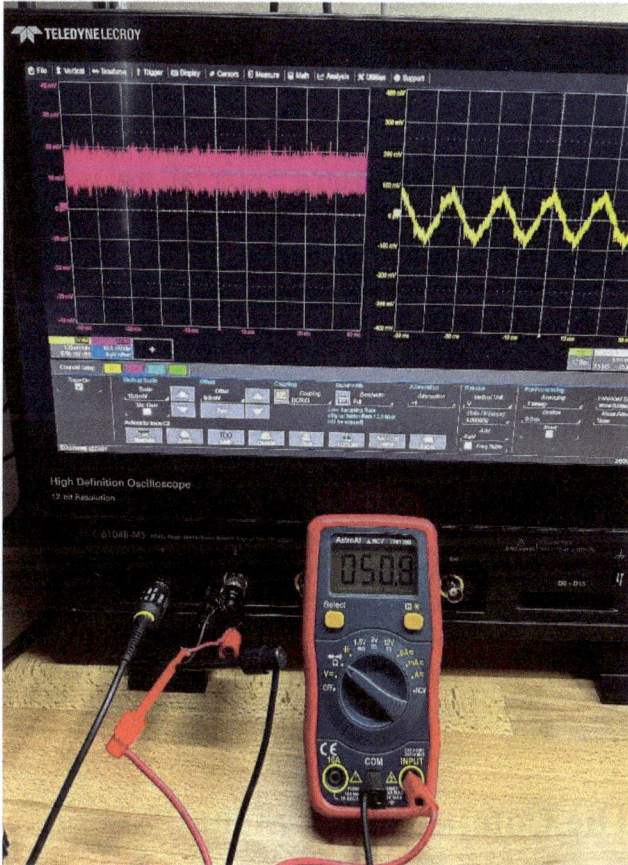

Figure 10.5 The measured input resistance when the input coupling is set for 50 ohms. The DMM reads 50.8 ohms, which is slightly above the 1% tolerance for the input resistance. However, the measurement uncertainty in resistance measurements of a DMM is always +/- the last two digits.

Never use the 50-ohm input resistance coupling unless you know your input signal very well. On this setting, the signal connects directly to a 50-ohm resistor to ground on a circuit board inside the scope. This internal 50-ohm resistor can safely dissipate up to 0.5 Watts. If it consumes more than 0.5 watts, it may get so hot as to melt its solder and fall off the board. This means you have to send the scope back for repairs, and the board has to be replaced. It only takes a 5 V rms signal at the input to consume 0.5 watts in a 50-ohm resistor:

$$P = \frac{(V_{rms})^2}{R} = \frac{(5V)^2}{50\Omega} = 0.5 \text{ watts}$$

If you measure the DC voltage on a power rail that is 7 V, for example, and the input coupling is set for 50 ohm, you will damage your scope. However, there is no damage if the input coupling is set for DC 1 megaohm input.

When measuring signals from general laboratory instruments with BNC connectors, a coax cable with BNC connectors will connect the instrument and the scope. As a general practice, use the 1 megaohm DC input coupling unless you have a compelling reason otherwise. This is a safe and robust setting.

If you are measuring a signal with a rise time shorter than about 10 nsec, and a source impedance other than 50 ohms, using the 50 ohm input to the scope will prevent reflection artifacts. This is the main purpose of the 50 ohm input setting and is described in great detail in the next chapter.

10.5 The Compensation Signal

The most important best measurement practice for a new instrument is to always start out measuring a signal for which you know the answer. This is one way to apply rule #9.

Two common signals readily available should be the first signals to measure: the compensation signal and a signal from a function generator. The measurement principles you learn from these two signals will be applied to all other signals.

The best signal source to start with, available on every scope, is the signal used to compensate a 10x probe. This is often labeled as the comp or cal signal and has lug connections available on the front of the scope.

10.5.1 Direct Connection with a Coax Cable

In this first example, we will use the compensation signal to explore some of the scope's measurement features. The probe we use initially is a mini grabber on the end of a coax cable. This way, we have higher confidence that we are looking at a signal that may have less distortion than a potentially uncompensated 10x probe. Alternatively, you can use a 10x probe after it has been compensated by this compensation signal.

In this example of measuring the comp signal with a coax cable, the input coupling to the scope is set for 1 meg. Whenever you use a 1 megaohm input resistance *and* a coax cable, *always* think about potential reflection artifacts that can arise. This topic is covered in detail in the next chapter. For now, this artifact will be ignored.

Coax cables come in a variety of specifications. The most common cable is the RG58. This label is usually written on the side of the cable. It is a very popular coax cable and is available on Amazon, for example, for about $10 for a 6 ft cable. On the ends are connectors compatible with most scopes, called bayonet or BNC connectors.

BNC stands for Bayonet Neill-Concelman, introduced to the industry in the 1940s in radio systems. It has become an industry standard for cables with signal bandwidths from DC to about 2 GHz. Some versions of a BNC connector can be used at bandwidths as high as 10 GHz.

These slip on the end of the scope's channel connection and rotate 90 degrees to lock in place.

Another popular cable is RG174. This is a small-diameter coax cable and is very flexible. These cables are also available on Amazon for a comparable price. They typically come with either BNC or SMA connectors on the ends. There are other comparable cable types, such as RG316. Just be aware that some cables use BNC connectors but are not 50 ohms. For example, RG6 cables are

75 ohms. Using a BNC connector is not a guarantee of a 50-ohm cable.

A coax cable with a BNC connector on one end and the other end opened up and connected to mini grabbers can be purchased on Amazon for about $4 each. These make excellent, general-purpose probes for any lab. **Figure 10.6** is an example of a BNC coax cable with a minigrabber on one end.

Figure 10.6 Example of a BNC cable with mini grabbers on the ends.

The BNC end of the cable connects to the scope channel and the mini grabber ends clip to the terminals of the compensation signal output on the scope.

An example of the connection to the compensation signal on two scopes is shown in **Figure 10.7**.

Keysight 4024 scope   Teledyne LeCroy HDO8108 scope

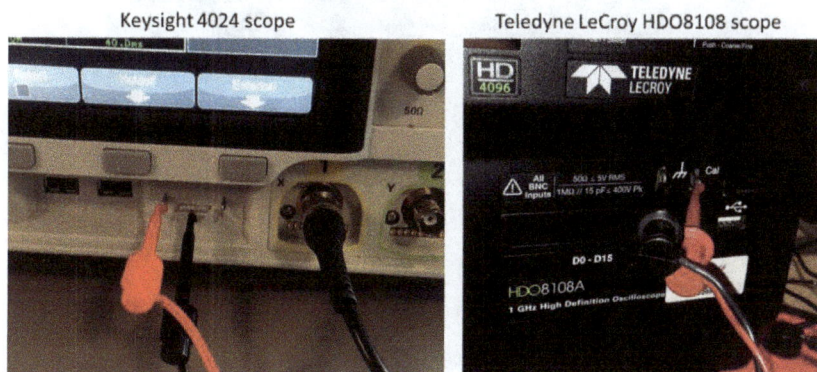

Figure 10.7 Connections to the comp signal on two different mid-level scopes.

Whether using a 10x probe or the mini grabbers on the end of a BNC cable, use an input coupling of DC and 1 meg input resistance. This should always be the default coupling to use unless you have a strong, compelling reason to do something different.

Before you look at the signal on the scope, apply Rule #9: What do you expect to measure? You should see a 1 kHz square wave signal with a period of 1 msec and an amplitude of about 2 V. Using this expectation, it is easy to set the scope scales to see the signal. A voltage scale of about 1 V/div and time scale of 1 msec/div will show the signal very easily. The trigger should be set for auto-mode, the channel being measured, rising edge, and a level of about 1 V.

Figure 10.8 shows a measurement of the comp signal on the Teledyne LeCroy WavePro HD scope. This particular compensation signal has a peak-to-peak value of 2.5 V.

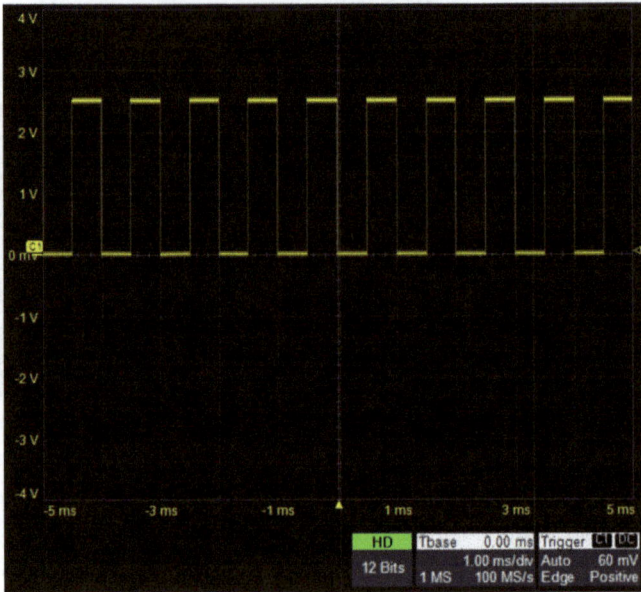

Figure 10.8 Example of the measured comp signal set up on the recommended scales.

This is a perfect signal to use to explore the three most important features of every scope: the vertical scale, the horizontal scale, and the trigger.

10.5.2 Adjusting the Scale Settings

Before you begin the measurement, identify where on the scope, the three most important controls are for the scope:

- ✓ *The vertical scaling and offset for each channel*
- ✓ *The horizontal time base controls*
- ✓ *The trigger controls*

For the Keysight DSOX3042 scope, these controls are identified in **Figure 10.9.**

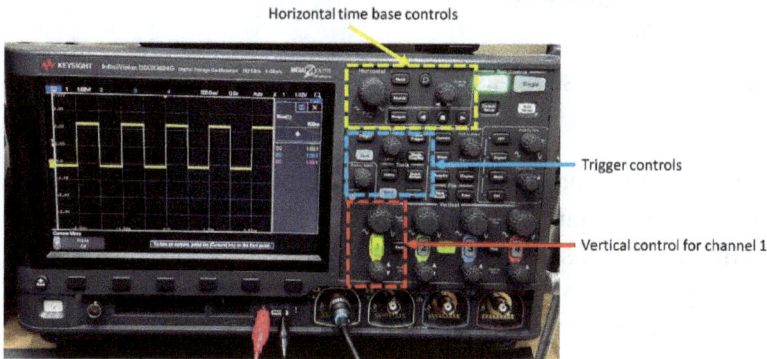

Figure 10.9 Example of the location of the three important controls for this Keysight 3024 scope.

Once you see the signal and it matches your expectations, you can adjust these settings to see the impact a change makes.

The compensation signal is a repetitive signal. This means that the trigger can be set for channel 1 as the source, on the rising edge, in auto mode, and adjust the trigger-level knob to center the trigger level roughly midway between the high- and low-level signals. On many scopes, you can push the trigger-level adjust knob to auto-center the trigger level midway between the peak-to-peak values of the signal on the triggered channel. This is a quick and simple way of finding a good trigger level.

It is important to build "muscle memory" by playing with the control settings and getting a feel for how the knobs affect the scales.

✓ *The vertical scale adjustment expands or contracts the trace vertically on the screen.*

✓ *The vertical offset moves the trace up and down on the screen.*

✓ *The horizontal scale adjustment expands or contracts the time base of the trace displayed on the screen.*

> ✓ *The horizontal offset moves the trace left or right on the screen. It effectively moves the location of the t = 0 position.*

Each button, on most scopes, also has a hidden feature when pressed in.

When the horizontal or vertical scale adjustment knobs are pressed in, the scale adjust changes from coarse changes, with multiples of 1x, 2x, and 5x changes with each click of the knob, to the fine adjust setting, with typically 2% increments in the scale adjustment with each click. Rarely should the fine adjust be used as this will result in scales that are difficult to read directly off the front screen.

When the horizontal or vertical offset control knobs are pressed, the location of the ground voltage is vertically centered on the screen, and the location of t = 0 is horizontally centered on the screen.

Note that in all these settings adjustments, the signal and its values are not changing, only its position displayed on the screen changes.

The most important feature to note is how easy or difficult it is to read important figures of merit off the front screen, like the period or peak-to-peak voltage, when the scale settings are adjusted.

The most important best practice when adjusting the scales is to end up with a scale setting that makes it easy to read figures of merit directly off the front screen with your Mark 1 eyeball. This is always the starting place in every scope measurement.

10.5.3 Adjusting the Trigger

The comp signal is a repetitive signal with a period of 1 msec. When setting up the trigger, be sure to select the channel on which the signal is connected, the simple edge triggering on the rising edge, and a DC level between the peak-to-peak values.

The initial mode should always be auto. This will ensure that even if the trigger level is set outside the signal's range, the scope will still trigger, and you will see something displayed on the screen.

With the trigger setting on auto, press the trigger-level button, and the trigger level will automatically adjust to midway between the peak values.

Alternatively, move the trigger level up and down until it is within the min and max values of the signal, and you will see the signal stationary on the screen. Move the trigger level above the signal level, and suddenly, the trace moves around and is not stationary on the screen.

In auto-trigger mode, when the scope does not see a trigger event, it waits about 50 msec and triggers the scope anyway. Whatever happens to be recorded in the acquisition buffer at this instant is displayed on the screen. Since the arbitrary time is asynchronous with the pattern of the signal, there is no correlation between the newly displayed signal and the previous one. This makes the display jump all over the screen.

It is tempting to use the stop acquisition button to freeze the screen at the last acquisition to see the signal. While this will display only the previous acquisition, you will lose the information about how this signal from the DUT might be changing over time.

If your signal is periodic, you should be able to make it appear stationary by adjusting the trigger level. If you can't go through the trigger setup and verify you have the correct channel, mode, feature, and level, something else is going on.

Only if you are unable to see the repetitive pattern displayed stationary on the front screen should you use the stop button. In this event, it is worth exploring why you can't use the trigger function to get a stationary pattern. You may learn an important feature about your signal, or how you are using your scope.

If you do use the stop acquisition button, there is another important artifact to always be on watch for. In the stopped mode, the scope is not acquiring new measurements. You see displayed on the screen the stored measurements in the acquisition buffer from the ADC. This means that if you change the display settings for the screen, you are only replotting the measurements; you are not changing the amplifier gain settings or the sampling rate of the scope.

Be aware of the fixed vertical resolution and the fixed sample rate from the ADC when in the stopped mode. If you are acquiring live measurements, you will often change the amplifier gain and the sampling rate when you change the scales. This is not the case in stopped mode.

10.5.4 Extracting Important Figures of Merit

When you perform a measurement of a signal from a DUT, you often times record and display as many as 1,000 to 1,000,000 individual V(t) measurements. The first task in understanding your measurement is to translate these raw measurements into useful information.

We will always compare the waveform we record to an ideal waveform, which is mathematically characterized by a few figures of merit. We will approximate the real waveform by the ideal waveform's figures of merit.

We first decide what ideal waveform this signal is closest to and identify the figures of merit that completely describe the ideal waveform. Then, we pattern-match the measured waveform to the ideal waveform and extract the corresponding values of each figure of merit.

The key step is that we are approximating the real, measured waveform with the ideal waveform. How well the ideal waveform matches the measured waveform is an indication of how useful the extracted figures of merit are.

For example, the compensation signal matches an ideal trapezoidal waveform. **Figure 10.10** shows the setup for simulating an ideal trapezoidal wave in a common simulation tool, LTSPICE. There are six figures of merit that define it. These are pattern-matched to the measured signal. This general waveform has a finite 0-100 linear rise and fall time and a specific duty cycle.

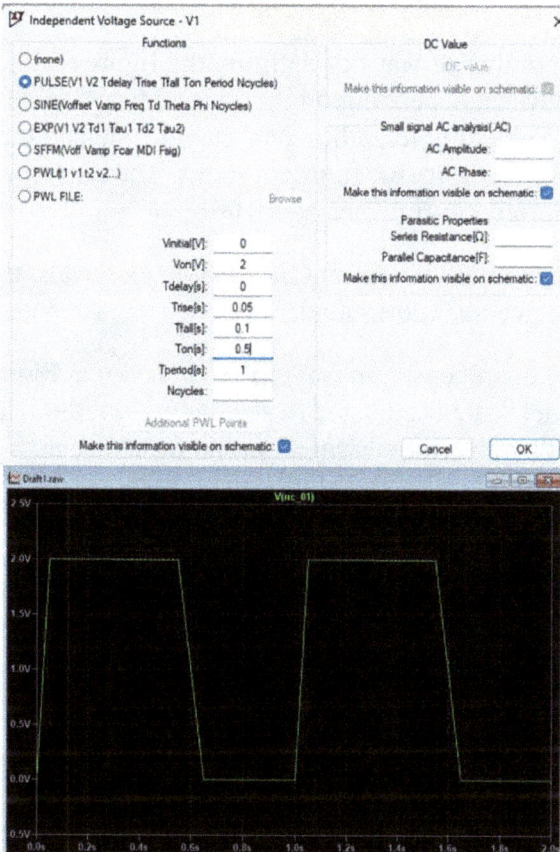

Figure 10.10 An example of an ideal trapezoidal wave as set up in LTSPICE, identifying the six important figures of merit and the resulting displayed ideal waveform.

Note that an ideal square wave is really a special case of a trapezoidal waveform with a 0 rise and fall time and a 50% duty cycle.

The simulated ideal trapezoidal waveform is purposefully adjusted from an ideal square wave to show off some of these features. We sometimes refer to the trapezoidal waveform as an ideal square wave with its various figures of merit.

Every simulated waveform is, by definition, an ideal waveform in that it is mathematically defined and has exactly the features that define it. We sometimes refer to the trapezoidal waveform as a square wave with finite rise time and non 50% duty cycle.

There are a number of ways of defining the figures of merit of an ideal square wave, depending on the information you need in your application. For example, rather than the on-time and the period, the duty cycle is a valuable figure of merit. The duty cycle in percent is literally on-time/period x 100.

Instead of the initial voltage and the on-voltage levels, the peak-to-peak and the average are equivalent.

The measured compensation waveform is shown in **Figure 10.11**. When the time scale is adjusted to see the period, the rise and fall times cannot be directly measured on the front screen. To get these values, the time scale was expanded enough to see the rising and falling edges.

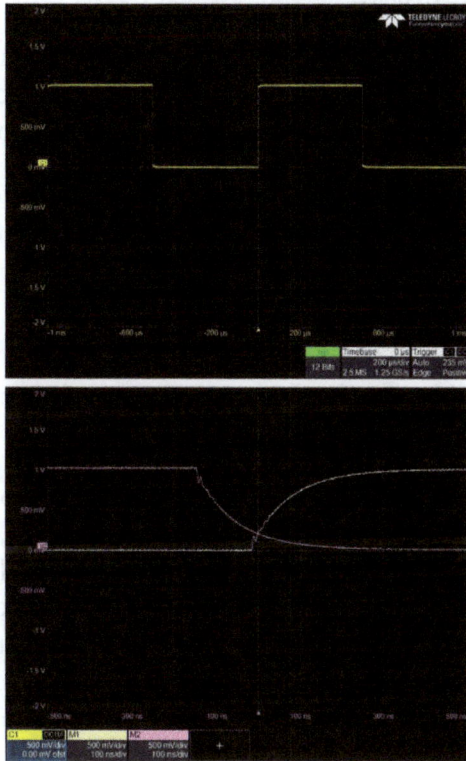

Figure 10.11 Measured comp signal with the time base adjusted to see the entire waveform in the top screen and then expanded to show the rising and falling edges in the bottom screen. The rising and falling edges were saved into memory traces to show on the same screen. The rising and falling edges are both displayed in the 100 nsec/div scale.

To show the falling edge, it is tempting to offset the time scale all the way to the far left until the falling edge is visible. While this may work in principle, it is very awkward to implement.

The best practice to see the rising and falling edges is to use the trigger function. Setting the trigger on a rising edge brings the rising edge to the $t = 0$ position. Setting the trigger on the falling edge brings the falling edge to the $t = 0$ position. In these settings, it is easy to expand the time base to see the details of the rising and falling edges.

When measuring a time parameter, there are three important questions to always ask:

1. If you are using a 1 meg input coupling *and* a coax cable, do you have potential reflection artifacts affecting the signal edges? This is covered in the next chapter.

2. What is the rise time of the signal compared to the bandwidth of the scope and measurement system?

3. What is the time resolution of the scope compared to the resolution required for the measurement?

For now, we will ignore the first question, as this is covered in the next chapter.

As a rough rule of thumb, if the bandwidth of the scope and measurement system is less than 1/ (10-90 rise time), there may be an artifact in the measurement due to the bandwidth limitations in the system and the bandwidth of the signal. In this example, the rise time is about 200 nsec. The required measurement system bandwidth to reduce the artifacts is about 1/200 nsec = 5 MHz. The measurement system bandwidth is well above this value and is not an issue.

The time resolution of the scope is related to the time interval between samples of the scope's ADC. While some interpolation is possible to resolve a time interval shorter than the sample interval, the sample interval is a good starting place for the resolution.

When the time base was set for 2 msec full scale to see the entire waveform, the sample rate, as seen displayed on the lower right of the screen, was 1.25 GSps. This is a sampling interval of 1/1.25 GSps = 0.8 nsec per point. When the scale was expanded to 1 usec full scale, while the scope was acquiring new measurements, the sample rate of the scope increased to 10 GSps. This is 0.1 nsec as the sampling interval and the time resolution.

Of course, if the time base had been expanded on a stopped screen, the time resolution would not have changed. Even though the time base could have been expanded to 1 sec/div, the sampling interval would have stayed at 0.8 sec/div. This is another important reason to avoid using the stopped acquiring mode unless you have a strong, compelling reason otherwise.

The time resolution of the sample interval sets a fundamental limit to the uncertainty of any time-related measurement. Be aware of this uncertainty in every time measurement, such as a period, an interval, or a rise or fall time.

When reading a numerical value off the front screen, there is also inherent uncertainty just from the coarseness of the scales and the displayed trace. Each square division is divided into five tick marks. There are 8 divisions in a vertical measurement. This means each tick mark is about 20% /8 = 3% of full scale. There are 10 horizontal divisions. This means each tick mark is about 20%/10 = 2% of full scale.

As a rough rule of thumb, unless there is a compelling reason otherwise, assume a typical measurement uncertainty using your Mark 1 eyeball is about 3% of the full-scale setting. This uncertainty is related to the full-scale setting. This suggests that to reduce the measured uncertainty in a signal's figure of merit, you should expand the scales so the signal takes up as large a fraction of the screen as practical, consistent with easily and quickly measuring the value from the screen.

To reduce the measurement uncertainty in a figure of merit below this level, the built-in functions should be used. Care should be taken when interpreting these values. Ultimately, there is a limit to the uncertainty of any figure of merit based on how well the measured waveform matches the ideal waveform from which the figure of merit is defined. Always be aware of this limitation.

As part of Rule #9, we expect to see a period of 1 msec, a voltage of about 1-2 V, and a 50% duty cycle. The rise and fall time and the source impedance are unknowns. This is why we will measure them with the scope.

In the comp signal from the Teledyne LeCroy HDO6104 scope, we can literally read off the front screen the six important figures of merit:

Figure of Merit	As Measured Off the Front Screen	Measurement Uncertainty (+/- 3% Full Scale)
Period (frequency)	1 msec (1 kHz)	+/-0.06 msec
Duty cycle	50%	+/- 6%
Max value	1 V	+/- 0.1 V
Min value	0 V	+/- 0.1 V
Rise time	200 nsec	+/- 30 nsec
Fall time	200 nsec	+/- 30 nsec

Once you estimate the measurements with your Mark 1 eyeball, it is appropriate to turn on measurement functions to get a more precise reading.

10.5.5 Using Measurement Functions

Measurement functions can be set up for all the important figures of merit, such as period, duty cycle, min and max values. These are sometimes called measurements or parameters in different scopes.

To set up a measurement function, you can select the type of measurement from a predefined list. They are usually separated by a horizontal measurement, which is a time value of some sort, or a vertical measurement, which is a voltage value of some sort. The wide variety of measurement functions available under the vertical category in the Teledyne LeCroy HDO6104 scope is shown in **Figure 10.12**.

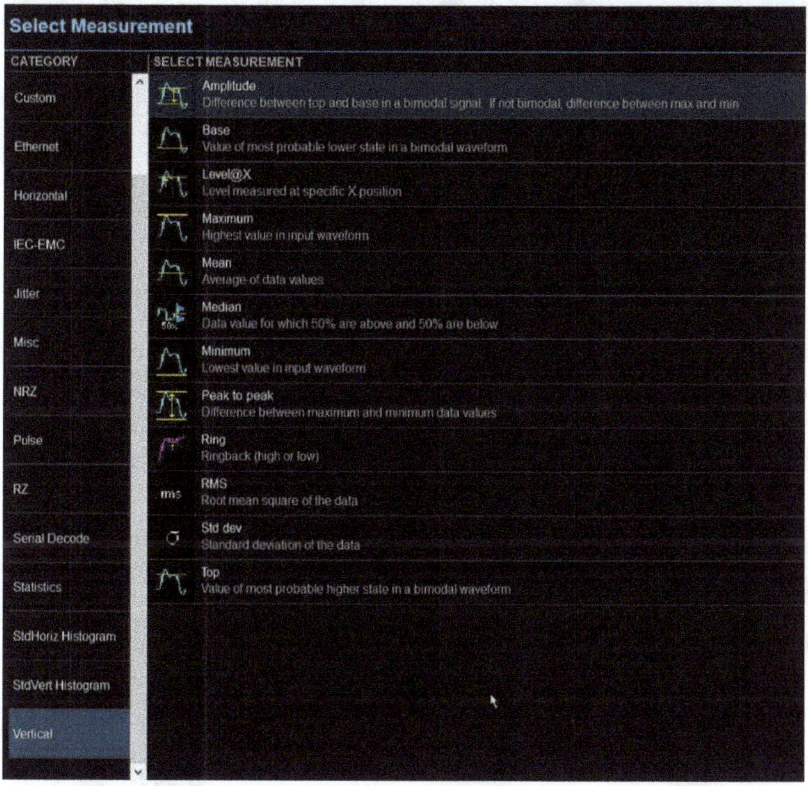

Figure 10.12 Examples of some of the measurement functions on the Teledyne LeCroy HDO6104 scope under the vertical category.

The six figures of merit can be selected from the list of parameters available to measure. A seventh parameter, the mean or average, was added to the list.

In addition to the figure of merit measured from the current acquisition window, some scopes can collect the statistics of the variation in the measured parameter over many consecutive acquisitions and can even plot a histogram of the distribution of the parameter over the many acquisitions. For a meaningful histogram, at least 1000 acquisitions should be analyzed. Figure 10.13 shows the seven parameters, their statistics, and the histogram of each measurement.

Figure 10.13 An example of using the measurement function to directly measure each important figure of merit and their statistics using the Teledyne LeCroy HDO6104 scope.

The uncertainty in each measurement is limited by the specified accuracy of 1% for the vertical scale and the time resolution of 0.8 nsec, with an acquisition rate of 1.25 GSps. In addition, the statistics reveal the relative uncertainty from the stability in each measurement. This is about the drift over the approximately 5 minutes of the acquisitions and the random noise. The fact that each histogram looks relatively Gaussian suggests the drift is small and the standard deviation, a measure of the measurement uncertainty, is due to random noise.

Figure of merit	As Measured Off the Front Screen	Measurement Uncertainty (+/- 3% Full Scale)	Measurement Function With +/- 1 Standard Deviation
Period (frequency)	1 msec (1 kHz)	+/-0.06 msec	1.000 +/- 0.0000005 msec
Duty cycle	50%	+/- 6%	50.00% +/- 0.00004%
Max value	1 V	+/- 0.1 V	1.037+/- 0.003 mV
Min value	0 V	+/- 0.1 V	-0.021 V +/- 0.003
Rise time	200 nsec	+/- 30 nsec	209.9 +/- 2.2 nsec
Fall time	200 nsec	+/- 30 nsec	209.1 +/-2.7 nsec

The horizontal uncertainty is limited to the 0.8 nsec of resolution of the ADC sampling. The standard deviation in the period of 0.5 nsec is lower than this sample interval, an example of the stability of the clock frequency of the source and that some interpolation is used in the measurement function. The standard deviation of about 2.5 nsec in the rise and fall time intervals is probably a measure of the small variation in this parameter.

The vertical measurements are at 12-bit vertical resolution, and a 4 V full-scale range is about 4 V/4095 values = 1 mV. The standard deviation reported for the min and max value is about 2.5 mV, or about 2 bit levels.

The average values and their uncertainty taken as 1 standard deviation are compared with the Mark 1 eyeball measurements:

There are two important observations. First, each measurement obtained from the built-in functions is within the range of what was measured with a Mark 1 eyeball, given the uncertainties. This is an important consistency check. Second, the uncertainties in the measurements from the built-in functions are orders of magnitude smaller than what can be measured directly off the front screen.

This exercise points out two important results: the values obtained both ways are consistent. Neither result may be correct, but together, your confidence level is higher. When a more precise value is required, using built-in measurement functions that are verified to be consistent with your visual observation can unlock your scope as a precision data acquisition instrument.

10.6 Measuring the Source Resistance of a Voltage Source

The source impedance is a very important figure of merit for any voltage source. The simplest model of a real voltage source is an ideal voltage source. The output impedance of an ideal voltage source is 0. This means it can deliver the specified voltage independent of the load attached. While this may be a simple starting model for a real voltage source, it is not a very good approximation. Instead, a better, first-order model for a real voltage source is a Thevenin circuit as shown in **Figure 10.14**.

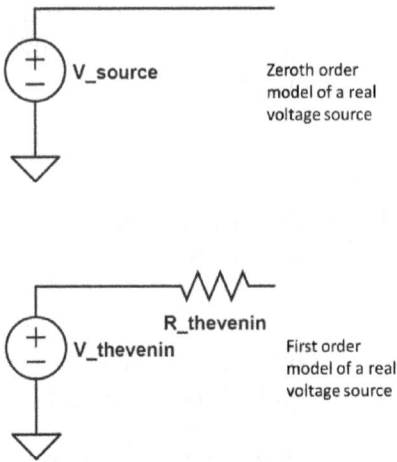

Figure 10.14 Examples of a zeroth and first-order equivalent circuit model of a real voltage source.

This model has two figures of merit that describe it: the Thevenin voltage and the Thevenin resistance. The Thevenin voltage is literally the output voltage of the source as measured with no load. This is what the scope would measure with an input resistance setting of 1 meg.

The Thevenin resistance requires a little more effort to measure. The voltage source should be loaded with a known resistance, and the voltage across this load should be measured. The Thevenin resistance is the voltage drop across the Thevenin resistor divided by the current through it. This is:

$$R_{thevenin} = \frac{V_{thevenin} - V_{load}}{I_{load}} = R_{load}\frac{V_{thevenin} - V_{load}}{V_{load}}$$

In principle, any load resistor can be attached to the source, and the loaded voltage across it is measured. In practice, not all voltage sources are linear. As the current draw increases, sometimes the output resistance dramatically increases. This means this simple first-order model is not always a good approximation.

If you are going to measure the Thevenin resistance with just one load, try to select a high enough resistance so the loaded voltage is within 10% of the unloaded Thevenin voltage. If the voltage drops more than 10%, use a higher-value resistor. A discrete resistor could literally be added as an external component across the scope probe. Resistor kits that are specified with a 1% tolerance can be purchased from a variety of vendors such as Amazon for less than $15.

As long as the Thevenin voltage is less than 5 V, so the power consumption is less than 0.5 watt, one simple process is to use the internal 50 ohm input resistance of the scope's channel as the load resistor. Measuring the Thevenin resistance just requires two measurements of the output voltage of the source: with the 1 meg input resistance of the scope and with the 50 ohm input resistance of the scope. Of course, this cannot be done with a 10x probe; it can only be done with a direct coax connection. An external resistor could be used with a 10x probe connection.

Measuring the output resistance of the compensation source requires two measurements: the voltage with a 1 megaohm input resistance of the scope and then with the 50 ohm input.

Figure 10.15 shows the saved waveform with the 1 meg input. In this example, cursors were used to measure the low and the high-voltage levels of the compensation signal. The open-circuit Thevenin voltage can be read directly from the cursors as the voltage difference between the high and low levels as 1.016 V + 0.004 V = 1.020 V. The uncertainty using cursors is about 1% of full scale or +/- 0.016 V.

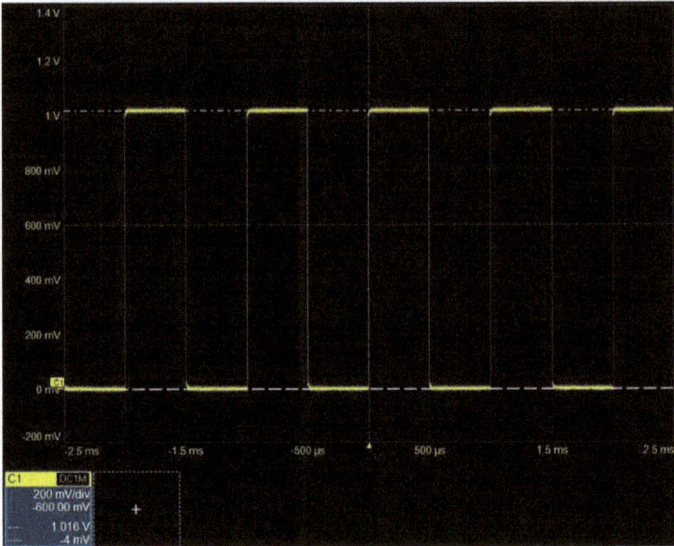

Figure 10.15 Measured voltage from the comp signal with a 1 meg input using a coax cable connection to the comp signal. The values of the horizontal cursors are in the small icon box in the lower left of the screen.

The input to channel 1 was changed to 50 ohms. The measured voltage immediately dropped when there was a 50 ohm load on the comp signal. The horizontal cursors were moved to align with the top and bottom of the trace. This new measurement of the compensation signal with a 50 ohm input to the scope as the load resistor is shown in **Figure 10.16**.

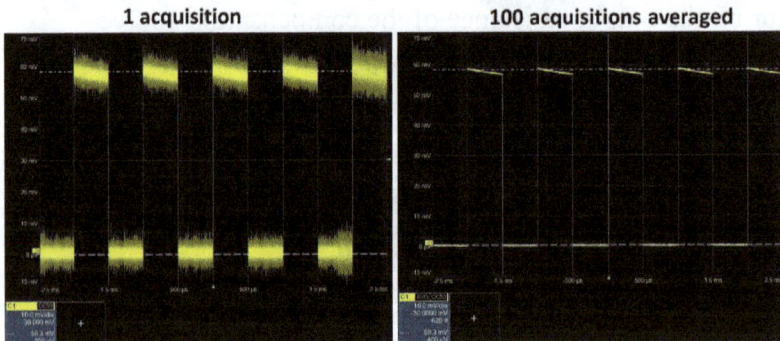

Figure 10.16 The measured compensation signal with 50 ohm input to the scope. The noise seen on the left was reduced by averaging 100 acquisitions in the right screen capture.

With a 50-ohm load on the compensation signal, the measured peak-to-peak voltage dramatically dropped. In this case, it is 58.5 mV, compared with 1.02 V. The random noise of almost 10 mV peak-to-peak makes the measurement uncertainty of the loaded voltage almost +/-5 mV. This is almost 10% uncertainty. In this case, the noise appears to be random, and the signal is periodic.

To reduce random noise on a synchronous signal, we can apply a simple technique of averaging consecutive acquisitions, all synchronous with the trigger signal. The signal will stay the same, but the random noise will decrease in amplitude with the square root of the number of averages. With 100 consecutive acquisitions averaged, the noise should drop by a factor of 10 to about +/- 0.5 mV.

The screen capture in this figure shows the measured compensation signal with 100 averages. The random noise has dropped significantly, easily below 1 mV.

Also apparent in this measurement is a large overshoot signal at the transitions. This means that using a peak-to-peak measurement function would result in a wrong measurement of the compensation signal into a 50 ohm load. The cursors show a loaded voltage of 58.7 mV.

Using the simple second-order model for the compensation source, we can calculate the output resistance as

$$R_{thevenin} = R_{load} \frac{V_{thevenin} - V_{load}}{V_{load}} = 50\,\Omega\,\frac{1.02\text{ V} - 0.059\text{ V}}{0.059\text{ V}} = 814\,\Omega$$

This is a very high resistance.

Different scopes have different output resistances for the comp signal. It varies from 150 ohms to more than 800 ohms. When used as a compensation signal with a 10x probe, the output resistance of the source is not an important performance figure of merit.

However, a common user problem is to accidentally connect the ground lead of the 10x probe to the signal terminal of the compensation source. This shorts the source to the chassis and ultimately to earth-ground. A high resistance is used as the output resistance of the compensation signal source to limit the current and prevent any damage to the source, the probe, or the scope.

This is a surprisingly common accident.

All voltage sources can be modeled to second order as Thevenin sources with some Thevenin output resistance. Generally, voltage sources fall into two categories:

- ✓ Those designed to deliver a signal.
- ✓ Those designed to power a circuit.

When used as a signal source, little current needs to be supplied to the circuit. Signal generator signal sources generally have output resistances that are high: 50 ohms and above.

When used to deliver power, sometimes large currents are delivered. To provide a nearly constant output voltage under large and sometimes changing currents, a low output resistance is required. This is generally on the order of 1 ohm or less. If 0.1 A is drawn from a source that has a 1 ohm output impedance, there would be a 0.1 V drop to the output voltage under the full load. If this is too large a drop in an application, a lower output impedance power source would be required. Knowing the impact on the supplied voltage under different current loads is partly why knowing the output Thevenin resistance of a voltage source is so important.

Given the much higher output resistance of a signal source, it is rarely a good idea to use a signal source as a power supply for a circuit. It will rarely be able to supply enough current without a significant drop in its output voltage.

10.7 Function Generators

Some scopes have a function generator built in. This is the case for the Analog Discovery 3 and most mid-range scopes. When there is no built-in function generator, an external function generator can be purchased.

The lowest cost function generator, based on the XR2206 IC, that is still capable of sine, square, and triangle waves of variable frequency, can be purchased as a kit for as low as $4 from eBay, for example, or even Amazon for $10, as shown in **Figure 10.17**.

Figure 10.17 The lowest cost, $4 function generator, as a kit and after assembly, is seen at the top of the figure.

For as little as $29, an assembled, slightly higher-quality function generator based on a direct digital synthesis (DDS) chip, probably the AD9837 or equivalent, can be purchased from Amazon.

This device is capable of higher frequency, and since it is digitally synthesized and then converted to an analog signal by a DAC, it has better signal quality. An example of this simple $29 function generator is shown in **Figure 10.18**.

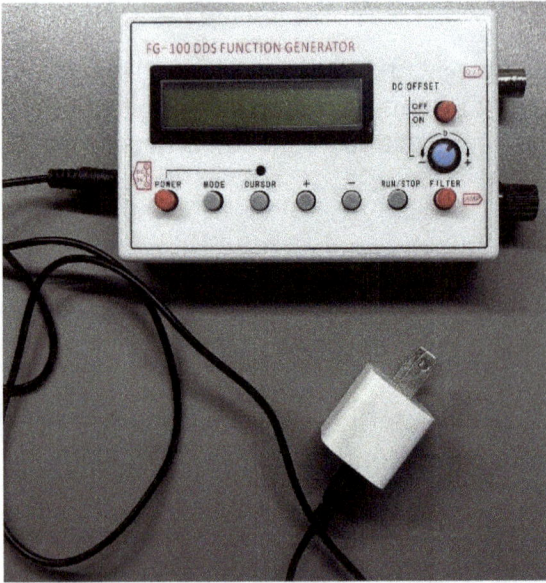

Figure 10.18 A simple $29 function generator as a source of common signals.

With these very low-cost function generators available, there is no reason not to have one in your lab. The downside is that these are not the highest quality. Sometimes, they do not work, or the assembly quality is poor. They have limited functionality, offering just a DC, sine, square, and triangle wave with a limited frequency range.

Beyond these very low-cost function generators are a wide range of higher-performance units ranging in price from $100 to $5,000. They differ in their frequency range, quality of signals, and additional features such as signal modulation and arbitrary waveform generation (AWG or Arb).

All function generators can produce a DC value, as well as sine waves, square waves, and triangle waveforms. They are excellent as a starting signal for practicing measuring repetitive signals.

10.8 Built-In Waveform Generators in a Scope

Most mid-range scopes have a built-in function generator. This is the case with the Digilent AD3 scope, the Keysight DSOX3024, the Rohde and Schwarz RTC1000 and the Teledyne LeCroy HDO6104.

The features of the function generator are all very similar. If a more versatile function generator is required, a high-end external function generator can be purchased for under $1000.

10.8.1 Using the Internal Function Generator of a Mid-Range Scope

Every scope has a slightly different location for the BNC connection to its function generator. On the Keysight and Rohde and Schwarz scopes, the BNC connection is on the front. For the Teledyne LeCroy scopes, the connection to the internal function generator is in the back.

The features of the function generators range from simple, basic functions to complex modulation patterns and the ability to play .wav files, such as those found on the Digilent AD3 scopes.

In the following examples, we will look just at sine wave sources. They are repetitive signals with a very well-defined behavior. The reason they are so common is that they are naturally occurring in many electronic circuits. A sine wave is the solution of a second-order, linear differential equation. This means they appear in the natural or step response of circuits. These naturally arise in circuits with R, L, and C elements.

There are four terms that define a sine wave:

✓ *The amplitude*
✓ *The period or frequency*

✓ The phase

✓ The DC offset

The phase of the sine wave is the location in the cycle at t = 0. In a scope measurement, t = 0 is defined by the trigger event. The part of the sine wave at t = 0 is arbitrarily defined by the trigger setup and not intrinsic to the sine wave, so it is not a term by itself that is important. The relative phase to other signals is important.

The DC component of an ideal sine wave is always 0 V. This is the definition of a sine wave. However, when synthesized as a voltage waveform, a DC component can be added to the signal. This is why phase is not a parameter to set for generating the sine wave, but DC offset is. **Figure 10.19** shows the setup of the sine wave in the Teledyne LeCroy HDO6104 scope for a sine wave.

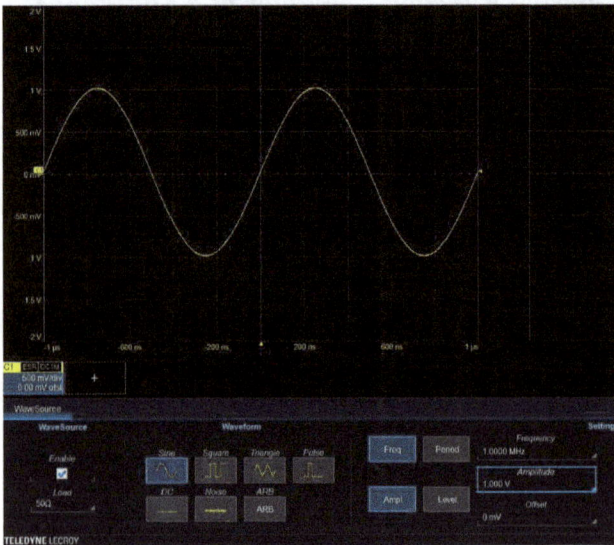

Figure 10.19 An example of the setup for a sine wave and the measured output of the function generator in a Teledyne LeCroy HDO6104 scope.

Even a $29 DDS function generator with a limit of no higher than 500 kHz, can produce a very nice-looking sine wave, as shown in **Figure 10.20**.

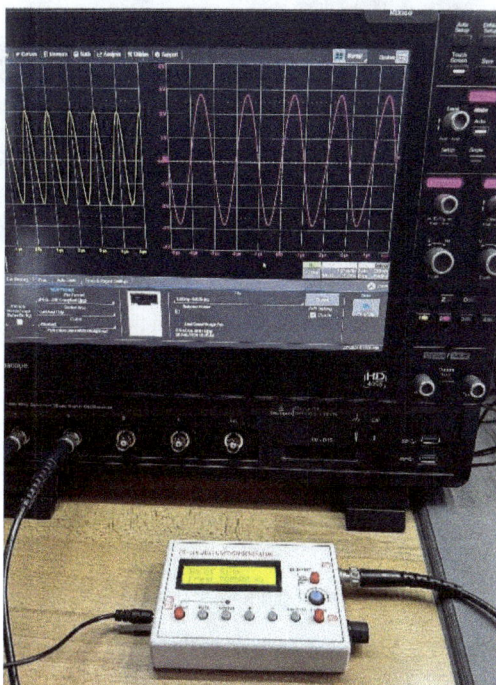

Figure 10.20 A low-cost function generator produces a 500 kHz sine wave with nearly 3 V amplitude.

Even a $29 function generator provides a range of signals that can be used to exercise the scales, cursors, and built-in functions of any scope.

As an example of the power of the measurement functions, the three important figures of merit of a sine wave are the peak to peak, the DC offset, and the period were measured using measurement functions for the $29 DDS function generator at 3 V amplitude and 500 kHz and for the internal function generator in the Teledyne LeCroy HDO6104 scope. The measurements are shown in **Figure 10.21**.

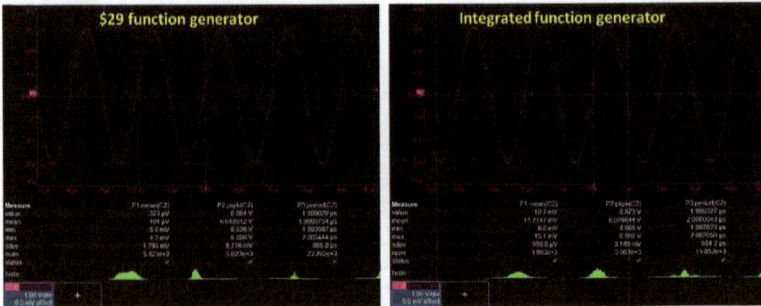

Figure 10.21 Comparison of the sine wave figures of merit a sine wave from a low-cost function generator and the built-in function generator in the Teledyne LeCroy HDO6104 scope.

These values are:

Figure of merit	$29 function generator	Integrated function generator
Average DC offset	-0.4 mV +/- 1.7 mV	11.7 mV +/-0.96 mV
Peak-to-peak	6.048 V +/- 0.009 mV	6.077 V +/- 0.003 mV
Period	1.994 usec +/- 0.0009 usec	2.000 usec +/- 0.0005 usec

The absolute values are really about how well the knob on the function generator can be adjusted. The real metric of the performance of the sine waves is in the standard deviations in their measurements. The low cost $29 DDS function generator has a performance just as good as that of a built-in function generator in the mid-range scope.

The amplitude variation or drift over a one-minute period is about 0.15%. The period variation is about 0.045%, or 450 ppm.

10.9 DC Validation of the Scope

An important question is always the absolute accuracy of the voltage measurements with a scope. All scopes are rated with an absolute accuracy of about 1%, with some uncertainty based on the scale settings. When the accuracy is limited by digitizing noise, the average of consecutive points or from acquisition to acquisition can reduce this random digitizing noise.

Some scopes offer a specific function to average an acquisition buffer and prominently display this average voltage so the scope can be used as a DMM.

Using a function generator as a DC source, the absolute voltage measured by the scope can be calibrated against a DMM with known absolute accuracy. **Figure 10.22** shows the voltage measured by a Teledyne LeCroy HDO6104 12-bit scope and two benchtop DMMs rated with better than +/- 0.004% absolute accuracy.

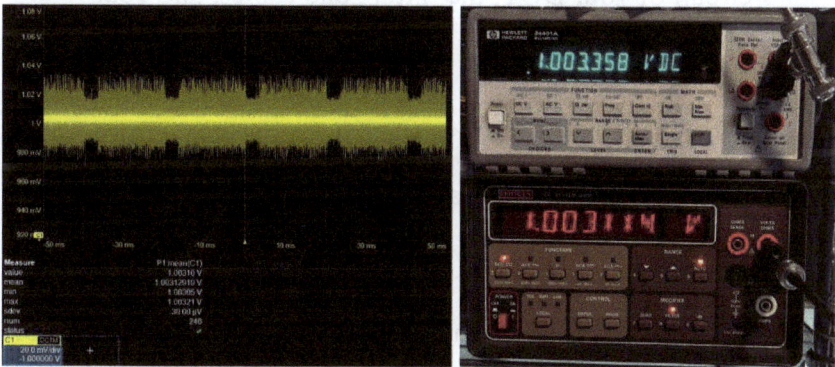

Figure 10.22 Three instruments measuring the same DC voltage source. The scope measured 1.003,13+/- 0.000,03 V. The HP 34401A DMM measured 1.003,36+/- 0.000,04 V. The Keithly 196 DMM measured 1.003,11+/- 0.000,04 V. Note the large AC noise introduced by the sampling by the two DMMs.

Both DMMs measured a voltage of 1.003V to within 200 uV of each other. This is a difference of 0.02%, outside the 0.004% absolute accuracy of either DMM. The scope measured a voltage of 1.003,13 V within 100 uV of the average of the two DMMs. This is within 0.01% absolute accuracy, far exceeding the scope's rated 1% absolute accuracy.

10.10 The Bottom Line

1. When starting out with initial measurements on a new instrument, always measure something for which you know the answer. This is the only way of applying Rule #9.

2. Follow Rule #9.

3. Perform all the consistency tests you can think of.

4. Apply situational awareness to identify possible measurement artifacts by understanding the equivalent circuit of your DUT, the fixture and cables, and your instrument.

5. With a new instrument, start by measuring something for which you know the answer.

6. Always try to estimate a measurement directly from the front screen before you apply built-in, complex measurement functions.

7. Turn the raw measurements into information in the form of a few figures of merit.

8. Use the information to answer the "So What?" question.

9. Use the built-in compensation signal to all scopes as a signal to practice using all the vertical, horizontal, and trigger controls of your scope.

10. Become familiar with a function generator, either built-in or external. If you do not have one, purchase even a low-cost one so you have these simple, well-defined signals available to you.

Chapter 11
Scopes, Transmission Lines, and Reflections

The 10x probe is specially designed to reduce the artifact from reflections of the signal between the DUT source and the scope input impedance. When other probes, such as a coax cable, are used to connect between the DUT and the scope, an important artifact can arise that is easily avoided once understood.

In this chapter, the principles behind signals propagating on interconnects are introduced, and how reflections of signals on transmission lines connected between the DUT source impedance and the scope input impedance create a measurement artifact and can be eliminated.

11.1 A Misleading Observation

It is a common misconception that the length of a coax cable influences the rise time of the signal to which it is connected. The longer the cable length, the longer the signal rise time at the receiver will be.

After all, the coax cable has some capacitance, and the signal source impedance will be charging up the cable capacitance. The longer the cable, the larger the capacitance and the longer the rise time. Right?

Alternatively, it is commonly believed that the losses in the cable cause rise time degradation. The longer the cable, the larger the rise time degradation. In principle, this is correct. However, the impact on the signal rise time in RG58 or RG174, common coax cables, is easily measured when the scope is set up correctly and is on the order of 2.1 nsec for 10 m long cables, not 900 nsec.

This misconception that the rise time increases with cable length is reinforced in a simple experiment that anyone can do.

In this example, the comp signal from the front terminals of a scope is the signal source, and a simple RG58 coax cable is used as the connection between the comp source and the scope's channel 1. A BNC to minigrabber adaptor connects the scope's terminal lugs to the BNC connector of the coax cable. As before, the scope's input coupling is set to 1 meg.

We measure the rise time of the signal as we increase the length of the coax cable. To measure the rise time, the time base is expanded so that the highest sample rate of the scope is used and the rise time can be read off the front screen. The measurement function is used to extract a more precise value of the 10-90 rise time.

The initial cable length was 3 feet. The measurement setup is shown in **Figure 11.1**.

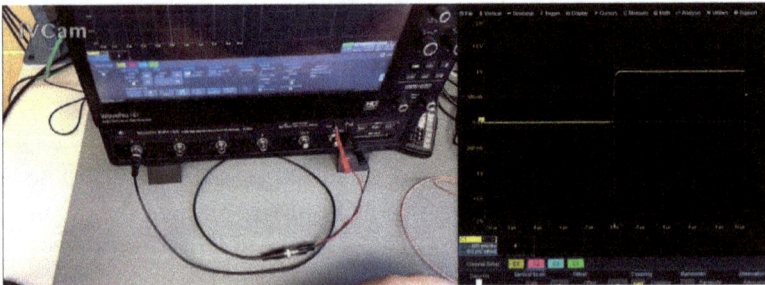

Figure 11.1 Connection to the comp signal with mini grabbers and the measured rise time by the scope of about 203 sec. The coax cable was 3 feet long.

As the cable length is increased, the rise time also increases. The extreme case of a total length of 12 feet is shown in **Figure 11.2**. In this case, the rise time has increased to 590 nsec.

Figure 11.2 An example of a 12-foot-long coax cable with a measured rise time of 600 nsec.

As the cable length increased, so did the measured rise time. The length and measured rise time of the compensation signal is listed below:

Coax cable length	Rise time
6 ft	264 nsec
12 ft	590 nsec
18 ft	920 nsec

These observations are summarized in **Figure 11.3**, comparing the rising edges of the comp signal with three different cable lengths.

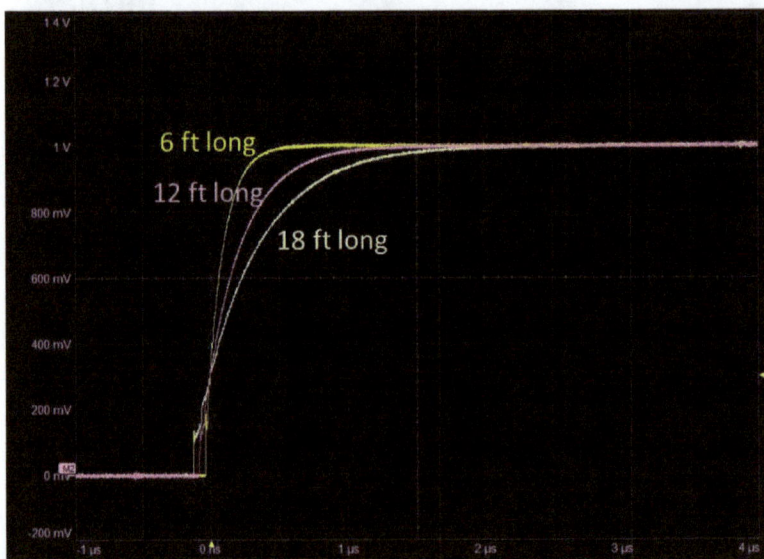

Figure 11.3 The same comp signal was measured with three different lengths of coax cables, showing a longer rise time with cable length.

This observation supports the commonly held notion that the longer the cable, the longer the rise time of the measured signal. While this is a correct observation, it is the result of a measurement artifact. If you take these measurements as the correct behavior, you will develop incorrect engineering judgment about how signals propagate on interconnects and would misinterpret all future measurements.

When the scope is set up correctly, the rise time of the signal into the scope is nearly independent of the cable length. There is a very minor rise time degradation due to the frequency-dependent losses in the cable, which is on the order of 1 nsec for 10 feet, not 600 nsec. This specific effect is measured, and an example is shown later in this chapter.

The origin of this artifact of the apparent rise time increase with cable length is due to the properties of signals propagating on transmission lines, the nature of reflections, and the impact of terminations.

11.2   Five Principles of Signal Propagation

There are five important principles at the foundation of the behavior of signals on interconnects. These principles apply not just to the effects related to signals between a DUT and the scope but also to any signal from any source propagating on any interconnect to any receiver.

11.2.1 Principle #1

Principle #1 is that all interconnects are transmission lines.

The first principle is that all interconnects, no matter their size, length, or geometry, are transmission lines. A transmission line is composed of two conductors. We pick one and call it the signal conductor and we call the other the return conductor.

The signal that propagates on a transmission line is the voltage difference between the signal and the return conductor. A voltage difference between the signal and return conductors is applied at one end of the transmission line and this voltage difference propagates between the two conductors to other parts of the transmission line.

It is tempting to call the return conductor "ground." While the return conductor is usually also connected to the net in a circuit labeled as ground, it does not play the role of a reference point in the circuit to which all voltages are referred. Instead, the return conductor plays the very important role of carrying the return current for the signal. Get in the habit of referring to the second conductor as the return path, rather than the ground conductor so that you train your intuition into thinking about the important role of the second conductor in a transmission line. This is illustrated in **Figure 11.4**.

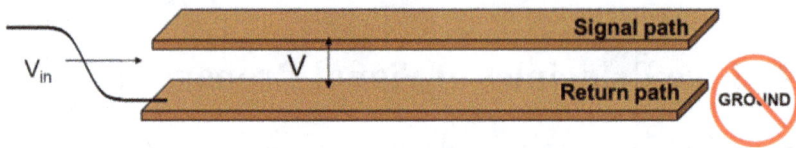

Figure 11.4 Every interconnect is a transmission line composed of a signal path and the return path.

In a coax cable, the center conductor is the signal conductor, and the outer shield is the return conductor. In a circuit board, we typically only see the top, surface traces of the microstrip. These are the signal traces. The return conductor for these signal traces is usually the plane on the layer directly below the signal layer.

Since we cannot see this conductor with our eyes, we must learn to see it with our engineer's mind's eye. **Figure 11.5** shows an example of the signal traces on two circuit boards. We only see half the transmission lines. The return conductor in each case is the plane on the layer directly below the signal layer.

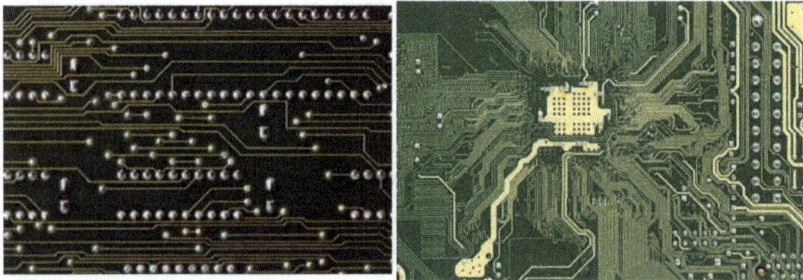

Figure 11.5 The signal traces on a circuit board's outer layer have their return conductor on the layer directly beneath them.

Without exception, there is always a return conductor associated with every signal conductor. This pair makes up the interconnect.

If you do not engineer the return conductor as carefully as the signal conductor, the electrical properties of the transmission line may not be optimized for good signal transmission. One of the most important design features of an optimized transmission line is a uniform cross section designed to achieve a specific target characteristic impedance.

11.2.2 Principle #2

Principle #2 is that signals are dynamic.

A signal is the voltage difference between the signal conductor and the return conductor measured between adjacent points on a transmission line. Once a voltage is launched on a transmission line, for example, by turning on a driver, the signal will propagate. This is a fundamental property of electromagnetic fields.

The voltage is turned on, creating a changing electric field between the signal and the return conductor. This changing electric field creates a changing magnetic field, which creates a changing electric field, and so on. This is the principle of the propagation of an electromagnetic field, a form of light. The speed at which the signal propagates down the transmission line is the speed of light in the dielectric between the signal and the return conductor and is given by

$$v = \frac{c}{\sqrt{Dk}}$$

where

v = the speed of the signal in the transmission line

c = speed of light in vacuum, about 11.8 inches/nsec

Dk = the dielectric constant of the material between the signal and return conductors

In an FR4 or similar polymer material, with a dielectric constant of about $Dk = 4$, the speed of the signal is about

$$V = \frac{c}{\sqrt{Dk}} = \frac{11.8\,{}^{in}\!/\!_{n\,sec}}{\sqrt{4}} \sim 6\,{}^{in}\!/\!_{n\,sec}$$

In a coax cable with polyethylene as the dielectric, the speed of a signal is closer to

$$V = \frac{c}{\sqrt{Dk}} = \frac{11.8\,{}^{in}\!/\!_{n\,sec}}{\sqrt{2.4}} \sim 8\,{}^{in}\!/\!_{n\,sec}$$

This means that the time delay, TD, of a 12-inch coax cable, for example, will be TD = 12 inches/8 inch/nsec = 1.5 nsec. For a 6 ft coax cable, the time delay would be about 1.5 nsec/ft x 6 ft = 9 nsec.

In a 6-inch-long transmission line with an FR4 dielectric, the time delay, TD, of the transmission line will be 6 inches/6 inch/nsec = 1 nsec.

This is an important property of any transmission line.

The signal refers to the propagating voltage difference between the signal and return conductor. A signal propagates in a specific direction. This is slightly different from the voltage that is measured between two points on the signal and return conductor.

When a voltage is measured with a scope probe, for example, there is no information about the direction of propagation of the signal. For this reason, we sometimes refer to the voltage signal a scope measures as a *scalar voltage* in that it has no direction information, just magnitude.

A scope probe only measures the instantaneous voltage difference between the signal and return conductors at the location to which it is connected. There is no way to tell what direction the signal that created the voltage may be propagating. Yet, there is a voltage wave that is propagating down the transmission line. This can be demonstrated in a simple experiment.

The fast edge signal source from Bodnar Electronics was used as a source to launch a signal into a coax cable. The coax cable was created by stringing together three, six-foot-long coax cables with a tee connector between them. At the source of the driver, a 10x probe was used to measure the voltage between the signal and return conductors. This signal triggered the scope and defined t = 0. A second 10x probe was used to measure the signal along the transmission line at different distances from the source. This setup and measurements are shown in **Figure 11.6**.

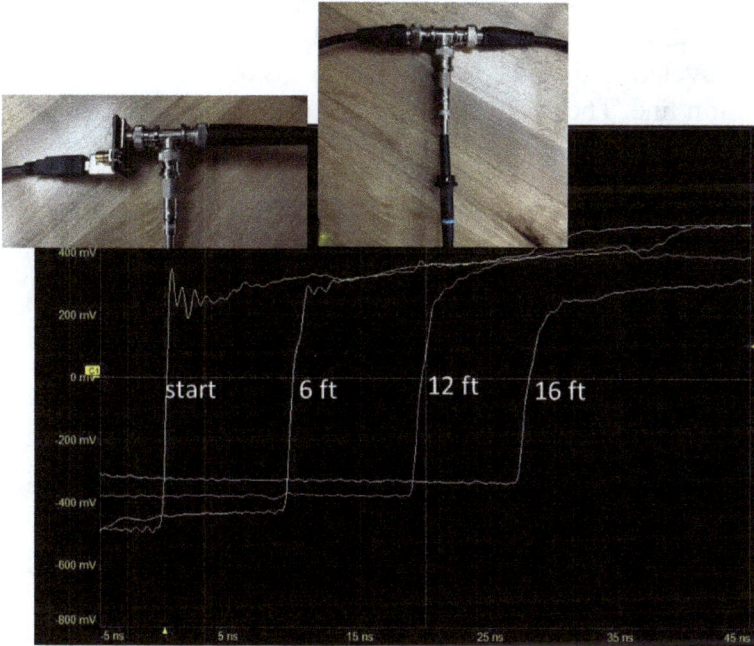

Figure 11.6 The measured signal as it propagates down a coax cable is measured at 6 ft intervals.

The signal measured at the three locations, spaced by roughly 6 feet, was measured by the scope. The scope measured the voltage at the Tee connector at different time intervals, as the signal propagated down the transmission line.

With a length of 6 feet, or 72 inches between Tee connectors, and a speed of about 8 inches/nsec, the expected time delay, TD, between each junction is 72 in /8 in/nsec = 9 nsec. The bandwidth of the 10x probe sets the fundamental bandwidth of the measurement system and the shortest rise time that would be measured from the source. On this scale of 5 nsec/div, the signal rise time at the source is seen to be less than 1 nsec.

The time interval between the rising edge centers is about 10 nsec, close to the expected 9 nsec, based on our rough estimate of the speed of a signal in the coax.

This experiment illustrates that it takes a finite time for a signal, once launched in a transmission line, to propagate down the line. This time is directly related to the speed of propagation and the distance traveled. This same behavior happens on every transmission line. The shorter the line, the shorter the time delay.

In this example, the rise time of the signal measured at each interval is nearly the same. Even after propagating 18 feet of coax cable, the rise time had increased from about 0.7 nsec to only 1.5 nsec. This is the impact from the frequency-dependent losses in the cable. This is a very different behavior than observed with the comp signal.

The time delay on an interconnect is independent of the rise time of the signal. These are two different, independent time intervals.

11.2.3 Principle #3

Principle #3 is that signals see an instantaneous impedance as they propagate.

The most important electrical property of an interconnect the signal sees as it propagates down the transmission line is the instantaneous impedance. Impedance is always the ratio of a voltage to a current:

$$Z = \frac{V}{I}$$

In the case of the propagating signal, the instantaneous impedance the signal sees each step along the way is the ratio of the voltage of the signal to the current associated with the signal, flowing between the signal and the return path.

As the signal propagates down the transmission line, there is conduction current flowing down the signal conductor and back on the return conductor. But how does the current make a complete path between the signal and return conductor? Aren't they separated by an insulating dielectric? Does the signal current have to make it all the way to the end of the transmission line before it connects to the return conductor and then makes its way back? What if the end of the transmission line is open and there is no connection between the signal and return conductors?

The resolution to this puzzle comes from a new type of current called displacement current. This was one of the innovations introduced by James Clerk Maxwell in the late 1870s. In his generalization of the equations Oliver Heaviside introduced that described electromagnetic fields, he realized that a changing electric field, a dE/dt, behaved exactly like a conduction current. But the dE/dt was in an insulating dielectric.

In Maxwell's view, displacement current is the current that flows along the changing electric field lines. It is not a flow of free charges like conduction current, but a fundamentally new type of current, just as real as conduction current. It is through the displacement current that the current flows between the signal and return conductors at the wave front where the electric field and the signal voltage are changing. This is illustrated in **Figure 11.7**.

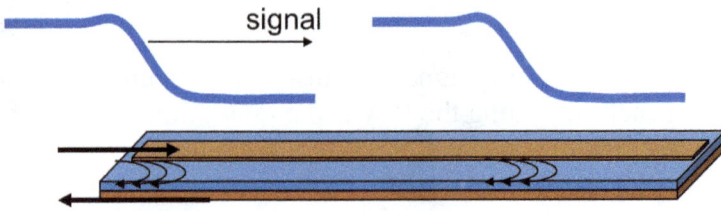

Figure 11.7 An illustration of the signal propagating down a transmission line at two locations shows the current flowing between the signal and return conductors where the dV/dt and the dE/dt are at the edge of the signal wavefront.

The displacement current flows wherever there is a changing electric field between the signal and return conductors. This is the same location as where there is a changing voltage between the signal and return conductors, where the signal is changing at the wavefront. Coincident with the changing voltage between the signal and return conductors is the displacement current wavefront.

Along the entire length of the transmission line, the *only* place there is a current flowing between the signal and return conductors is at the propagating wavefront coincident to the changing voltage and, hence, changing electric field.

The displacement current is also the current that flows through a capacitor when the voltage across it changes. The amount of current that flows through a capacitor is

$$I = C \frac{dV}{dt}$$

This is the basis of calculating the current flowing between the signal and return conductors.

As the signal propagates, the voltage swing is from 0 V to the full V. It has a rise time, Δt, and a spatial extent, Δx. The connection between the two is related to the speed of the signal in the dielectric medium, v, by

$$\Delta x = v \times \Delta t$$

The transmission line has some capacitance per length, C_{Len}. This means at any instant of time, the capacitance through which the return current flows is equal to

$$C = C_{Len} \times \Delta x = C_{Len} \times v \times \Delta t$$

The current flowing through the transmission line from the displacement current is calculated as

$$I = C \frac{dV}{dt} = C_{Len} \times v \times \Delta t \frac{V}{\Delta t} = C_{Len} \times v \times V$$

As the width of the signal conductor increases, for example, the capacitance per length increases, and more current flows between the signal and return conductor.

The impedance the signal sees is the ratio of the voltage of the signal, V, to the current carried by the signal, I, flowing between the signal and return conductors. It is calculated as:

$$Z = \frac{V}{I} = \frac{V}{C_{Len} \times v \times V} = \frac{1}{C_{Len} \times v}$$

This is a very simple form for the instantaneous impedance the signal sees. If the capacitance per length increases, the instantaneous impedance decreases. If the transmission line is uniform, the capacitance per length of the line is constant, and the speed of the signal is constant. This means the instantaneous impedance the signal sees, each step along the transmission line is also constant.

In a uniform cross-section transmission line, there is one value of instantaneous impedance. This one value of instantaneous impedance characterizes the transmission line. We call this value of instantaneous impedance the characteristic impedance of the transmission line and designate it with the letter and number Z0.

Every transmission line has a characteristic impedance, which is an intrinsic property of its cross-section and material properties and is independent of its length. The two most important electrical properties that determine how a signal interacts with the interconnect are the characteristic impedance and the time delay of a transmission line.

11.2.4 Principle #4

Principle #4 is that initially, when the driver looks into a transmission line, it sees a resistor equal to its characteristic impedance.

When a source launches a signal into a transmission line, the input to the transmission line looks like a resistor for a time equal to the round-trip time of the transmission line, 2 x TD. After this time, the input impedance of the transmission line changes and is difficult to estimate other than in extreme cases or using a SPICE simulator.

If the transmission line is 6 inches long and composed of an FR4 dielectric, the input to the transmission line will look like a resistor for a time equal to 2 x 1 nsec = 2 nsec.

This means that the circuit composed of the source voltage, V_{source}, the source impedance, R_{source}, and input resistance of the transmission line, Z0, looks like a voltage divider. This equivalent circuit is shown in **Figure 11.8**.

For t < 2 x TD

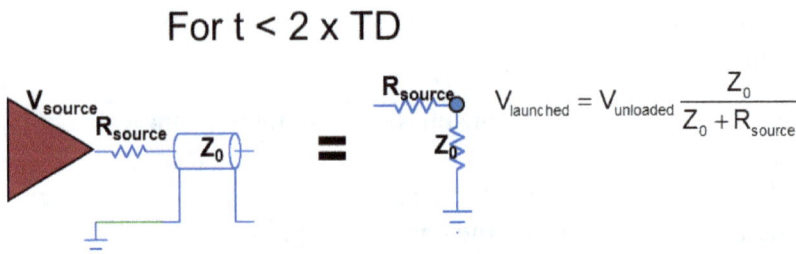

$$V_{launched} = V_{unloaded} \frac{Z_0}{Z_0 + R_{source}}$$

Figure 11.8 Equivalent circuit model of a driver with a source resistance connected to a transmission line.

The voltage launched into the transmission line, $V_{launched}$, that propagates to the end of the line, is the voltage division of the source voltage:

$$V_{launched} = V_{source} \frac{Z_0}{R_{source} + Z_0}$$

There are three special cases to consider.

Case 1: $R_{source} \ll Z0$

In this case, the initial voltage launched into the transmission line that continues to propagate is nearly the same as the source voltage. A source device with a low output impedance is sometimes referred to as a line driver, in that it can drive a signal on a transmission line nearly equal to its source voltage.

Case 2: $R_{source} \gg Z0$.

In this case, the voltage launched into the transmission line is small compared to the source voltage. This is an important distinction. There is a difference between the source voltage, which would be the voltage from the source measured in an unloaded condition, and the signal launched into the transmission line. These are two different voltages.

When the source resistance is very large compared to the transmission line's characteristic impedance, the voltage signal that propagates to the end of the transmission line is a small fraction of the source voltage.

Case 3: $R_{source} = Z0$

In the special case, the source resistance is equal to the transmission line's characteristic impedance, and the voltage launched into the transmission line is half the source voltage.

11.2.5 Principle #5

Principle #5 is that reflections occur when the instantaneous impedance a signal sees changes.

The instantaneous impedance of the line is so important because of what happens when its value changes. If the instantaneous impedance the signal sees is constant, the signal continues to propagate undistorted down the line. But if the instantaneous impedance changes, some of the signal reflects, and the signal transmitted changes.

When the signal comes from an impedance environment, Z_1, and enters an impedance environment, Z_2, the reflection coefficient, rho, the ratio of the reflected voltage to the incident voltage, is given by

$$\text{rho} = \frac{V_{reflected}}{V_{incident}} = \frac{Z_2 - Z_1}{Z_2 + Z_1}$$

The transmission coefficient, t, is given by

$$t = \frac{V_{transmitted}}{V_{incident}} = \frac{2 \times Z_2}{Z_2 + Z_1}$$

This is illustrated in **Figure 11.9**.

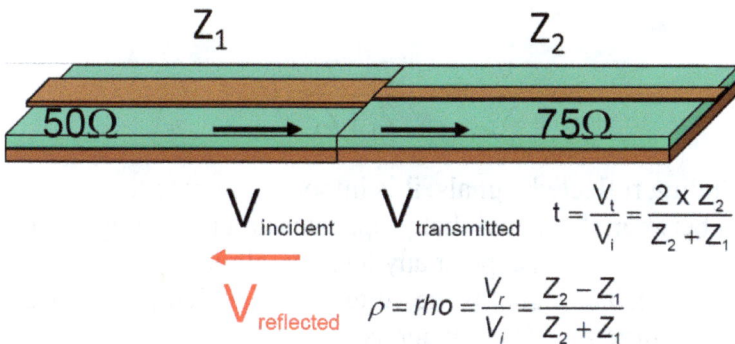

Figure 11.9 A reflected signal is created whenever the instantaneous impedance changes.

There are three special cases to note.

Case 1: When the second impedance is an open, the reflection coefficient is 1:

$$\text{rho} = \frac{V_{\text{reflected}}}{V_{\text{incident}}} = \frac{\infty - Z_1}{\infty + Z_1} = 1$$

This means all of the incident signal is reflected back in the direction of the source. When a signal propagates from a 50 ohm transmission line and encounters the 1 megaohm input resistor of a scope, the reflection coefficient is (1 − 0.000025), which is nearly 1.

Case 2: When the second impedance is a short, the reflection coefficient is -1:

$$\text{rho} = \frac{V_{\text{reflected}}}{V_{\text{incident}}} = \frac{0 - Z_1}{0 + Z_1} = -1$$

All of the signal is reflected back, but it is changed in sign.

Case 3: When the second impedance is the same as the first impedance, there is no reflected signal. The reflection coefficient is 0:

$$\text{rho} = \frac{V_{reflected}}{V_{incident}} = \frac{Z_1 - Z_1}{Z_1 + Z_1} = 0$$

When including reflected signals, it is important to note the distinction between the signal that propagates and the voltage that would be measured by a scope at any location along the transmission line. The signal is the voltage change that propagates in a particular direction. The voltage is the total net voltage difference between the signal and return conductor, which is what would be measured by a scope. The scope cannot distinguish the direction in which the voltage is traveling.

For example, a 1 V signal traveling from left to right in a transmission line would be measured exactly the same as a 1 V signal propagating from right to left.

When a 1 V signal is traveling down a transmission line and hits the nearly open impedance at the 1 megaohm input impedance of a scope, the 1 V signal will reflect, and the reflected signal will also be 1 V. However, at the 1 megaohm resistor, there are really two waves propagating. A 1 V signal travels into the 1 megaohm resistor and a 1 V signal travels away from the 1 megaohm resistor. There is a net 1 V + 1 V signal across the 1 megaohm resistor. This means the scope measures the scalar voltage of the 2 V signal across the 1 megaohm resistor.

This is illustrated in a simple experiment, shown in **Figure 11.10**. A 400 mV signal is launched into a 6 ft long, 50 ohm RG58 transmission line. This signal is measured initially as it is launched into the transmission line with a short-length coax Tee connector and the input coupling of the scope's channel set for 1 megaohm. About 10 nsec later, this 400 mV signal is detected at the input to the scope on channel 3. It is seen as a 400 mV incident signal + 400 mV reflected signal, or 800 mV total voltage.

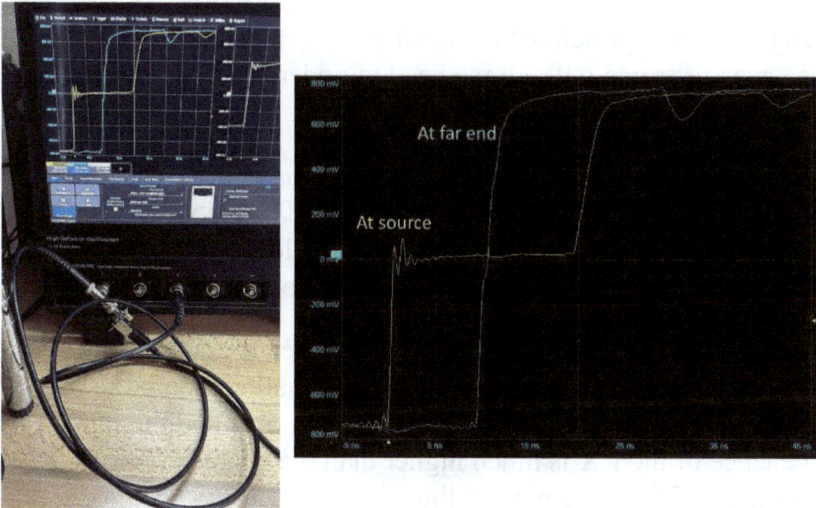

Figure 11.10 A fast source drives a 400 mV signal into a transmission line. 10 nsec later, the signal is measured at the far end, with 1 meg coupling. The signal at the far end is the incident 400 mV and the reflected 400 mV signal.

The 400 mV reflected signal makes its way back to the source another 10 nsec later and is seen as an additional 400 mV voltage at the source, on top of the already present 400 mV signal. This specific source had a 50 ohm source impedance, which is the same impedance as the coax cable transmission line, so there is no further reflection when the reflected 400 mV signal hits the source.

The scope measurements tell you nothing about the dynamic nature of the signals and their reflections. You must see the reflections in your engineer's mind's eye and interpret the scalar voltages measured by the scope in terms of your understanding of the behavior of the signals propagating on the transmission line.

These reflections happen at every interface where the instantaneous impedance changes. It is also important to keep track of the direction the signal is propagating. The direction of propagation determines the impedance from which the signal is coming and to which it is going. The challenge is keeping track of all the reflections at all the interfaces.

11.3 The Origin of Ringing Noise

Using these five principles of signal propagation on transmission lines, we are armed with all we need to understand the origin of ringing noise.

The way to analyze the behavior of signals from the DUT to the scope is by applying situational awareness. The physical design of the system needs to be turned into an equivalent circuit model and the circuit model used to analyze the behavior of the signal.

Ringing noise at the receiver, RX, arises when the source impedance of the transmitter, TX, is very small compared to the characteristic impedance of the transmission line, and the input impedance of the RX is much higher than the characteristic impedance of the transmission line.

This is one of three special cases of combinations of source impedance and termination impedance with three distinct behaviors to consider.

11.3.1 Case 1

Case 1 is when the source, the TX, is a low impedance and the far end, the RX, a high impedance.

When the driver source, the transmitter, TX, is connected by a transmission line to the receiver, RX, there are three interfaces to analyze: propagating from the TX into the transmission line, propagating from the transmission line into the RX, and propagating from the transmission line back into the TX.

When the RX has a very high impedance compared to the transmission line, the reflection coefficient at this interface will be nearly 1. When the TX has an impedance very low compared to the transmission line, the reflection coefficient will be nearly -1.

In this case, all of the signal will reflect from the RX, make its way back to the TX, and reflect back toward the RX with a sign change and small amplitude change. The resulting signal that makes its way to the RX after its first reflection from the TX, will be a negative signal, bringing the received signal at the RX lower. Each time the signal reflects from the TX, there will be a sign change and a reduction in amplitude.

When there is more than one interface, and the reflected signal continues to bounce between impedance changes, it is sometimes difficult to keep track of all the reflections. This is when a circuit simulation tool is valuable to keep track of all the reflections. Most versions of SPICE can simulate the reflections in a transmission line circuit.

One example of a simulation tool to simulate reflections in transmission line circuits is Keysight's ADS. An example of the circuit to simulate the voltage at the RX when the TX has a low impedance source and the simulated signal at the RX is shown in **Figure 11.11**.

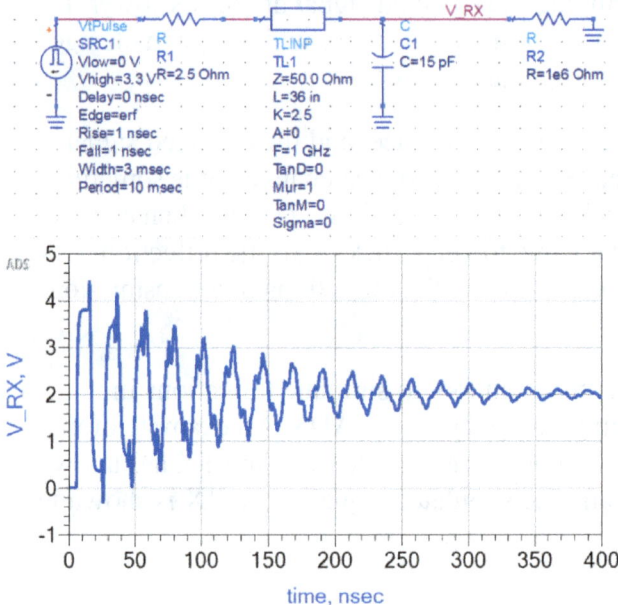

Figure 11.11 A simple circuit model in ADS to simulate the signal from a source through a transmission line to a receiver and the resulting simulation.

In this example, the signal launched into the transmission line hits the end of the line at the RX and reflects. The RX sees the initial incident signal plus the reflected signal, hence the first peak. A capacitor is added to the input of the scope in the model to simulate the impact of the scope's 15 pF input capacitance, in addition to its 1 megaohm input resistor.

The reflected signal propagates back to the TX's input, where it sees a low impedance. The reflected signal heading back to the RX is negative. After reflecting from the RX, this negative signal makes its way back to the TX, reflects, and changes sign.

At the RX, the combination of the initial positive incident signal from the TX and the negative reflected signal from the TX combine to bring the signal lower, hence the first dip.

The negative signal reflected from the RX hits the TX, reflects, changes sign, and heads back to the RX. When it hits the RX as a positive signal, it adds to the current, net voltage, hence the second peak. This process repeats. Each time the signal reflects from the low-impedance TX, the reflected signal has a smaller and smaller amplitude.

The time between consecutive peaks is 4, one-way time delays, 4 x TD. If the signal rise time >> 4 x TD, the reflections still happen but they are smeared out over the rising edge and are not visible. But if the rise time is < 4 x TD and the source impedance of the TX is lower than the transmission line's characteristic impedance, a ringing signature will appear at the scope input.

11.3.2 Measurement Simulation Correlation

In order to demonstrate this effect, a low impedance TX is needed with a rise time that is short compared with 4 x TD of the interconnect. Using a cable that is 36 inches long, the time delay is TD = 36 in/8 in/nsec = 4.5 nsec. This would require a source with a rise time shorter than 4 x 4.5 = 18 nsec.

An example of a source is a SN74AHC14 hex inverter, with a rise time of about 2 nsec. Each output has an output impedance of about 15 ohms. With six of them in parallel, the output impedance is about 15/6 = 2.5 ohms.

The system is the hex inverters in parallel driving a 36 inch RG174 coax cable connected to the 1 meg input of a scope. The system and measured signal at the scope is shown in **Figure 11.12**.

Figure 11.12 Measured signal into the scope with 1 meg input coupling from a low impedance, fast source and a 3 ft 50 ohm coax cable transmission line. The ringing is due to the reflections between the cable and the RX, and the cable and the TX.

This ringing signature is very similar to the predicted ringing behavior of the ideal model. There are just seven parameters that define this simulated system:

1. The unloaded voltage of the source, V_{source}

2. The rise time of the source, RT

3. The impedance of the source, R_{source}

4. The characteristic impedance of the transmission line, Z0

5. The time delay of the transmission line, TD

6. The input resistance of the scope, R_{scope}

7. The input capacitance of the scope, C_{scope}

The measured and simulated response can be directly compared in the Keysight ADS simulation environment. The measured signal at the scope was saved as a CSV file and imported into ADS. The nominal values of the system were used as starting values for this simulation using these values:

$V_{source} = 3.3$ V

$RT = 2$ nsec

$R_{source} = 2.5$ ohms

$Z0 = 50$ ohms

$TD = 4.5$ nsec

$R_{scope} = 1$ meg

$C_{scope} = 15$ pF

The match between the measured response by the scope and the simulated response at the RX is very close. **Figure 11.13** shows the match using the nominal values.

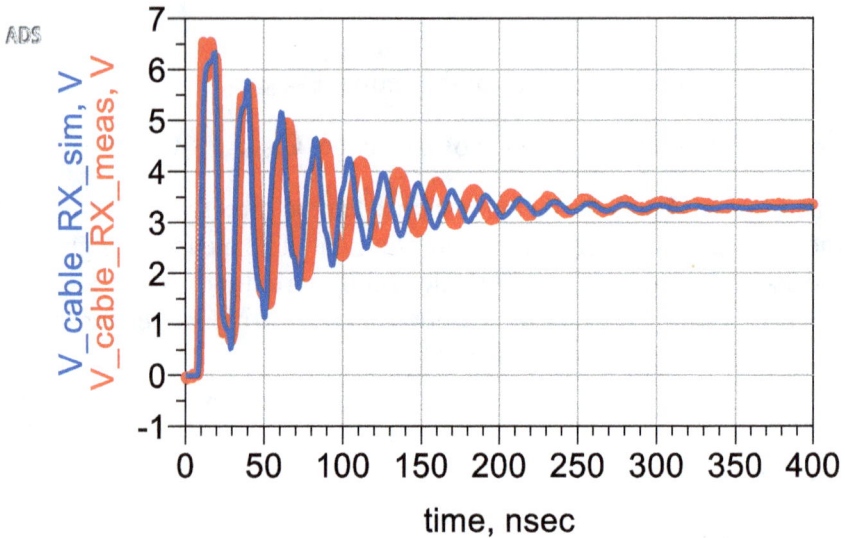

Figure 11.13. Comparing the measured and simulated signal at the scope using nominal conditions.

By slightly adjusting the parameter values from the nominal values, the agreement between the measured and simulated ringing response using this simple ideal circuit model, as shown in **Figure 11.14**, is very close.

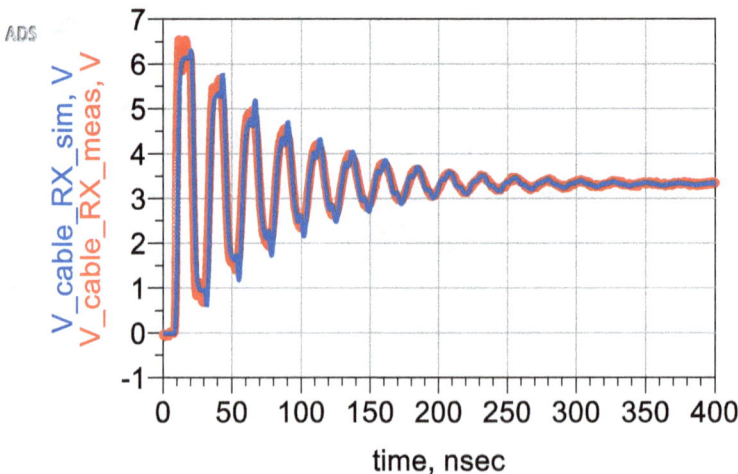

Figure 11.14 Match between the measured and simulated signal at the scope input using slightly optimized parameter values.

Adjusting the model's parameter values until the simulated result matches the measured result is sometimes called "hacking" the model.

This gives confidence that this simple model is a close representation of the actual behavior of this system. In this model, the transmission line is an ideal lossless transmission line and the source is an ideal voltage step source with a finite rise time. The adjusted parameters are shown in the table below:

Parameter	Nominal Value	Adjusted Value
RT =	2 nsec	2 nsec
R_{source} =	2.5 ohms	4.2 ohms
Z0 =	50 ohms	50 ohms
TD =	4.5 nsec	5.0 nsec
R_{scope} =	1 megaohm	1 megaohm
C_{scope} =	15 pF	15 pF

Ringing-like behavior will be seen at the RX when three conditions are met:

1. The driver output impedance is low compared to the characteristic impedance of the transmission line.

2. The rise time of the driver is shorter than about 4 x TD of the transmission line.

3. The input impedance of the RX is very high compared to the characteristic impedance of the transmission line.

11.3.3 Reducing the Ringing Artifact

Preventing the reflections can eliminate the ringing behavior. This is accomplished by fooling the signal into not seeing a change in the instantaneous impedance at either or both ends of the line. This is accomplished by strategically adding resistors, referred to as terminating the line.

A resistor can be added at the far end, so the signal sees a 50 ohm instantaneous impedance when it hits the far end. This is called far-end termination. This is the purpose of the 50 ohm input impedance of the scope. It is designed to terminate any signal propagating down the 50 ohm transmission line connecting to the DUT.

When the terminating resistor used in the scope is selected, there is an explicit assumption made that the coax cable feeding the signal to the scope will be a 50 ohm transmission line. Any other impedance cable, such as a 75 ohm RG6 cable, will not be terminated by the 50 ohm resistor built into the scope and even when the scope is set for 50 ohms, there may be reflection artifacts in the measurement.

Alternatively, a resistor can be added in series with the output source impedance of the driver so the sum of the added resistor and the source resistance adds up to the characteristic impedance of the transmission line. This makes the impedance the signal sees as it transitions from the transmission line to the input of the TX also 50 ohms, with no change in the instantaneous impedance and no reflections. This is called source series termination.

This implicitly assumes the characteristic impedance of the coax cable connecting the DUT to the scope is 50 ohms. Before changing the input coupling of the scope from 1 meg to 50 ohms, of course, being sure the RMS voltage is less than 5 V, the reflections in the cable connecting the DUT to the scope can be completely eliminated. **Figure 11.15** shows the resulting signal measured by the scope.

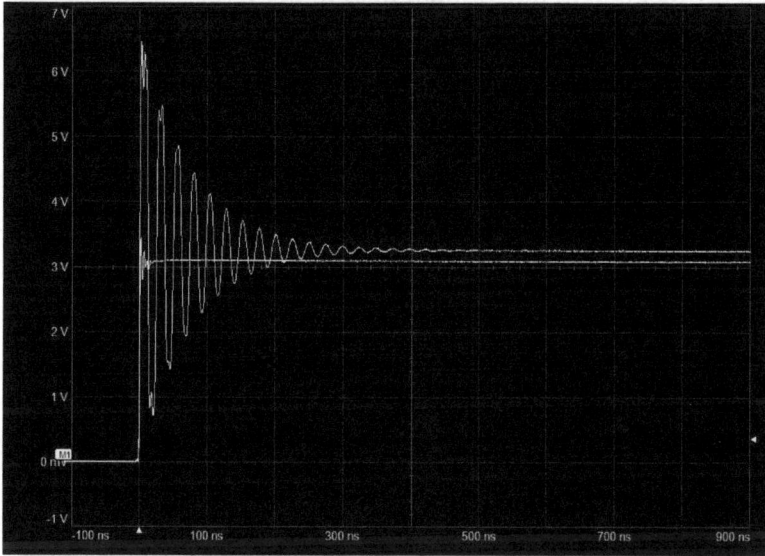

Figure 11.15 The measured signal from a fast, low-impedance driver with a 3 ft RG174 coax cable and a 1 meg input to the scope showed ringing, while a 50 ohm input coupling to the scope showed no ringing artifact.

In this measurement, the slight difference in the DC voltage after the ringing has died down is due to the 50 ohm DC load on the source Thevenin resistance. The slight ripple at the start of the measurement in the case of the terminated line is the actual noisy waveform from the source's output signal due to the switching noise inside the hex inverter's SOIC package. The time interval for the ringing noise is much shorter than the time interval of the reflections due to the long delay of the cable.

Two methods are commonly used to terminate the signal and prevent ringing artifacts when using a 50 ohm cable connection to the DUT.

Method 1: Use the input coupling of 50 ohms to the scope. This completely eliminates the ringing. Two important considerations are to ensure the RMS voltage is under 5 V to prevent damage to the scope and loading the DUT with a 50 ohm DC resistance.

Depending on the type of source, this method will introduce a potential artifact due to the 50-ohm load to the DUT.

Method 2: Use 1 meg input coupling to the scope and add a series resistor between the tip of the cable and the DUT. This is sometimes called a transmission line probe. The equivalent circuit model and two implementations are shown in **Figure 11.16**.

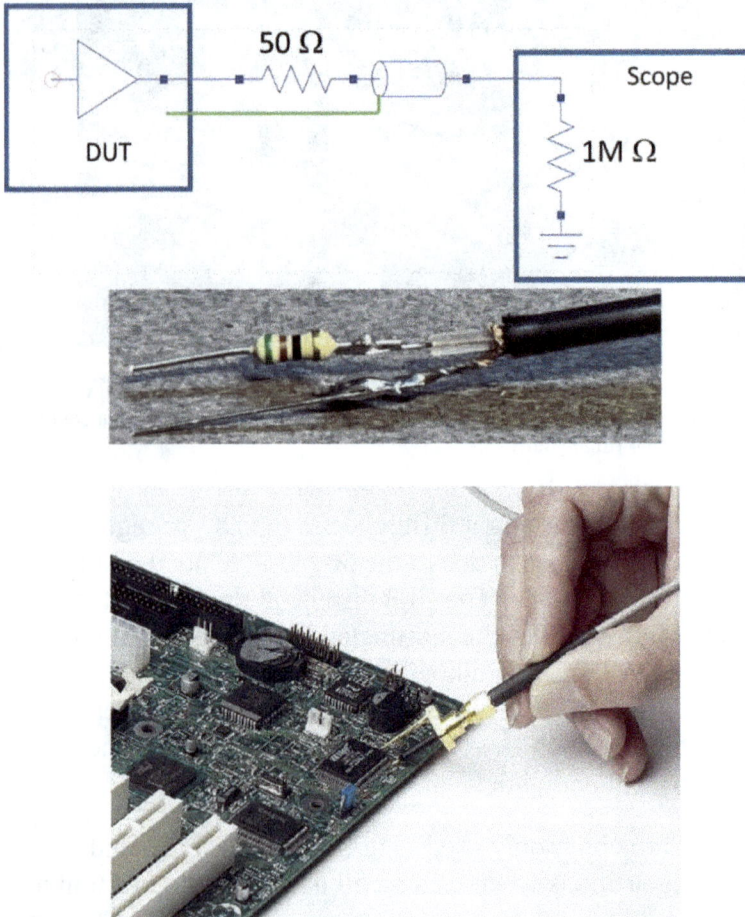

Figure 11.16 A transmission line probe is created using a 50 ohm series resistor at the tip of the coax cable and 1 meg input coupling to the scope.

The downside of this approach is that you need to know the output source impedance of the DUT to select the resistor value to add.

If the source resistance is less than 5 ohms, such as a power source or a low output impedance driver, a 50 ohm resistor is a good approximation to a series terminating resistor. This is a common resistor to add when probing power rails.

The bandwidth of this probe is limited by the bandwidth of the 1 meg input amplifier, typically 1 GHz or less, and the tip loop inductance.

11.3.4 Case 2

The second case of a combination of source impedance and receiver impedance is when the TX impedance = Z0 and the RX impedance is very high.

In this case, the voltage launched into the transmission line is just half the source voltage. When this signal makes its way to the RX, it will reflect. The scalar voltage measured at the RX will be one-half the source voltage propagating into the RX plus this same voltage reflecting from the RX, making the voltage at the RX equal to the source voltage.

This reflected signal will make its way to the TX where it will see an impedance equal to the characteristic impedance of the transmission line. There will be no reflection. Instead, the signal will be terminated.

The consequence of this circuit is that the full source voltage will be seen by the RX, one TD after it is launched into the transmission line. This is the special case of a source series terminated transmission line.

When the source impedance is 50 ohms, the input coupling to the scope can be 1 meg. The DC loading is the 1 meg input to the scope. The load the DUT sees is the length of the coax cable open at the far end. From the scope's perspective, the signal measured by the scope is the source signal with no distortion and no loading.

11.3.5 Case 3

Case 3 is when the source impedance, TX, is much higher than the transmission line Z0 and the TX impedance is very high.

In this case, the initial voltage launched into the transmission line will be a small fraction of the source voltage due to the voltage divider of the source resistance and the input to the impedance of the coax cable transmission line. This small voltage will make its way to the high impedance of the RX and reflect.

When this reflected wave hits the high impedance of the TX, it will again reflect, with a slightly smaller amplitude but the same sign. This additional signal will make its way to the RX and again reflect.

Each time the smaller amplitude voltage hits the high impedance of the RX, it will add to the previous signal. Eventually, all the small amplitude waves will add together and equal the source voltage, but it will take many round-trip time delays. It will look a lot like an RC charging of the transmission line but it is actually due to many small reflections.

Measured examples of these three cases of different source impedances and the resulting signal measured at the high impedance of the scope input are shown in **Figure 11.17**. They show the three different behaviors for each case.

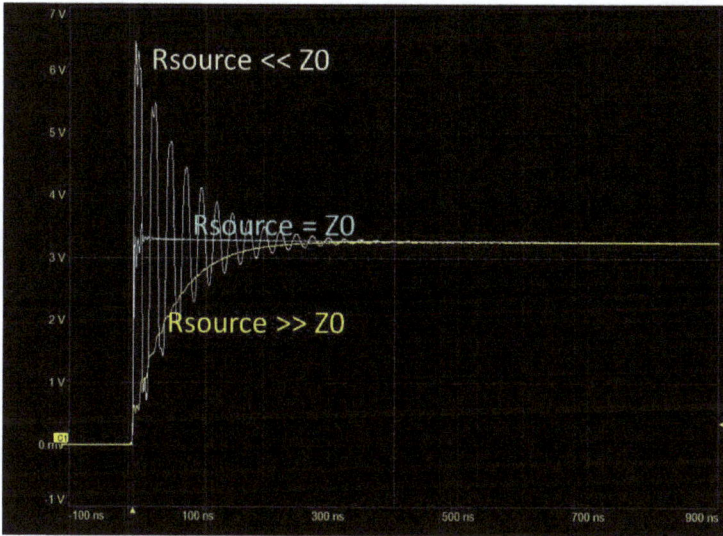

Figure 11.17 The three cases of source resistance all with the scope input 1 Meg coupling and a 3 ft long RG174 transmission line from the DUT to the scope.

11.4   Selecting the Right Scope Input Impedance

Using 50 ohms as the input coupling to the scope eliminates the reflection noise between the DUT and the coax cable connecting them. If a reflection noise artifact is possible, use 50 ohm coupling. Of course, this assumes the transmission line connecting the DUT and the scope has a 50 ohm characteristic impedance.

This is a common assumption and is why 50 ohms is the universal input impedance of a scope and most other instruments. If all interconnects have a characteristic impedance of 50 ohms, reflections at the ends will be eliminated.

The downside of using 50 ohms in the scope is that it will load the DUT with 50 ohms. And, of course, it is not an option if the input rms voltage from the source is > 5 V.

Use the 1 meg input coupling to the scope when using:

- The 10x probe, suitable for signal bandwidths < 500 MHz

- The 1x probe or a direct coax connection to the DUT for signal bandwidths < 10 MHz

Use 50-ohm input coupling to the scope for:

- Input RMS voltages < 5 V

- A 10x transmission line probe for signal bandwidths < 5 GHz

- Direct coax cable connection with the highest bandwidth and loading the DUT with a 50 ohm resistive load

- Most active probes

11.5   Resolving the Paradox of the Compensation Signal

When the transmission line connecting the comp signal to the scope is terminated by the 50 ohm input resistance of the scope there are no reflections at the scope end of the transmission line. Whatever signal is launched into the transmission line at the source makes its way to the end of the transmission line and is measured across the 50 ohm internal resistor and displayed by the scope.

The scope measures two important signal features: the voltage launched into the transmission line due to the voltage divider of the DUT source impedance and the cable's 50 ohms, and the signal's rise time.

The unloaded voltage of the comp signal in the Teledyne LeCroy HDO6104 scope is measured as 1 V into a 1 meg resistor. This is the Thevenin source voltage.

The voltage launched into the RG174 transmission line, which makes its way to the 50 Ohms of the input coupling to the scope, is measured as 60 mV. This is due to the very high output Thevenin resistance of the comp signal source and the input impedance of the 50 ohm transmission line. This voltage-divided signal makes its way to the scope input, where it is terminated by the 50 ohm scope impedance, preventing any further reflections. The voltage and the rise time of the signal from the comp signal is measured as 2.2 nsec. These features are shown in **Figure 11.18**.

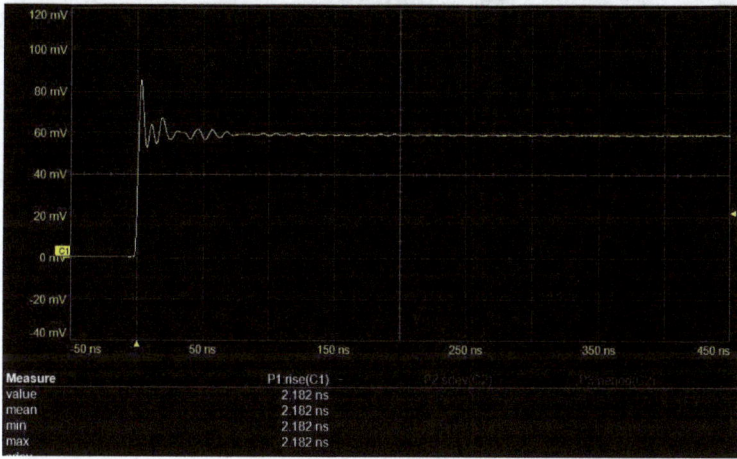

Figure 11.18 Measured comp signal from the Teledyne LeCroy HDO6104 scope into 50 Ohm input coupling of the scope. The rise time, with 0.1 nsec resolution, is measured as 2.2 nsec.

This 2.2 nsec rise time is intrinsic to the source of the comp signal. There is a little ringing structure to the signal from the comp signal. This is part of the comp signal and is related to the noise on the driver that creates the comp signal.

The 2.2 nsec rise time is independent of the length of the transmission line. **Figure 11.19** This shows the measured signal into the scope from the comp signal after a length of 6 ft and 24 ft of RG174 cable. The signal's rise time is hardly changed as the cable length increases.

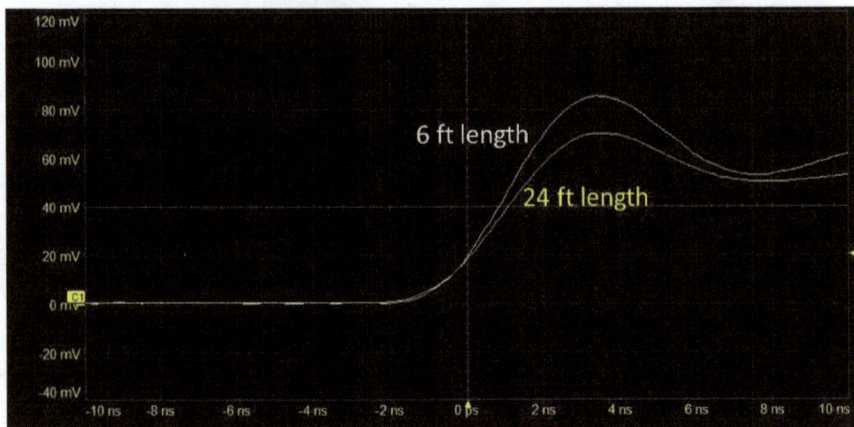

Figure 11.19 The measured signal from the comp source into 50 ohm coupling of the scope after 6 ft of coax cable and after 24 ft of coax cable. The rising edges have been shifted to align together. Note the less than 1 nsec increase in rise time due to the losses in the cable.

It is not true that a longer cable will increase the rise time of a transmitted signal. There is a slight contribution from the cable losses, which increases the rise time by less than 1 nsec, but this is very slight. This observation is contrary to the observation when the input to the scope is 1 meg, where the measured rise time of the comp signal depended on the cable length. What is going on?

The increasing measured rise time of the comp signal with cable length is an artifact of the reflections in the cable due to the combination of the comp signal's high output resistance and the 1 meg coupling to the scope. This behavior can be understood in the context of the equivalent circuit model of the system composed of the comp source, the RG174 coax cable, and the input resistance of the scope. This equivalent circuit model is shown in **Figure 11.20**.

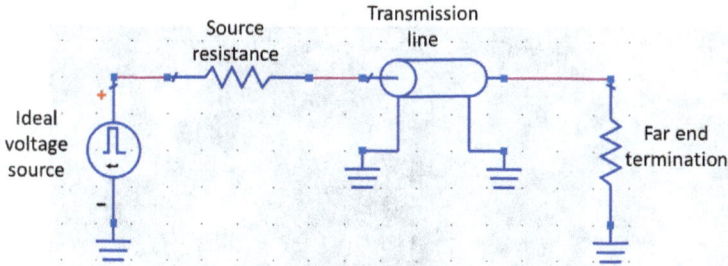

Figure 11.20 Equivalent circuit model of the comp signal driving a transmission line and scope termination.

In the Teledyne LeCroy HDO6104 scope, the source impedance of the comp signal is 785 ohms. This means the 1 V unloaded voltage of the source is divided down to a smaller voltage that is launched into the transmission line:

$$V_{launched} = V_{source} \frac{Z_0}{R_{source} + Z_0} = 1V \frac{50\Omega}{785\Omega + 50\Omega} = 0.06V$$

This 60 mV voltage propagates to the end of the RG174 cable and is reflected by the 1 meg input to the scope. The expected voltage measured at the scope is 60 mV going into the 1 meg resistor and another 60 mV reflected, for a total voltage of 120 mV measured.

The 60 mV reflected signal will propagate to the source, where it will reflect with a reflection coefficient of

$$rho = \frac{785\Omega - 50\Omega}{785\Omega + 50\Omega} = 0.88$$

This reflected voltage, 60 mV x 0.88 = 53 mV, will head to the scope, the RX. For a 6 ft long coax cable, the time delay is 10 nsec. After a round trip time of 2 x 10 nsec = 20 nsec, the next wave will be received at the scope. This process is repeated, and multiple steps are reflected from the TX and added together to recompose the full 1 V signal from the TX eventually. The initial voltage and the first few reflections from the TX can be measured by the scope by expanding the time scale to resolve each step, as seen in **Figure 11.21**.

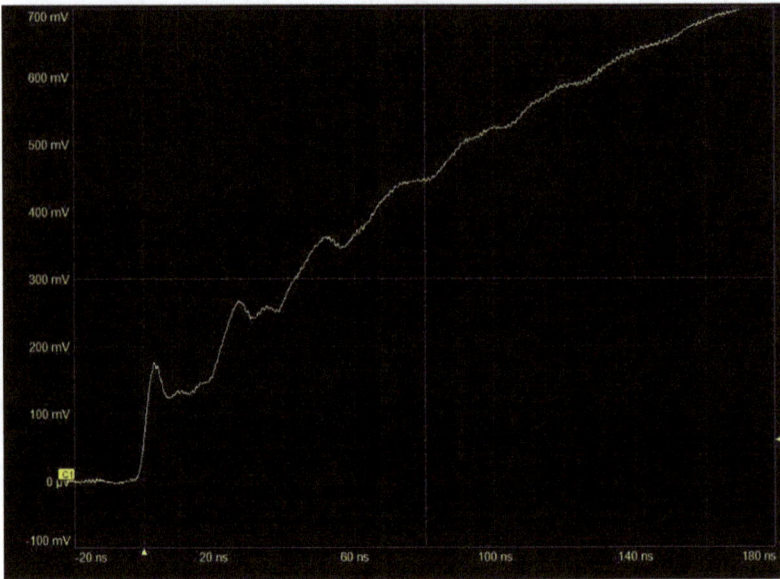

Figure 11.21 The initial voltages measured by the scope from the comp signal with 1 meg coupling into the scope, on an extended time base.

The initial voltage measured is about 120 mV. About 20 nsec later, the next step of about 100 mV is detected. The later steps every 20 nsec are clearly seen. After about six reflections, the 15 pF of input capacitance on the 1 meg input to the scope increases the rising edge, and the individual reflections are smeared over the rise time. Nonetheless, the reflections still happen.

The increase in time with the cable length is seen as an artifact of the multiple reflections on the transmission line. This is an artifact of how the signal is measured by the scope. The longer the cable, the longer the time interval between the steps and the longer the apparent rise time of the signal.

The origin of the anomalous rising edge can be understood by analyzing the measurement system in terms of the equivalent circuit model. Translating the real system of the DUT, the cables, and the scope into an equivalent circuit model and analyzing the behavior of signals in this circuit is an example of applying situational awareness.

The reflections can be eliminated by using 50 ohms as the input coupling to the scope to measure the intrinsic rise time of the comp signal.

However, another problem arises. The low impedance of the cable characteristic impedance loads the comp signal source, and the voltage measured by the scope is the loaded voltage, not the source voltage.

A 10x probe should be used to measure the rise time and the source voltage. Due to the distributed lossy nature of the special 10x probe cable, this will not load the comp signal and will not be subject to the ringing artifact. **Figure 11.22** shows the measured comp signal on an expanded time base, showing the rising edge and the full unloaded amplitude of 1 V.

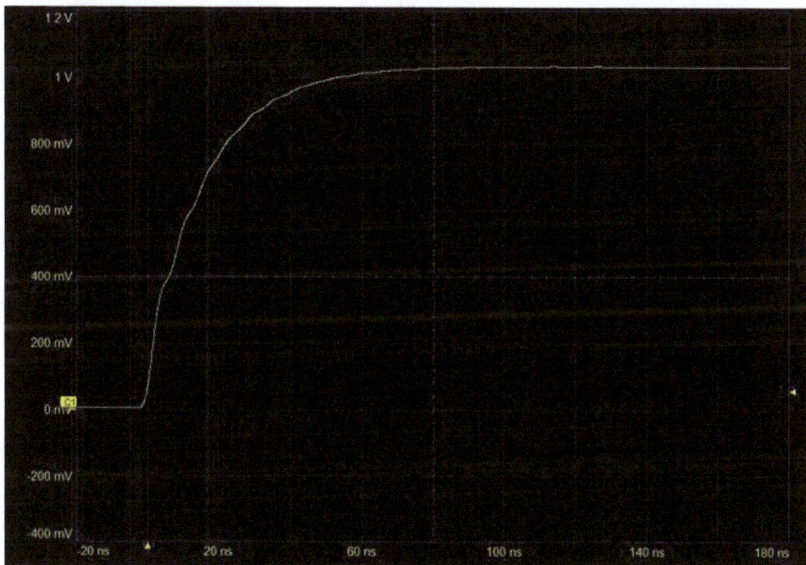

Figure 11.22 Measured comp signal using a 10x probe on the 1 meg input coupling.

It should be noted that the 10x probe also introduces an artifact. Electrically, it is equivalent to a 10 pF capacitor. The 785 ohm output impedance of the comp signal, when connected to a 10 pF capacitor, has an RC rise time of 785 ohms x 10 pF = 8 nsec. The 10-90 rise time is about 2.2 x 8 nsec = 18 nsec. The rise time measured by the 10x probe is approximately 30 nsec. This is much larger than the 2.2 sec signal rise time measured into the 50 ohm coupling of the scope and is another artifact due to the high bandwidth and high-impedance source.

Measuring a fast signal from a high-impedance source is surprisingly difficult without introducing artifacts. It requires a high bandwidth, low input capacitance, and high-impedance probe. Alternatively, it requires analyzing the system using an equivalent circuit model and reverse engineering the parameters that describe each circuit element.

11.6   The Bottom Line

1. Reflections play a critically important role when measuring a signal with a rise time shorter than about 40 nsec.

2. Understanding how reflections are created and suppressed enables you to manage them in any measurement system.

3. Use a 50 ohm termination into the scope when connecting a coax cable to a signal with a rise time shorter than 40 nsec.

4. Be aware of the loading to the DUT by the 50 ohm input coupling to the scope.

5. Never apply an input signal to the 50 ohm input of the scope larger than 5 V rms, or the scope may be damaged.

6. Model all interconnects as transmission lines with a time delay and a characteristic impedance. Be aware of your cable's impedance and time delay.

7. When transmission line reflection artifacts may be a problem, either terminate the scope input, add a source series resistor, or use a 10x probe.

8. Even a 10x probe, with its 10 pF of input capacitance, will show an increased rise time when the source impedance is large due to the RC charging.

9. In any measurement, note the source impedance, the cable characteristic impedance, the scope input impedance, the DUT signal rise time, and the cable's TD.

10. When reflection noise creates a measurement artifact, use a termination strategy to reduce the ringing noise and analyze the system using situational awareness to interpret the scope measurements to extract the DUT features.

Chapter 12
Alternative Probing Options

The most valuable general-purpose probe with a scope is the 10x probe. This is a good balance between

- Not loading the DUT with a low impedance
- Good bandwidth
- Easy connections to the DUT
- Higher voltage probing
- Low RF interference pickup
- Readily available
- Very low cost

But there are four properties of the 10x probe that affect how the probe interacts with the DUT:

- The signal-ended nature of the probe
- The 10x attenuation of the signal
- The input capacitance of 10 pF
- The limited bandwidth of the probe/scope system

12.1 Simple 10x Probe Alternatives

Generally, the first probe to use for all signals is the 10x probe. It is a good balance between performance, not loading the DUT, and accessibility.

But there are three conditions when it is not the best probe:

- When the signal level is < 10 mV

- When a true differential probe is needed

- When the signal bandwidth is > ~ 200 MHz.

When slightly better sensitivity is required and the signal bandwidth is less than 20 MHz, a simple alternative is to use the 1x setting built into some low-cost probes. This switch shorts the 9 Megaohm input resistor and the 10 pF capacitance. An example of this switch is shown in **Figure 12.1**.

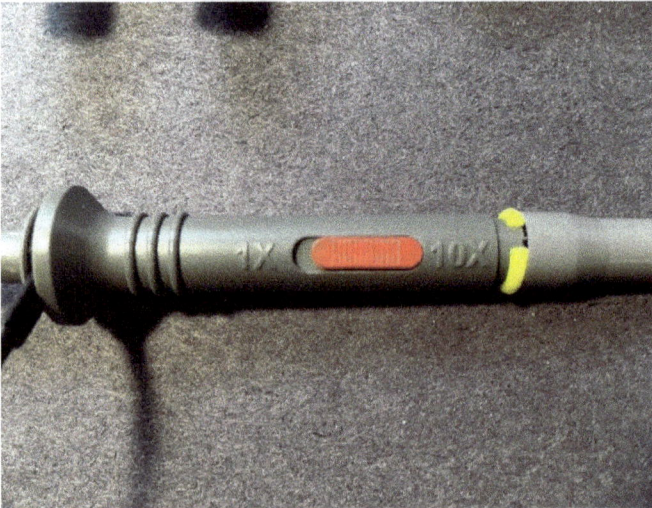

Figure 12.1 Some 10x probes have a switch that shorts the input resistor and capacitor, turning the probe into a 1x probe. This limits the bandwidth to about 10 MHz.

This setting turns the probe into a direct, though very high-resistance, coax cable connection between the DUT and the scope. While this is an easy connection to make and enables the scope's full sensitive range of 1 mV/div, it limits the measurement bandwidth to about 10 MHz.

The limited bandwidth is set by the roughly 300 ohm series resistance of the special cable that makes up the 10x probe, and the distributed capacitance of the cable and the input capacitance of the scope. This RC time constant can be on the order of 300 ohms x 0.1 nF = 30 nsec.

A second way around the 10x attenuation of a 10x probe is to use a low-loss coax cable, such as an RG58 or RG174 with BNC or mini grabber connectors.

Care must be taken to avoid reflection artifacts when using a direct coax cable connection. Generally, if the DUT output impedance is other than 50 Ohms and the input to the scope is 1 meg, there will be reflections in the coax cable. These artifacts limit the useful bandwidth to about 20 MHz. When this limitation is acceptable, a simple mini grabber adaptor can be added to the end of the coax cable to connect the DUT. An example is shown in **Figure 12.2**.

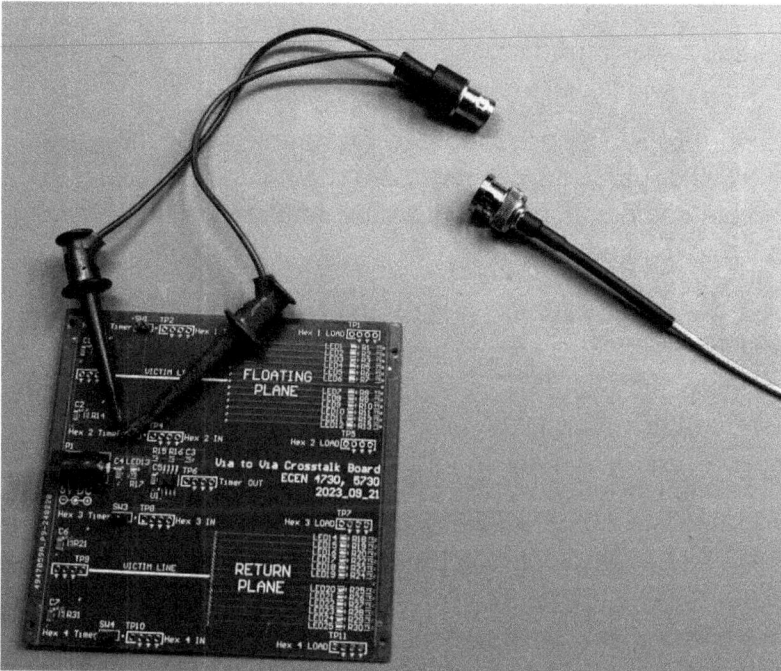

Figure 12.2 The fixture between the DUT and scope consists of a mini grabber to a BNC adapter and a coax cable.

Reflection artifacts in a coax cable connection can be eliminated if the DUT's source impedance is 50 ohms or the scope's input impedance is set for a 50 ohm input. Either of these conditions will terminate the reflections.

12.2 Probing Low Bandwidth, Low Level, or Differential Signals

When low-level signals require a more sensitive scale than 1 mV/div, such as some sensors or biomedical applications, or a true differential measurement is required, such as in power supply systems, there are generally two probing options:

- Use an external preamplifier between the DUT and the scope

- Use a specialized probe that is usually custom to a specific scope vendor's family of probes

A preamplifier will increase the signal level into the scope's range and buffer the source impedance to drive the scope cable and channel but not load the DUT with a low impedance. They can generally be used on any scope.

The simplest and lowest cost option is to build your own opAmp circuit. OpAmps with a gain-bandwidth product higher than 50 MHz are generally considered high-speed. Many have input impedance higher than 1 Megaohm and can drive the 50 ohm input to scopes. The typical price is < $5.

Two examples of suitable opAmp buffers are the ADA4807 with a GBW product of 180 MHz and the OPA858 with a gain bandwidth product of 5500 MHz. The downside of this approach is that it requires building your own custom PCB.

A differential amplifier can be designed with an op-amp, but achieving a high common-mode rejection ratio is sometimes difficult. This is because a careful balance of the gain and offsets of the p and n sides of the amplifier is required. Instead, an instrumentation amplifier, such as the LT110 and the INA849, could be used. While these are true differential amplifiers, they are limited in bandwidth to about 35 MHz.

Stand-alone instruments providing this same capability are available from multiple vendors. These are sometimes referred to as preamps or signal conditioners. Their gains are typically 10x to 10,000x but are sometimes limited in bandwidth to < 50 MHz.

An example of this type of preamp is the Stanford Research Systems SR560, shown in **Figure 12.3**. It has a gain of 1 to 50,000x and bandwidth filtering to reduce the high-frequency noise. Its bandwidth is limited to 1 MHz.

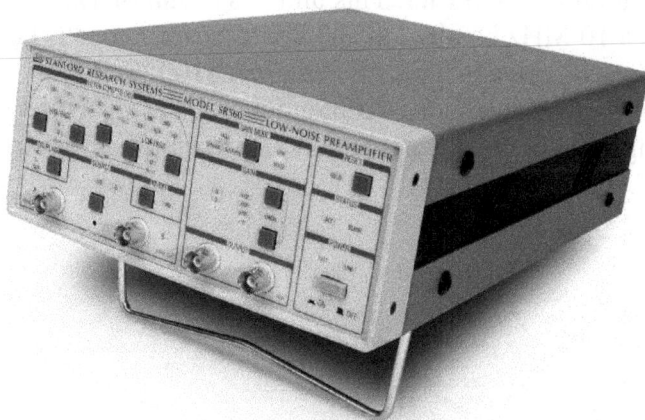

Figure 12.3 An example of a low-noise preamp with a 1 MHz bandwidth is courtesy of Stanford Research Systems.

Another example is the Teledyne LeCroy DA1855A differential signal conditioner. It has a gain setting of 10x, 1x and 0.1x, with a bandwidth of 100 MHz. An example is shown in **Figure 12.4**.

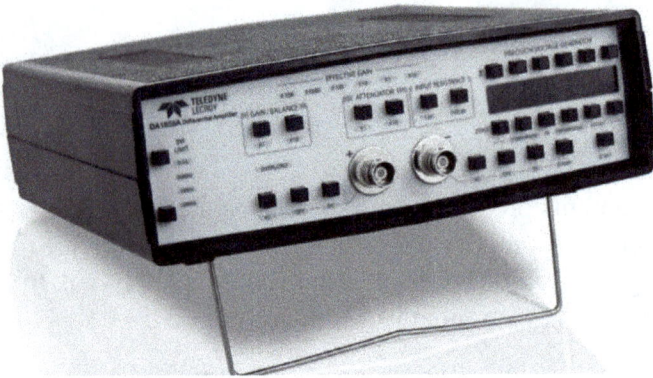

Figure 12.4 The Teledyne LeCroy DA1855A differential signal conditioner. It has a gain of up to 10x and 100 MHz bandwidth.

Multiple commodity differential preamps exist at the low end, such as the T100 differential amplifier. This unit has a gain of 1x, 10x and 100x with a 10 MHz bandwidth. While its performance is not as good as others, its price is typically less than $100. Some versions of this sort of standalone differential probe are rated for 1 kV input voltages with built-in attenuators. An example is shown in **Figure 12.5**.

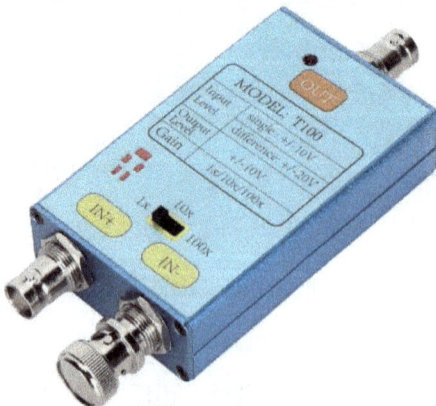

Figure 12.5 The T100 low-end stand-alone differential amplifier with a 10 MHz bandwidth.

A second probing approach uses a proprietary active probe that interfaces to a specific scope vendor's family of scopes.

For example, Keysight and Teledyne LeCroy both have active differential probes with similar performance. **Figure 12.6** shows examples of these two scope probes with bandwidths higher than 500 MHz. They are powered by proprietary sockets built into the scopes.

Figure 12.6 Examples of Keysight and Teledyne LeCroy active differential probes.

The amplifier at the tip of the probe has an input impedance of 1 megaohm. These probes convert the differential signal at the tip to a single-ended signal and buffer the DUT signal to drive the cable's 50 ohm impedance and the scope's input impedance.

12.3 Commodity LNA for Higher Bandwidth

Another class of signal conditioners is used primarily with spectrum analyzers and RF systems but can also be used with scopes. These are low-noise amplifiers (LNA) with a low-frequency cutoff, typically about 1 MHz, but a high-frequency limit over 4 GHz. Their input impedance is 50 ohms.

These preamps have two significant limitations. Their input impedance is 50 ohms, which means they cannot be used with high-output impedance DUTs. They are also AC coupled with a low-frequency cutoff generally higher than 1 MHz, which means they are unsuitable for a DUT when the DC levels are also important.

These amplifiers are designed for radio frequency communications applications, such as antenna preamps or software-defined radio applications. However, they are very useful when measuring low-level radio frequency signals, such as from near-field E or B probes.

Because so many vendors offer stand-alone LNA modules with similar performance, they are commodity parts. For example, **Figure 12.7** shows two low-cost generic modules, a Goozeezoo 6 GHz LNA and one that uses the Qorvo TQP3M9037 LNA chip. Both of these amplifiers cost less than $30.

Figure 12.7 Examples of two commodity LNA examples with a price under $30.

12.4 Directly Measuring Signals with a BW > 20 MHz

When the DUT can drive a 50 ohm load, the best connection between the DUT and the scope is with a coax cable using BNC or SMA connectors. This will always be the highest bandwidth measurement, limited by the bandwidth of the connector type used. Of course, the most important recommendation is to use a scope with a bandwidth rating of at least 3x of the signal bandwidth, or at least as much as you can afford.

These methods only work if the signal is probed at the receiver with a 50-ohm direct coax cable connection. Any other interconnect topology will result in connection stubs and artifacts of reflections and ringing.

The following are some of the limitations just from the connector:

- A BNC connector is limited to about 2 GHz
- A U.FL connector is limited to about 6 GHz
- A type N connector is limited to 11 GHz
- An SMA connector is limited to about 18 GHz
- A 3.5 mm connector is limited to about 26 GHz
- A 2.92 mm type K connector is limited to 40 GHz
- A 2.4 mm connector is limited to 50 GHz
- A 1.85 mm connector is limited to 67 GHz
- A 1.35 mm connector is limited to 90 GHz
- A 1 mm connector is limited to 110 GHz

The cost of a connector increases nearly exponentially with bandwidth. For example, an SMA connector is < $1, while a 2.92 mm connector can be $100, and a 1.85 mm connector can be $300.

An example of the measured transfer function of two coax cables is shown in **Figure 12.8**, an RG58 cable with BNC connectors and an RG174 cable with SMA connectors. Their -3 dB bandwidth is measured as about 5 GHz.

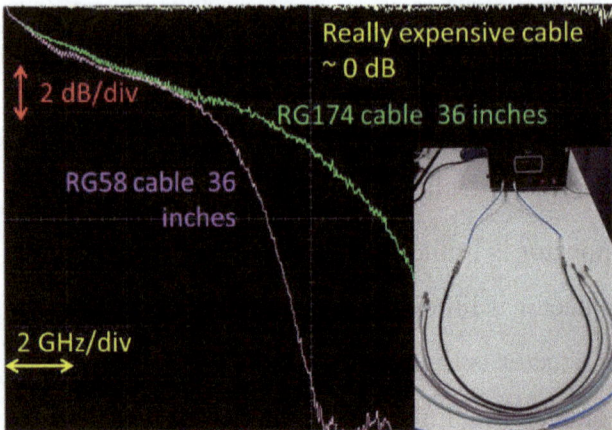

Figure 12.8 The measured transfer function of a 36-inch long RG58 and RG174 coax cable shows a -3 dB attenuation at about 5 GHz.

The highest bandwidth connection to the DUT will always be with a high bandwidth connector soldered directly into the DUT. An example of a 12.5 Gbps signal measured from an FPGA board, with an SMA connector soldered directly on the board, is shown in **Figure 12.9**.

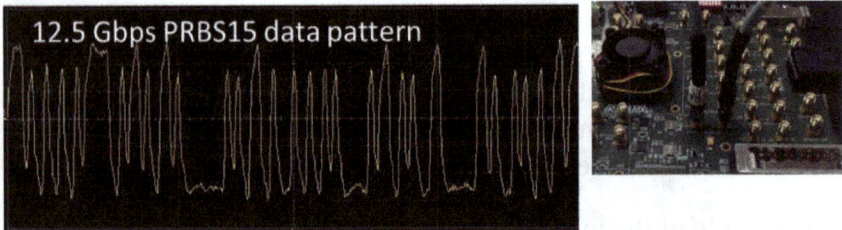

Figure 12.9 The highest bandwidth connection always has a coax connector soldered directly on the circuit board.

Many boards are designed with SMA connectors integrated throughout the board specifically to enable high bandwidth connections either to other boards or to a scope to measure the signals. Examples of test boards with SMA connectors soldered to the board are shown in **Figure 12.10**.

Figure 12.10 Examples of production boards with integrated SMA connectors to facilitate high bandwidth scope measurements.

Often, soldering a large coaxial connector to a circuit board is impractical. Alternatively, a smaller U.fl connector can be added to a board with a pigtail connecting to an SMA. **Figure 12.11** shows an example of a U.fl connector soldered to a board. The footprint can be left on the circuit board with little impact on the signal. The connector is soldered to the board only when it is needed.

Figure 12.11 Closeup of a circuit board showing the small footprint U.FL connector on the board and, to the right, pigtails connected to the board to SMA connections. Courtesy of Adinath Phene.

When integrating a high bandwidth coax connector into the board is not practical, pads on the board can be probed with a 50 Ohm microprobe, just requiring adjacent signal and return pads on a pitch from 0.5 mm to 0.2 mm. These probes can have > 20 GHz bandwidths limited by the pitch of the pads on the circuit board. An example of a microprobe touching a circuit board is shown in **Figure 12.12**.

Figure 12.12 Microprobes contacting PCB traces with a connection bandwidth of 20 GHz. Courtesy of MPI.

The downside of a microprobe is that it requires a probe station to align the microprobe to the board carefully. These can easily cost above $25k.

Yet another type of connection to the DUT is possible, which also has high bandwidth and does not require a probe station. Multiple vendors offer a compression mount coaxial probe that connects to signal pads with surrounding ground pads on the DUT board. These probes are assembled manually and aligned with pins on the board, so they do not require a probe station. Examples of these probes are shown in **Figure 12.13**.

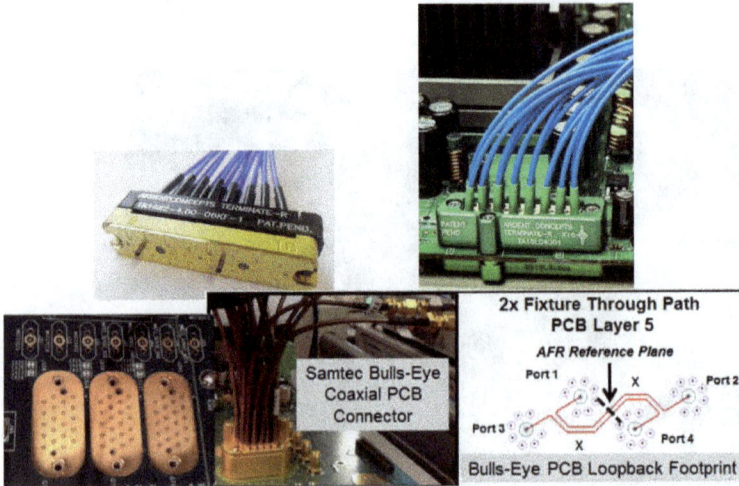

Figure 12.13 Examples of compression mount coax connector arrays from the circuit board to coax pigtails.

The lowest-cost probing method, which can be implemented after the design is complete, is to solder a coax cable to the board. A significant challenge is connecting the coax cable to the DUT. This connection should be engineered as close to coaxial as possible to maintain a high bandwidth. When the signal and return paths are pulled apart, such as when terminating a cable to the connector or contacting a circuit board, the loop inductance at the tip significantly limits the connection bandwidth.

For example, **Figure 12.14** shows how a minor disruption at the cable connector interface drops the cable's bandwidth from 5 GHz to less than 1 GHz.

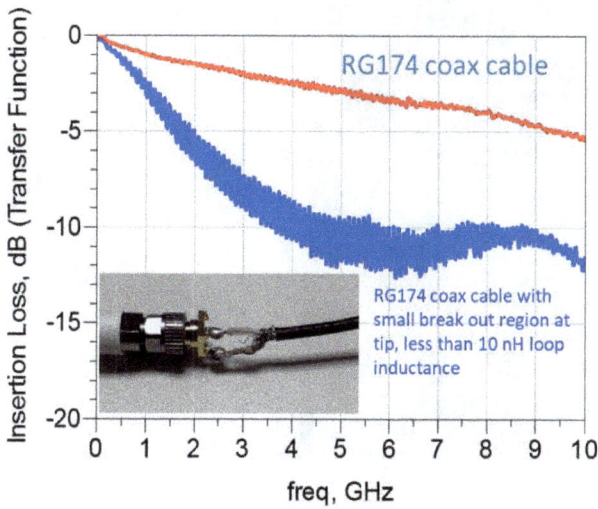

Figure 12.14 Measured insertion loss of a good SMA coaxial connection and one in which the signal and return have been separated.

If a coax cable is soldered directly to a board, care should be taken to reduce this loop inductance at the connection. Examples are shown in **Figure 12.15**.

Figure 12.15 Here are some examples of direct solder connections of a coax cable to a circuit board. Keeping the tip loop inductance low is critical to high bandwidth applications.

In all of these examples, the signal from the DUT must be probed at the RX. This minimizes any interconnect stub in the signal path and reduces the high-speed signal's measurement artifacts. The scope's 50-ohm input impedance terminates the signal and prevents reflections in the cable.

This method cannot be used if the signal in the DUT to probe is not at the RX. In this case, it must be measured with a high-impedance probe that does not load the signal. This is the case when the high-speed signal, either digital or RF, must be "sniffed" somewhere between the TX and RX in the DUT. In this case, a high-impedance, high-bandwidth probe must be used.

12.5 Sniffing Signals with a BW > 200 MHz

When a high-speed signal with a 50 ohm source impedance can be directly connected to the scope channel, the scope can measure the signal directly. To avoid reflection artifacts, make sure either the output resistance of the source is 50 ohms or the scope input is set to 50 ohms.

In many signal integrity and RF communications applications, high-speed signals are over 100 mV but can have very high bandwidths, and the DUT cannot tolerate a load much lower than a few hundred ohms. Generally, these signals are wideband and extend to DC. Below about 200 MHz, a conventional 10x probe can be used to sniff the signal. Other probes are needed above this bandwidth.

Two different probing options are available, each with different impacts on the signal bandwidth, signal-to-noise ratio, acceptable DUT loading, and acceptable cost.

The first version is a passive probe, very similar to a 10x probe. In this case, the input impedance of the probe needs to be much higher than 50 ohms. How high should the impedance of the probe be compared to the characteristic impedance environment of the DUT? This is a delicate balance between what is acceptable signal distortion and the cost of the final solution.

As a rule of thumb, a goal should be to keep the load on the DUT > 500 ohms. This means there is about a 10% decrease in impedance over the 50 ohm environment the signal may see when the probe is connected.

The simplest approach is to "roll your own" 10x probe, but designed for a 50 ohm environment. This means adding a 450 ohm series resistor right at the signal pin of the DUT. This makes the initial connection to the DUT, which is then connected directly into a 50 ohm coax cable and into the scope.

The scope is set for 50 ohm input impedance. This effectively makes a 10 to 1 voltage divider with the 450 ohm source resistor in series with the 50 ohms of the scope. If there is a sufficient signal level, a 950-ohm series resistor can be added, making a 20x attenuating probe with a 1k ohm input impedance. The equivalent circuit is shown in **Figure 12.16**.

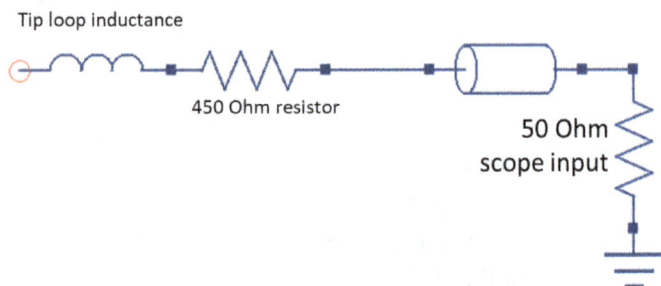

Figure 12.16 Equivalent circuit of the 10x voltage divider using the scope's 50 ohms as the load.

When the 450-ohm resistor is connected to the end of a manual probe, it is usually called a transmission line probe or a browser probe. An example of a commercial transmission line probe with a series 450-ohm resistor is shown in **Figure 12.17**. It can be implemented using a discrete resistor in the probe tip or with a resistor mounted on the circuit board.

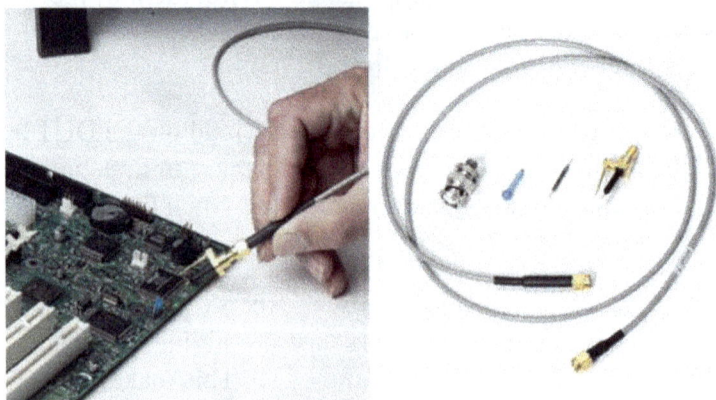

Figure 12.17 Example of a Teledyne LeCroy transmission probe with a 450-ohm series resistor.

The tip inductance limits the bandwidth of this connection to about 5 GHz. For higher bandwidth, the 450-ohm resistor can be integrated directly on the board.

An example of adding a 450-ohm resistor to a transmission line under test, which then connects through a U.FL connector to a 50-ohm transmission and into the 50-ohms of a scope is shown in **Figure 12.18**. This is one way of "sniffing" a high-speed signal using a passive transmission line probe.

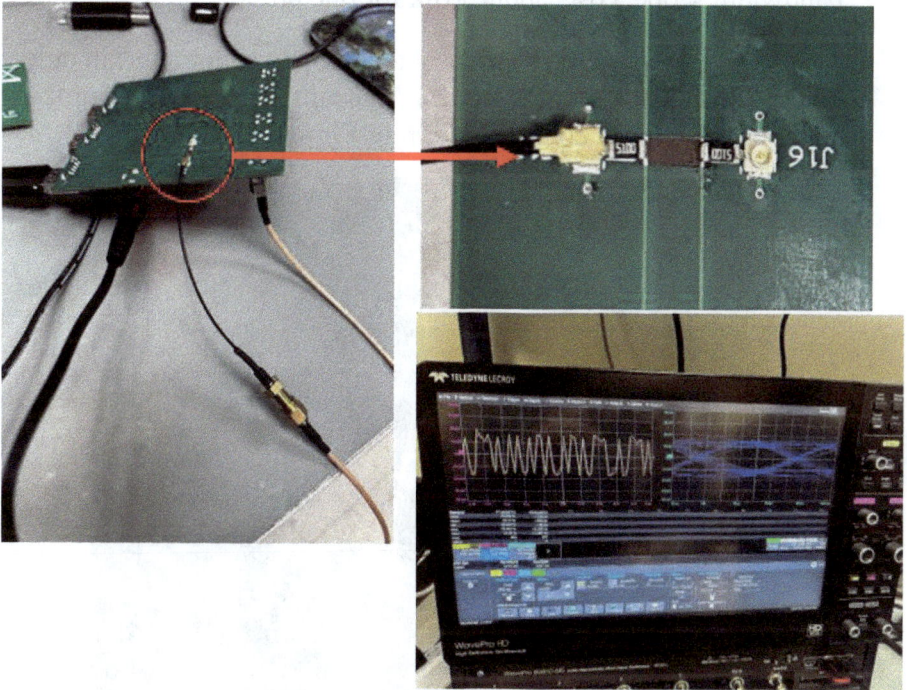

Figure 12.18 A transmission line probe using a 450-ohm resistor connected to a transmission line in series with a 50-ohm transmission and into the scope. The signal measured by the scope is a scaled version of the signal on the transmission line under test. The scope shows the real time 5 Gbps signal measured by the scope and by the transmission line probe, and the resulting eye.

The other limitation of this passive probe is that it attenuates the signal by 10x. An alternative is to use a high-bandwidth active probe. These probes are the general method used to sniff high-speed digital signals on circuit boards.

Active probes are carefully engineered to provide the lowest possible capacitive load while also providing a flat frequency response. These probes will offer the highest bandwidth with the lowest impact on the DUT system to which they are measuring.

The downside of active probes is their cost. An active probe costs about $1k per GHz of bandwidth, as a rough rule of thumb. This means a 10 GHz active probe, which should be used with a 5 GHz bandwidth scope, costs about $10k for one probe.

Most active probes are proprietary to a scope vendor and can only be used with that vendor's scopes. An example of an active probe soldered to a circuit board is shown in **Figure 12.19**.

Figure 12.19 A 5 GHz bandwidth active differential probe soldered to a DDR memory device, courtesy of Teledyne LeCroy.

One metric of the performance of the active probe is that it reproduces the waveform of the signal and that it does not degrade or affect the quality of the received signal in the system for which it is sniffing the signal.

By measuring the signal directly into the scope and then observing this signal with the probe attached, it's straightforward to verify the probe is not loading the signal. **Figure 12.20** compares a 10 Gbps high-speed serial signal measured at the receiver without the active probe connected to the middle of the interconnect and then with it connected. The loading of the probe on the signal is minimal.

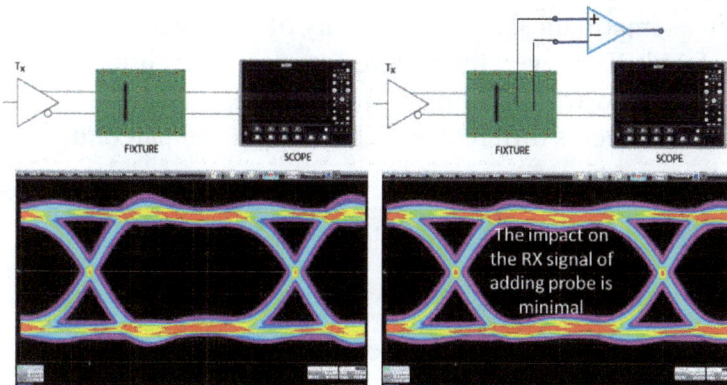

Figure 12.20 A 10 Gbps eye was received into the scope without the active probe connected to the single line and with the probe connected. It has no impact on the signal as it passes by.

12.6 Probing Power Rails

One of the most important applications of oscilloscope measurements is power integrity analysis. This relates to the part of a product designed to deliver a stable, low-noise voltage to power all active devices under all use conditions. The power distribution network (PDN) is the entire system, from the voltage regulator module (VRM), which is the source of the DC voltage, to the power rails on-die, which connect to the actual transistors on the die.

Some applications require measuring voltages well above the 400 V maximum voltage for which a 10x probe is rated.

Generally, any voltage above 50 V should be considered dangerous and attempted only after safety training.

This book does not cover the safe practices for measuring voltages higher than 50 V. Such measurements should only be attempted with adequate safety training and the right equipment.

While the purpose of the PDN is to deliver a DC voltage, the fluctuations or noise on the PDN with bandwidths of up to 1 GHz are sometimes important. This makes many of their features similar to signals. However, there are three special conditions generally associated with PDN measurements that make power integrity measurements different and challenging compared to signal measurements:

- The source impedance is low, less than 1 ohm

- There is a small voltage change on top of a large DC bias

- There are more RF interference sources in the vicinity of the PDN DUT

Even though the interconnects on the power rails are usually all the same net, the series resistance and inductance, plus the distributed capacitors throughout the PDN, mean different voltage levels may exist at different locations. Generally, the noise on the die is different than the noise on the circuit board power rails. Sometimes, it is higher on the die than the board, and sometimes, it is lower, depending on the source of the noise.

The typical PDN features to characterize at various locations include

- The DC voltage

- The absolute accuracy of the voltage

- The 60 Hz noise component

- The ripple noise from the switching supply

- The voltage noise spectral content

- The synchronous noise from transient events

- The VRM response to a step current change

- The VRM output impedance

There are three major types of VRMs: batteries, AC to DC converters, and DC to DC converters. Each type of VRM will show different behaviors.

Among DC-to-DC converters, switch mode power supplies (SMPS) and linear regulators are the two most common types. SMPS outputs either a lower voltage than what is input, as with a buck converter, or a higher voltage than the input, as with a boost converter, or both, as in a buck-boost converter. They all operate by passing the input DC voltage through a modulated switch and filtering the modulated voltage. This means their output voltage will have some amount of "switching" noise.

A linear regulator uses a transistor or MOSFET effectively as a variable resistor to maintain the voltage on a load as it may change.

An SMPS converts an input DC voltage into an output DC voltage with 85% to 95% efficiency. A linear regulator is less efficient. Depending on the voltage headroom between the input and output voltage and the current draw, it will generally be less than 80% and sometimes as low as 50% efficient. Its primary advantage is much lower noise.

12.6.1 AC or DC Coupling In The Scope

A chief feature of a voltage regulator module (VRM) is that low-level noise rides on top of an otherwise larger DC voltage. The low source impedance, large DC component that can exceed 5 V, and the need for wide bandwidth measurements make measuring power rails challenging.

Once the DC voltage is measured, it is tempting to use AC coupling to remove the DC component and zoom in on the AC component.

When using AC coupling input to the scope, a high-pass filter is connected in series between the DUT and the scope's amplifier inside the scope. It should be noted that the AC coupling feature is only available with the 1 megaohm input to the scope. It is not a built-in feature with 50 ohm input to the scope.

The AC coupling is created with a DC blocking capacitor in series, usually on the order of 1 uF. The equivalent circuit is shown in **Figure 12.21**.

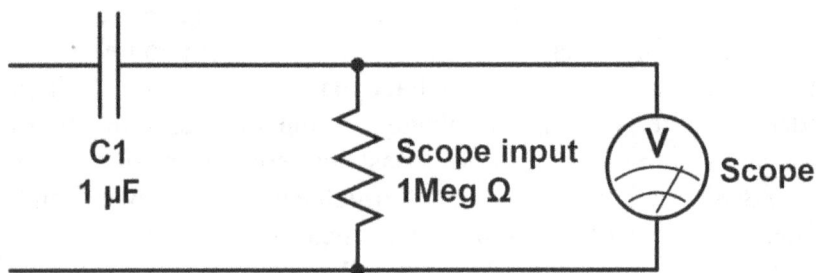

Figure 12.21 The equivalent input circuit of a scope set for AC coupling.

The cutoff frequency of this high-pass filter is,

$$f_{pole} = \frac{1}{2\pi(\,1uF \times \,1M\Omega)} = 16Hz$$

This passes any frequency component above 16 Hz but blocks the DC component and any slowly changing voltage associated with drift.

When you use AC coupling, you sometimes hide important behaviors related to the slow drifting of DC levels. When debugging the PDN, slow variations on the DC level, such as due to load fluctuations or temperature, can be important. This is why AC coupling should be avoided if practical.

Most scopes allow a large DC offset to be subtracted from the input signal so the scale can be adjusted to display small changes. There is a limit to how large a DC component can be subtracted. Generally, the vertical scale can be expanded to display less than 1% change per division. **Figure 12.22** is an example of a 5 V power rail expanded about the DC value, showing its less than 50 mV peak-to-peak noise out of 5 V or about 1% noise.

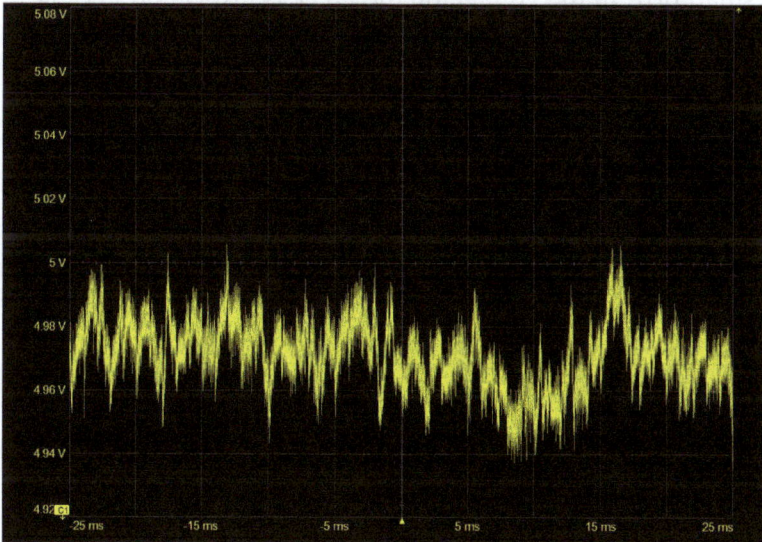

Figure 12.22 The measured voltage on the 5 V rail powered by a USB port shows less than 1% peak-to-peak noise on top of 5 V.

The AC scope setting is often suggested because when the scope's vertical scale is expanded, the signal is moved off-screen, and the offset has to be continuously adjusted each time the scale changes to bring the trace back on the screen. This is very inconvenient.

All scopes have a special setting that avoids this problem, but it is often not the default setting. When a scale changes, the scope has two options: expanding about ground or expanding about the center value on the screen. To use DC coupling to observe a signal with a large DC component and expand about the signal, set the scope to the "expand about the center value" option.

Sometimes, this option is buried under three or four menu trees. Once set, it becomes the scope's default setting. In this mode, position the DC value in the center of the screen. Then, when the scale expands, the center voltage value stays in the center of the screen. This is perfect for using DC coupling of the scope and expanding the scale to the appropriate level of sensitivity.

Unless you have a strong, compelling reason, *always* use the DC setting and an offset to zoom in on the small signal level. This way, a small drift in the voltage on the DUT is always observable.

12.6.2 *Active Power Rail Probes*

A 10x probe is always the first probe to consider for all applications. If nothing else is available, it is always a viable alternative. However, three almost contradictory features of the PDN limit the usefulness of a 10x probe:

- Noise less than 10 mV is sometimes important
- Signal bandwidths may exceed 100 MHz
- A high DC load impedance to the DUT is important

These special conditions have stimulated the introduction of a new class of specialized probes, called rail probes, which are optimized for the PDN. Every major scope vendor offers its own version of a rail probe. Since it is an active probe, it often connects to the scope with a proprietary connector, which provides power and digital bus connections to adjust specific settings in the probe. **Figure 12.23** shows an example of a rail probe from one vendor.

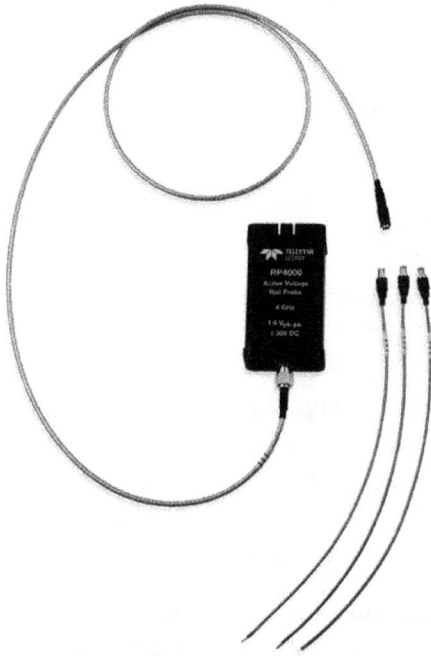

Figure 12.23 An example of a rail probe that uses a proprietary connection to the scope is Each major scope vendor has its own version of a rail probe.

All rail probes have four essential features that make them a better choice than a 10x probe when it is available:

- Providing a 1x attenuation to maintain a high SNR

- Capable of subtracting a large, stable DC offset

- High DC impedance for low-loading

- High bandwidth, up to 4 GHz

The rail probe achieves this using two parallel amplifier circuits, illustrated in **Figure 12.24**.

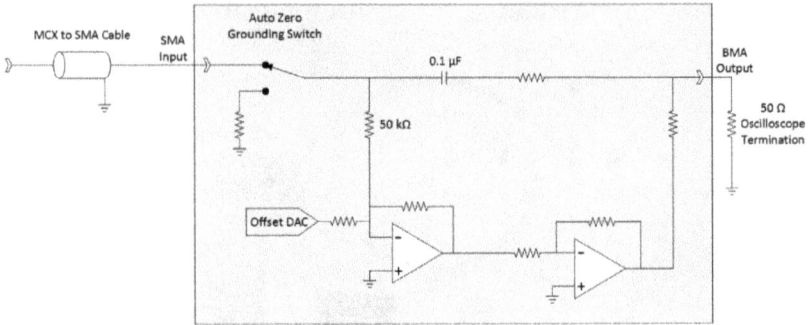

Figure 12.24 The equivalent circuit model of a typical rail probe shows the high bandwidth and DC signal chain paths—courtesy of Teledyne LeCroy.

At low frequencies, the signal path sees an inverting amplifier with an input impedance of about 50k in this example. A DAC provides a stable, programmable DC offset to the summing amplifier to subtract up to a +/- 30 V DC value. This means the residual voltage can be as small as 1 mV.

A second inverting amplifier inverts the signal again and provides additional gain. This amplifier circuit's frequency response is typically about 100 kHz. However, the output stage of the amplifier can buffer the input signal and drive the scope's 50-ohm input resistance.

In parallel is a direct connection to the scope's input amplifier and 50-ohm load. This is AC coupled with a high bandwidth DC blocking capacitor. The high-pass roll-off frequency is

$$f_{pole} = \frac{1}{2\pi(0.1uF \times 50\Omega)} = 30kHz$$

The amplifier signal-path pole frequency is optimized to overlap with the DC blocking capacitor's pole frequency to produce a flat response over a wide frequency. The typical frequency response is shown in **Figure 12.25**.

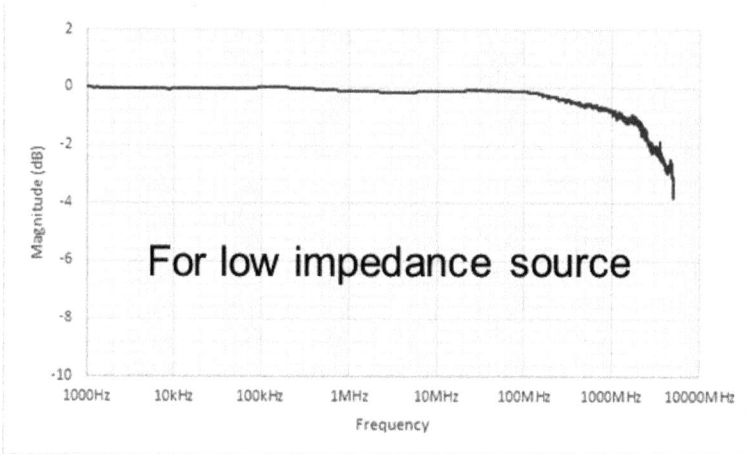

Figure 12.25 The measured frequency response of a rail probe with a low- and high-pass signal path in parallel—courtesy of Teledyne LeCroy.

An important limitation of a rail probe that makes it suitable only for low-impedance sources is that its input impedance changes with frequency. At low frequencies, the input impedance is due to the 50k input resistance of the inverting amplifier. This is perfect for not loading a power rail with a resistive load.

At high frequencies, the signal path is directly into a 50-ohm resistor, which is perfect for terminating high bandwidth signals so they do not reflect and cause artifacts. The transition of the input resistance between these two limits is shown in **Figure 12.26**.

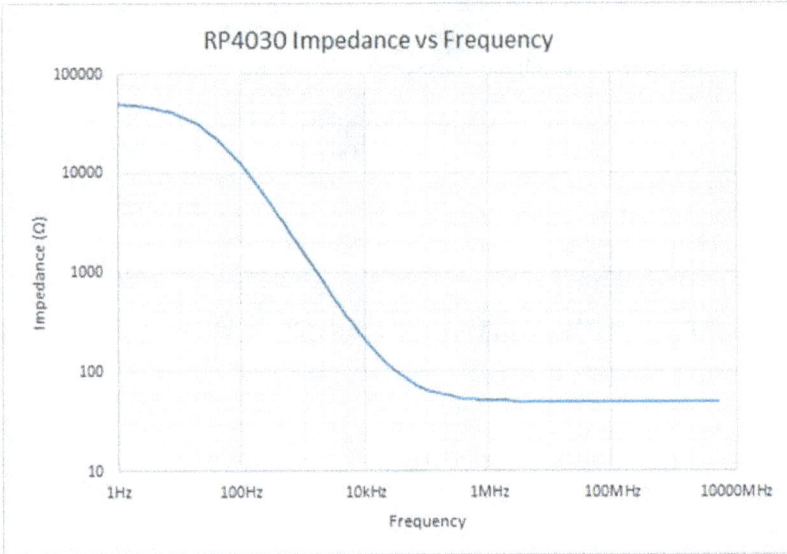

Figure 12.26 The input impedance of a rail probe shows the high impedance at low frequency and the 50 Ohms at high frequency—courtesy of Teledyne LeCroy.

If the source impedance is always much less than 50 ohms, the response of the rail probe is consistently flat. However, if the source impedance is on the order of 50 ohms, which might be the case for some low-current power rails, the voltage divider created by the output impedance of the PDN and the input impedance of the rail probe will make the gain frequency dependent. It will be 0 dB at low frequency but less than -6 dB at higher frequencies. This is an important artifact to watch out for when using a rail probe to measure a PDN with an output impedance higher than about 5 ohms.

Given this condition, a rail probe is a 1x probe with high bandwidth that does not load the PDN. **Figure 12.27** shows an example of a rail probe's better SNR because it does not reduce the SNR by 20 dB as in the case with an attenuating 10x probe.

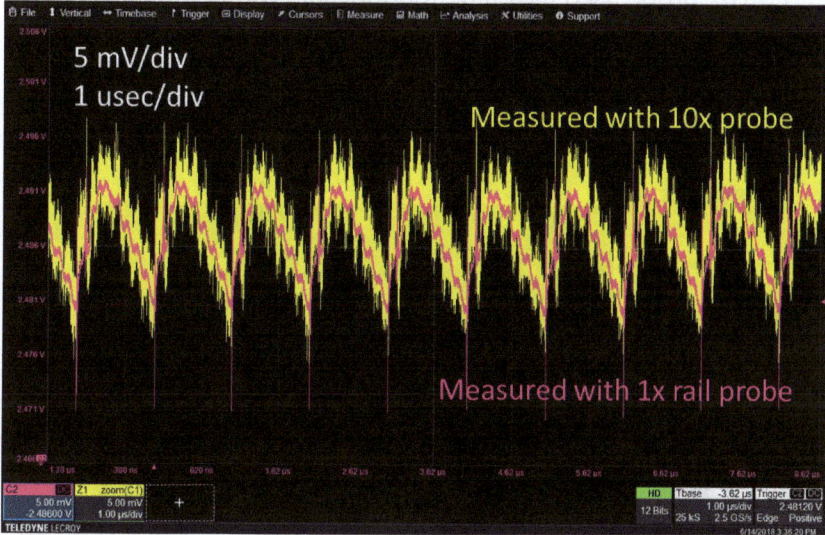

Figure 12.27 The same SMPS output signal was measured with a 10x probe and a rail probe. Note the improved SNR with the 1x rail probe.

12.6.3 Use Coax Connections to Reduce RF Pickup

An important consideration with a power rail measurement is that the near field in the vicinity of the DUT is sometimes noisy. If the board or the components are not designed well, the switching currents and poor interconnect design can contribute to RF noise that can be picked up in the probe as interference.

Nothing can be done if the switching noise contaminates the PDN conductors' voltage. After all, this is part of the PDN's signature, which is the signal to measure. However, to avoid the possibility of picking up any interference from the noisy near-field emissions, the connections to the board should be made as close to coaxial as possible. This will reduce the loop area of the probe and reduce the pickup of mostly changing magnetic fields.

For example, **Figure 12.28** shows how the connection to the PDN dramatically affects the RF interference. When a 10x probe is used with even short leads to the power rail, there is significant ringing noise in the measurement. This same signature is apparent from a simple 10x probe used as a pickup loop, suggesting most of the voltage on the PDN is really interference. When a coaxial connection is soldered to the board, and the same 10x probe is used, the resulting voltage measured on the board does not have any of the ringing noise.

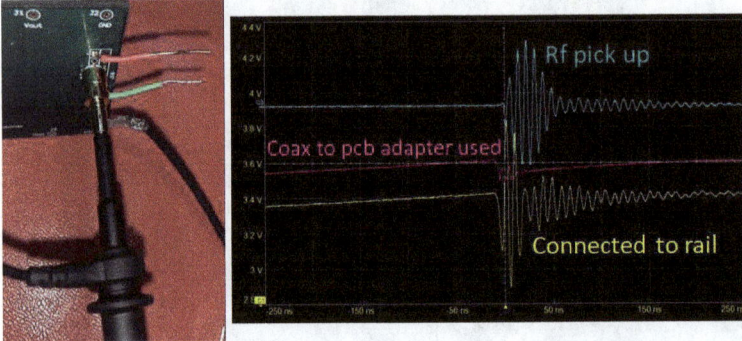

Figure 12.28 Measurements of the board-level PDN with wires and with a coax connector soldered to the board show how to eliminate the artifact of RF interference.

As a best practice, coax connections should be made to the board-level PDN, and the probe should be connected with a coax connector, such as a BNC, SMA, or even U.FL.

12.6.4 A Low-Cost Passive Rail Probe

An alternative method for probing a low-impedance, fast-switching source without the added expense of an active rail probe is to use a handmade passive rail probe.

This involves inserting a 50-ohm resistor in series between the DUT and the coax cable connection to the scope. The coax cable is then connected to the scope, set for 1 meg input. **Figure 12.29** shows an example of the equivalent circuit model and a simple implementation.

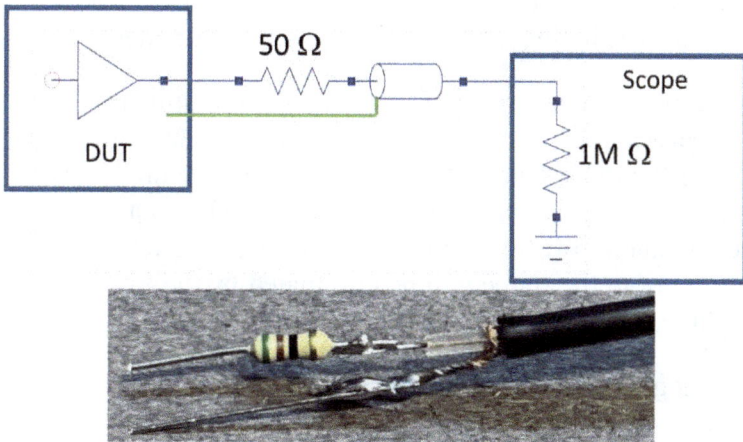

Figure 12.29 The equivalent circuit model of a passive, handmade rail probe and an example using an axial lead resistor.

Usually, to measure high-bandwidth signals, we should use 50 ohms as the input to the scope. This will terminate high-speed signals at the scope. But this is not well suited for power rails. A 50-ohm input to the scope creates two important problems when measuring power rails.

With 50 ohms input to the scope, the maximum voltage we can probe is 5 V. Any higher voltage will cause too much power consumption in the 50 ohm resistor, damaging the scope.

The second problem with using 50 ohms in the scope is that it will introduce a DC load to the power rail. If the DUT is a 3 V source, the 50-ohm load will draw 60 mA. If the DUT can source 100 A, a 60 mA draw from the probe is negligible. However, if the DUT is a 100 mA low-drop-out (LDO) power source, a 60 mA draw from the scope will dramatically affect the LDO performance.

Without an active rail probe, an alternative is a 10x probe. The limitation is the 20 dB reduction in the SNR due to the attenuation of the probe.

If a direct coax connection is made between the power rail and the scope set for 1 meg input, any fast transient signal on the power rail will result in a ringing artifact noise in the measurement.

The solution is to use source series termination to eliminate reflections between the high impedance of the scope and the low impedance of the PDN source. By adding a 50 ohm source resistor at the tip of the probe, we are assuming that the PDN source impedance is much lower than 50 ohms. The signal reflected from the 1-MHz input to the scope will be terminated by the PDN with a 50-ohm source series resistor.

With the 1 meg input to the scope, voltages of ±40 V can be measured with a negligible DC draw.

The impact of source series resistance in reducing ringing artifacts in a scope is illustrated in *Figure 12.30*. In this example, a 5 V SMPS with a 0.1 ohm source impedance switches on with a few nanosecond rise times. It is measured with the scope set to 1 meg termination. In the first case, there is a direct coax connection. The multiple reflections are evident. A 50 ohm source series resistor is added to the DUT in the second case. The reflections are gone.

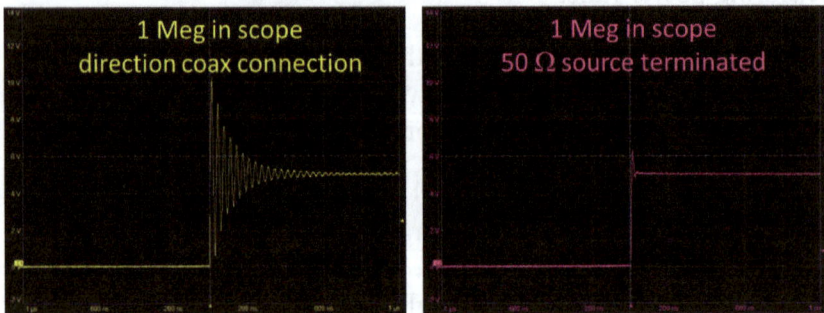

Figure 12.30 The transient voltage from a 5 V SMPS turning on was directly measured with the 1-meg input to the scope and with a 50-ohm series resistor added between the DUT and coax cable.

This simple approach of adding a 50-ohm source series resistance at the end of the coax cable is a low-cost alternative to probing power rails at high bandwidth. It is also an alternative to using a 10x probe.

This 50 ohm termination method has the advantage of being a 1x probe that does not attenuate the signal. This is important when voltages on the order of 10 mV are important, especially in low voltage, low impedance SMPS.

The example in **Figure 12.31** is a 3.3 V SMPS measured with the 10x probe and a 1x hand-made rail probe. The better SNR of the series terminated 1x probe is evident.

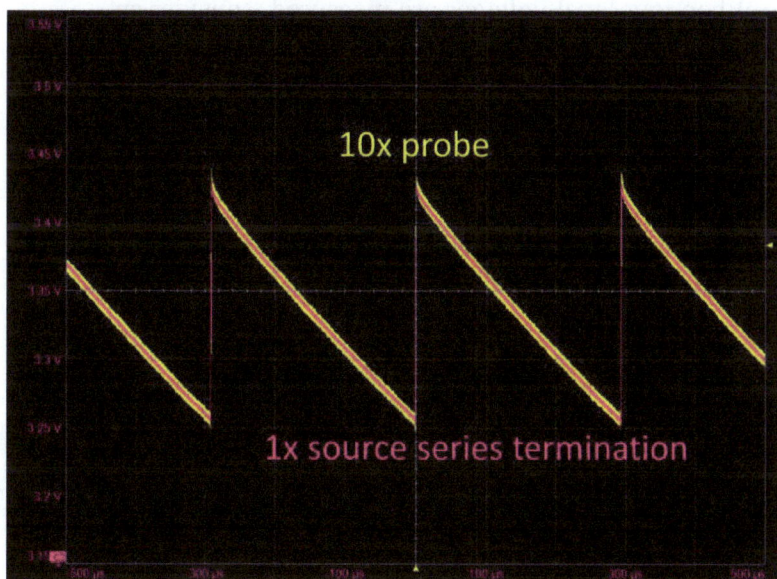

Figure 12.31 Measuring a 3.3 V SMPS with the two different probes.

This distinction defines the condition to decide which type of probing method to use: if your application is for a 20 V DC level or higher, the 10x passive probe will be a better choice. It has a higher voltage range, and its 10 mV typical noise level is less than 0.1% of the signal. Best of all, it is readily available.

If your application is for a 10 V DC level or lower, a 1x passive probe with source series termination may be a better choice. It will provide a better SNR.

However, if your application has a bandwidth above 500 MHz, neither approach may be acceptable. In these high-bandwidth applications, consider using an active rail probe.

12.7   The Bottom Line

1. The 10x probe should be your go-to probing solution in most cases.

2. When low-level voltages are measured, consider an alternative with better SNR.

3. When higher bandwidth is required, consider using a direct coax connection. This will always be the highest bandwidth.

4. If you need high bandwidth and do not want to load the DUT, use a 450-ohm series resistor to sniff the high-speed signal.

5. A high bandwidth 450-ohm resistor can be implemented as a transmission line probe.

6. Alternatively, consider using active probes. These are generally more expensive, costing about $1k per GHz bandwidth.

7. When probing a DC voltage rail, always use DC coupling when possible to see slow frequency drifts in the DC level.

8. When probing power rails, consider using a rail probe.

9. A low-cost alternative is to use a source series 50-ohm resistor.

10. Avoid RF pickup near the power rail by using a coaxial connection to the DUT.

Chapter 13
Spectral Analysis of Signals

So far in the exploration of signals measured by an oscilloscope, the signals have been measured, displayed, and analyzed in the time domain. While this is the real world and is often the domain in which to extract important figures of merit, sometimes further insight can be gained about the behavior or signature of signals by looking at their frequency components in the frequency domain. This is achieved by converting their time behavior into the frequency domain using spectral analysis.

13.1 A Spectrum

In the frequency domain, the only waveforms we are allowed to consider are sine waves. Unique combinations of sine waves in the frequency domain can describe any time-domain waveforms.

However, this property is not the only reason sine waves were singled out to be used in spectral analysis. There are other special waveforms, combinations of which can describe any time-domain waveform, such as Hermite polynomials, Laguerre polynomials, Jacobi polynomials, Legendre polynomials, or even wavelets. These functions are in the general class of complete, orthonormal basis functions, or eigenfunctions.

The reason we single out sine waves for a frequency domain description is that sine waves are solutions to second-order, linear, differential equations, the equations found so often in electrical circuits involving resistor, capacitor, and inductor elements. This means signals that arise or have interacted with RLC circuits are described more simply when using combinations of sine waves than any other function because sine waves naturally occur.

The only reason we would ever leave the real world of the time-domain to enter any other domain is to get to an answer or insight faster. Sometimes, a complex waveform in the time-domain can be more simply analyzed and understood by looking at its frequency domain description, composed of combinations of sine waves.

An ideal sine wave is described by only three figures of merit or parameters: its frequency, amplitude, and phase. A sine wave waveform, measured by an oscilloscope with one million voltage-time, V(t), data points in the acquisition buffer, is described by only three numbers in the frequency domain.

This is a dramatic simplification. This comparison between the time-domain measurement of a sine wave and its frequency domain description is shown in **Figure 13.1**.

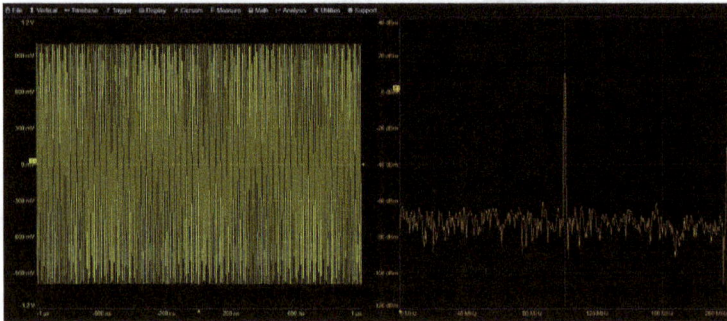

Figure 13.1 A 100 MHz sine wave in the time-domain and its spectrum in the frequency domain showing the one peak at 100 MHz.

In principle, any waveform in the time-domain over some time interval can be described by a unique combination of sine waves, each with its own frequency, amplitude, and phase. This combination of sine wave components is called its spectrum. It is a unique fingerprint describing the time-domain waveform.

The DC or average value of any ideal sine wave frequency component is 0. The DC offset of the time-domain waveform is stored as the 0 Hz frequency component of the signal. It is treated as just another frequency component.

The most common application of spectral analysis is to identify the frequency components in a time-domain waveform as a fingerprint to help identify the root cause of a problem.

What appears as a random, noisy waveform in the time-domain may reveal specific frequency components in the frequency domain. For example, the measured near field emissions from a digital system is a complex waveform in the time domain. The frequency domain decomposition into its spectrum reveals a pattern of specific frequency peaks, as shown in **Figure 13.2**.

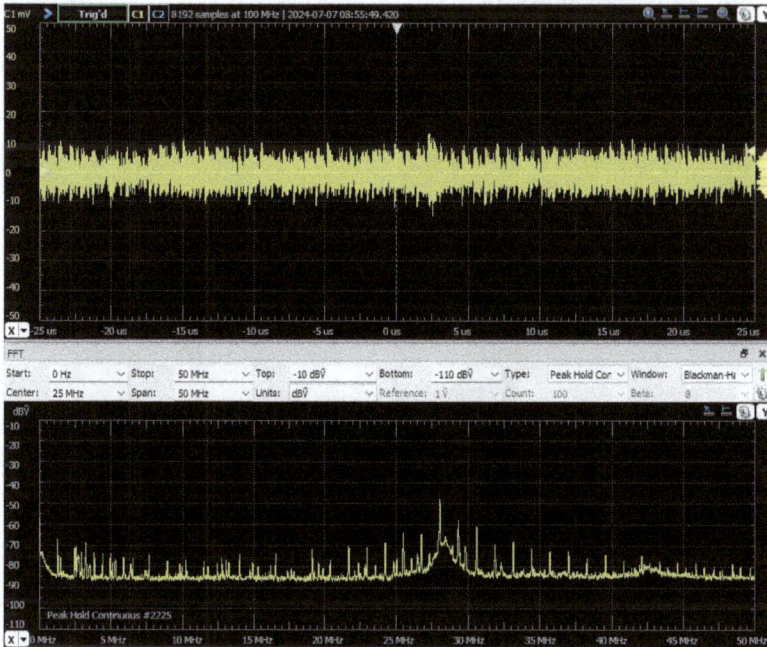

Figure 13.2 Top: time-domain waveform of the near field emissions from a digital device, bottom: the spectrum of this waveform, showing distinct peaks starting at 1 MHz with a broad peak at 28 MHz.

This spectrum shows the three common features found in most spectra:

- Sharp peaks, suggesting repetitive signals, such as clocks, carrier frequencies, or loops in the code running on the digital device.

- Broad peaks, suggesting periodic transient signals, or modulated carrier frequencies.

- Flat regions, suggesting broadband, random white noise.

The spectrum reveals a fingerprint identification of a possible source of noise or interference. In this example, searching in the performance of the DUT for clock frequencies that are 1 MHz or multiples, or repetitive operations at 28 MHz, might reveal the originating source of these features.

Another question is how the source of the noise then interacts with interconnects inside the DUT to create the near-field emissions. This is the start of the debug process.

Not all problems can be solved more quickly in the frequency domain, but when signals or noise are periodic and have specific frequency components, spectral analysis might be applied to the problem to gain insight faster.

13.2 Spectrum Analyzer or Real-time Spectrum Analyzer

Traditionally, the instrument used to measure the spectral content of a signal in the time domain is a spectrum analyzer. As a standalone instrument, a spectrum analyzer will scan across the frequency range with a tuned, narrow-band filter, measuring the voltage amplitude that gets through the narrow-band filter. This is a direct measurement of the power in the signal within the bandwidth of the narrowband filter.

The voltage through the filter, as the center frequency of the narrow-band filter is swept, is displayed on the front screen. There is no phase information extracted from the signal. This is why these instruments are sometimes referred to as scalar spectrum analyzers.

If the center frequency of the narrow-band filter is scanned quickly, it may appear like a continuous display of the spectrum on the screen, but in fact, the amplitude of the signal in each frequency band is really a snapshot measurement of the spectral content at the moment the filter happened to be centered at that frequency value.

If the spectrum of the signal is dynamically changing, such as in a spread spectrum clock, the dynamic nature of the spectrum of the signal may not be captured, or it would be an averaged response. Sometimes, this is enough from which to interpret the behavior of the signal source. If the source frequency changes more quickly compared to the sweep rate, information about the signal may be missed or misinterpreted.

Alternatively, if a narrow frequency region in the spectrum is of most interest, a spectrum analyzer can focus on just this narrow region with a narrow band filter and provide very high resolution on the frequency components about a carrier frequency.

A scope displays the spectral response of a signal in a fundamentally different way than a spectrum analyzer. A scope measures the signal in real time and records the V(t) buffer of measurements into a file. It then performs a calculation based on the FFT to convert the real-time voltage data into the spectrum. For this reason, the scope's spectrum is sometimes referred to as a *real-time spectrum analyzer*.

Using either hardware acceleration or an efficient software algorithm, the calculation of the FFT can be performed in a short enough time to appear to be instantaneous. Even though the time-domain signal and the spectral response are shown continuously on the screen, there is some dead time between successive time-domain buffer measurements, FFT calculation, and display as the spectral response. Some scopes allow multiple buffers to stream into memory and the FFT calculated after the total acquisition. This reduces the deadtime between consecutive buffer acquisitions.

The process of converting an arbitrary signal in the time-domain into its spectral components in the frequency domain is based on five fundamental principles outlined in the next section.

13.3   Principles of FFT Analysis

Every scope calculates the spectrum of a signal recorded in the time domain using five principles. By understanding these principles, we can identify and reduce potential artifacts created by the FFT process. In addition, the FFT process introduces some limitations in the features displayed in the spectrum.

Generally, all five principles are applied "under the hood" by oscilloscopes with an FFT function. While it is not necessary to understand these five principles, understanding these principles will eliminate the black box nature of the spectrum, and you will get the most out of your real time spectrum analyzer.

Every DSO available from the < $100 units to the >$1M units have spectral analysis built in. It is just as important a tool in analyzing signals and gaining insight into the root cause of problems as the real time voltage measurements in the time domain. If you wish to master the scope and the analysis of signals, you should master spectral analysis as well.

13.3.1 Turn Arbitrary Signals into Periodic Signals

When we take a waveform in the time domain and transform it into the frequency domain, we end up with a collection of sine waves, each with a frequency value, an amplitude, and a phase. Each frequency is separated into frequency bins. Alternatively, we can describe the waveform in the time domain as a collection of both sine waves and cosine waves, each with a frequency and amplitude value. The phase is contained in the relative sine and cosine amplitudes.

The amplitude of each frequency component bin is selected so that the combination of all the spectral component bins recreates the original waveform:

$$V(t) = \sum_n \left(a_n \sin(n2\pi f_0 t) + b_n \cos(n2\pi f_0 t) \right)$$

In this expansion, the a_n and b_n coefficients are the amplitudes of the sine and cosine components at each frequency bin, nf_0. In this formalism, f_0 is the initial, lowest repeat frequency bin in the acquisition buffer and n is the harmonic number of each frequency bin in the spectrum. Each harmonic frequency is an integer multiple of the fundamental frequency. The lowest frequency is related to the total acquisition time. The highest value of n is related to the Nyquist frequency of the sample rate, which is ½ the sample rate.

The coefficients and the frequency of each bin define the spectrum. They are calculated from the measured V(t) waveform using the discrete Fourier transform (DFT). This calculation is implemented with a special matrix math algorithm referred to as the FFT algorithm.

In the time domain, the measurements are taken and stored in an acquisition buffer with a total acquisition time, T_0, and a time interval between samples, ΔT. When we describe the same waveform in the frequency domain, we refer to the collection of all the sine and cosine wave components, each with a frequency and amplitude, as the spectrum.

Unfortunately, we can only use the DFT on a V(t) waveform that is periodic. If it is not periodic, we must artificially make it periodic. The trick we use to turn any arbitrary acquisition buffer of measured data into a periodic waveform is to take the acquisition buffer of total time, T_0, and concatenate identical, repeated acquisition buffers to the front and the back.

This creates an artificially repetitive waveform that repeats with a period equal to the acquisition buffer time, forever in the past and forever in the future. This is illustrated in **Figure 13.3**.

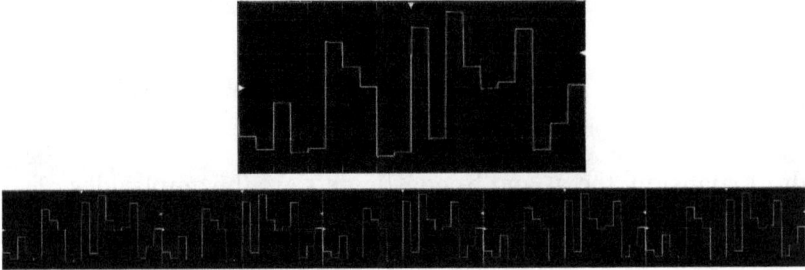

Figure 13.3 Top: a single measured acquisition buffer of some arbitrary waveform is turned into a period waveform by, (bottom) repeating the acquisition buffer forever in the past and forever in the future.

When we have this artificially repetitive waveform, we can apply the power of the DFT to mathematically calculate each frequency component in the spectrum. Each frequency bin in the DFT is an integer multiple of the fundamental, $f_0 = 1/T_0$.

The amplitude and phase of each frequency bin are calculated using:

$$a_n = \frac{1}{T_0} \int_0^{T_0} V(t) \sin\left(\frac{2\pi}{T_0} nt\right) dt$$

$$b_n = \frac{1}{T_0} \int_0^{T_0} V(t) \cos\left(\frac{2\pi}{T_0} nt\right) dt$$

The amplitude of each frequency component bin is given by

$$\text{Amplitude}_n = \sqrt{a_n^2 + b_n^2}$$

The spectrum calculated using the DFT is literally the collection of each frequency bin with its amplitude of each sine and cosine wave.

13.3.2 The Lowest Frequency

In the calculated spectrum, only discrete frequency values appear. The lowest frequency component bin is called the *fundamental*. It is the lowest frequency sine wave that will fit in the acquisition buffer time. The period of this lowest frequency sine wave is the total acquisition time, T_0.

$$f_0 = \frac{1}{T_0}$$

Each frequency component in the spectrum is a frequency that is an integer multiple of the fundamental, $f_n = n \times f_0$.

Multiples of the fundamental are the only frequency components that appear as bins in the spectrum. This means that the frequency spacing between each frequency component bin, or the *resolution*, is the fundamental frequency.

If you want a higher resolution to distinguish narrower frequency features in the spectrum, you need to use a longer acquisition time in the oscilloscope. This means a longer full-scale time interval.

The average value of an ideal sine wave is always 0. This means that when we use a collection of sine waves to describe a real waveform, the average value of the recreated time-domain waveform is always 0.

But real waveforms have an average value, or DC offset. To account for this, we store the DC component in the 0 Hz frequency component, which is 0 x the fundamental frequency. When calculating the DFT, the sine term for n = 0 is zero, but the cosine term is 1. This means the b_0 coefficient is literally the average value of the signal over the time interval, which is an approximation of the DC component.

In most oscilloscopes, you can suppress plotting the 0th frequency component to increase the dynamic range of the display.

13.3.3 The Highest Frequency

The highest frequency component in the spectrum is related to the sample rate of the measurements. At a minimum, two sampled points are required during one cycle to measure the amplitude of a frequency component. This means that the highest frequency component that can be extracted from sampled measurements is ½ the sample rate. This is the basis of the Nyquist theorem. The Nyquist frequency of sampled data is ½ the sample rate. The highest sine wave frequency that can be extracted is the Nyquist frequency.

$$F_{max} = F_{Nyquist} = \frac{1}{2} \times F_{sample\ rate}$$

If the sample rate is 10 GS/s, the Nyquist frequency is 5 GHz. The highest frequency we can calculate in the spectrum is 5 GHz.

Do not confuse the digitizing sample rate to record a signal with acceptable fidelity, with the sample rate to identify a specific frequency component of a sine wave component.

When the signal is a stable frequency sine wave and the acquisition trigger is synchronous with the signal phase, the measured waveform will have any arbitrary amplitude from the full value, to nearly 0, depending on the relative phase between the signal and the trigger. This is illustrated in **Figure 13.4**.

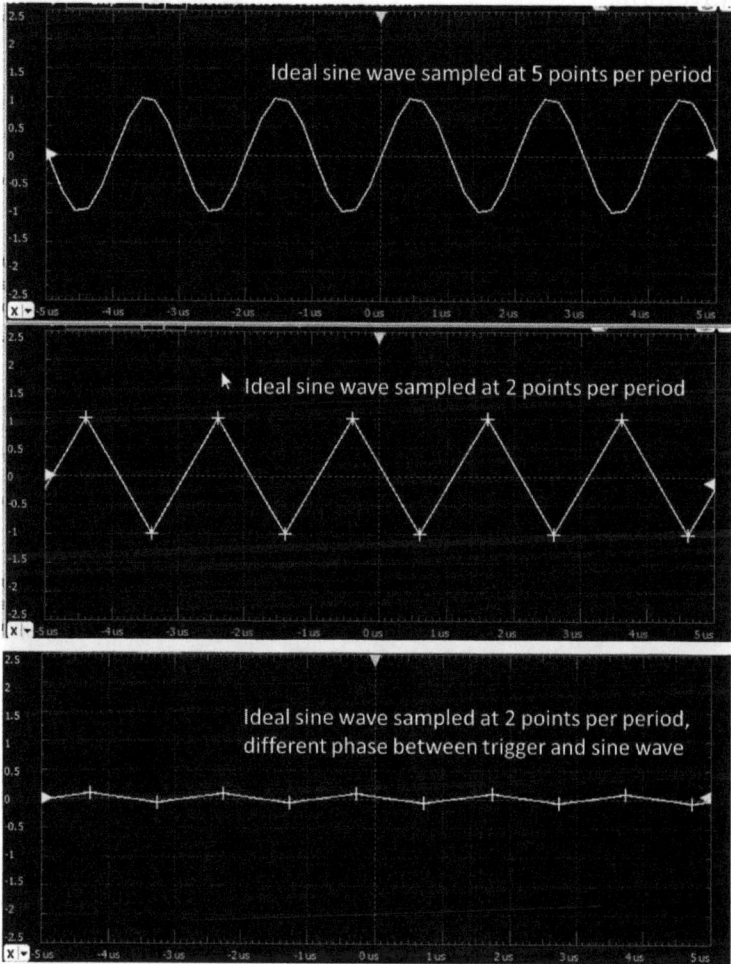

Figure 13.4 Impact of sample rate and measured signal for a sine wave. Top: acquisition rate = 5 x frequency. Middle and bottom, acquisition rate = 2 x sine wave frequency = Nyquist, but different phases.

If the frequency components of the signal are higher than the Nyquist frequency, the waveform is under-sampled, and the resulting recorded waveform is some arbitrary collection of points with no relationship to the original signal. Under sampling results in an artifact called aliasing. A signal will be recorded, and a spectrum created that is not an accurate representation of the original waveform.

As an example, **Figure 13.5** shows a 10 MHz sine wave signal, sampled at 1.37 MS/s. This means there is a voltage acquisition roughly every 7.299 cycles. Sometimes, the sampled point will catch a peak, sometimes a valley, and sometimes, somewhere in between. When the phase of the sine wave and trigger are both stable, their relative phases will be locked, and the sampled waveform will appear to be periodic due to aliasing. This will create artifacts in the spectrum at frequencies that are below the Nyquist frequency.

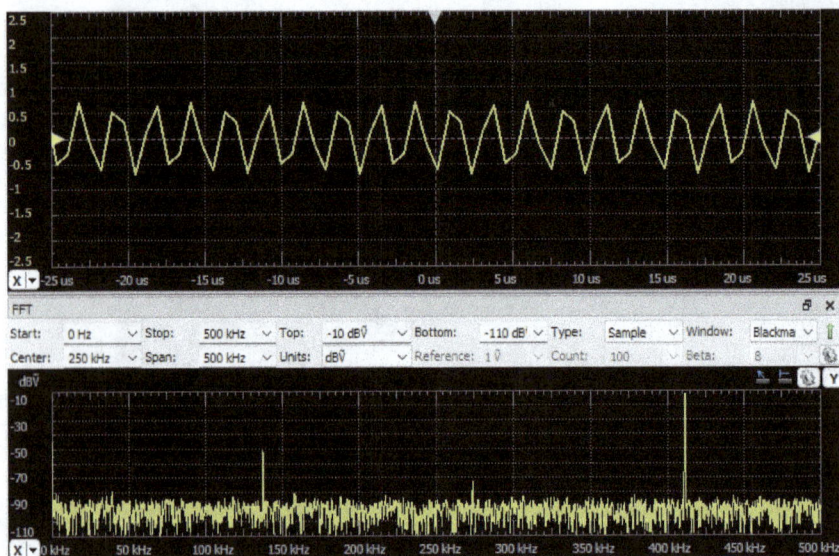

Figure 13.5 A 1 V amplitude, 10 MHz sine wave sampled at 1.37 MS/s. The sine wave and trigger are stable. The sine wave is undersampled. The resulting waveform and its spectrum are not representative of the 10 MHz waveform. Spectral components in the signal appear at 140 kHz, 275 kHz, and 410 kHz. These are aliasing artifacts.

This artifact is most apparent when the signal is composed of stable sine waves, which is common when measuring RF communications signals. It is less apparent when measuring analog or digital signals.

Two steps are important to avoid this artifact. First, the input signal measured should be filtered to remove frequency components above the Nyquist frequency. This eliminates the possibility of higher frequency components being aliased down to appear at lower frequencies. This is performed with an anti-aliasing filter. Sometimes, this is available and built into the scope.

Second, the signal should be over-sampled. At the very least, the Nyquist frequency should be 2.5x the frequency of the highest frequency component to measure. To preserve some fidelity of the signal features, the sample rate should be at least 5x the highest frequency component in the signal. This will allow at least 5 samples over one period.

This is why, in most scopes, the highest sample rate achievable on all channels simultaneously is usually about 2.5x the scope bandwidth. This is a cost-performance balance between the bandwidth of the scope, the ADC sample rate, the streaming rate for data into the memory buffer, and the impact from aliasing due to under-sampling the signal.

With all channels recording, frequency components above the Nyquist frequency will be filtered by the scope amplifier's bandwidth. If the scope is used at a lower sample rate than the scope's bandwidth, such as when the time base is reduced, aliasing artifacts can arise. This is why being aware of aliasing and adding a hardware filter is so important.

In some scopes that use one ADC for all four channels, the sample rate is 10x the scope bandwidth. When one channel is measured, it is over sampled by 10x. But when all four channels are measured, each channel is over sampled by 2.5x.

13.3.4 *Number of Points in the FFT Calculation*

Calculating the DFT of a time-domain waveform requires an integral over all the measurement points at each frequency bin value. With one million data points in the acquisition buffer, about one trillion calculations are required to create one spectrum. This may take too long and would not appear in real time.

To get around this problem, all scopes use a much faster version of the DFT which is the FFT. It calculates the same integrals as the DFT, but it applies matrix math to perform the calculations. The FFT matrix math can only operate on a total number of points that is a power of 2. If there are one million points in the buffer, the highest number of points that could be included in the FFT calculation would be $2^{19} = 524{,}288$ points. The scope throws out almost half the measured data to gain incredibly fast computation time.

The first step in performing an FFT is to define the region of the acquisition buffer that contains the 2^n points that will be computed. Most oscilloscopes allow you to pick either the central region of the time-domain screen or a count from the left edge. **Figure 13.6** shows the region of the acquisition buffer that will be included in the FFT calculation between the dashed lines on the screen.

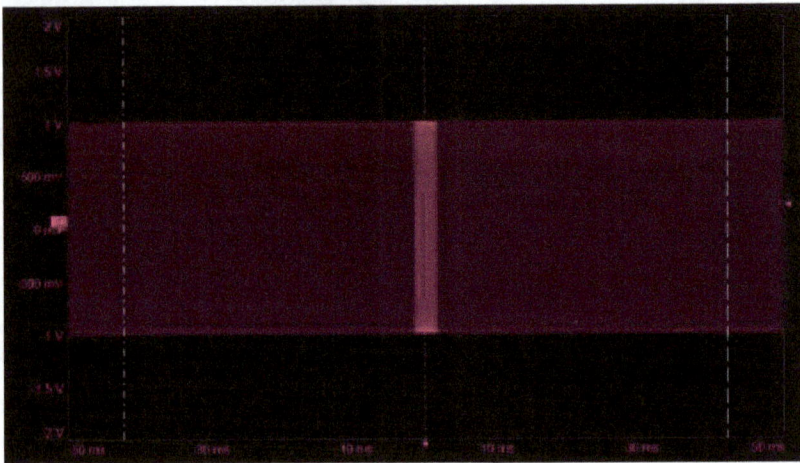

Figure 13.6 Between the vertical dashed lines is the region of the acquisition buffer that contains the 2^n points that will be used in the FFT.

When the acquisition buffer time is 1 μs, and we have one million points, we expect the fundamental frequency to be 1 MHz. In the spectrum, the FFT acquisition buffer is smaller than this, which means the actual resolution is slightly larger than 1 MHz. However, these estimates are still a good rule of thumb to use when thinking about the features of the spectrum.

13.3.5 Windowing Functions

To create a periodic waveform, we repeat the acquisition buffer indefinitely in the future and in the past. When using the FFT function, we further truncate the acquisition buffer and repeat the truncated buffer indefinitely. This means that at the boundaries of each appended acquisition buffer, there may be a discontinuity in the waveform corresponding to the end of one buffer and the beginning of the next one. This is illustrated in **Figure 13.7.**

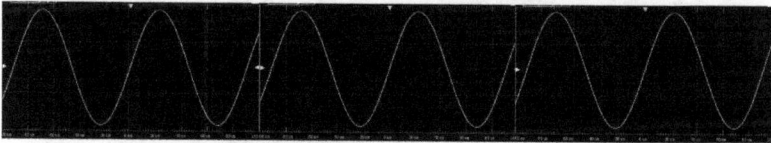

Figure 13.7 Example of a repeated buffer that ends at a different value than it starts. When concatenated, there is a discontinuity at each boundary in this infinitely repeated waveform.

In principle, the spectrum of a sine wave should be a single peak at the sine wave's frequency. If there are an integral number of cycles in the acquisition buffer, the end of one cycle matches up with the beginning of the next buffer, and there is no discontinuity. The spectrum will have a sharp peak and no other components in the spectrum.

However, if there is not an integral number of cycles in the buffer, then there is a discontinuity at the boundaries of the buffer's beginning and end. This means there will be a distortion in the spectrum. The discontinuity spreads spectral information from one frequency component into nearby components, causing *spectral leakage.*

An example of the spectrum of a measured sine wave signal with an integral number of cycles and a slightly shifted frequency that does not have an integral number of cycles is shown in **Figure 13.8**. In this example, the sample rate was 100 MS/s which is about 10 samples per cycle. The time base was 50 usec full scale so the frequency resolution was 1/50 usec = 20 kHz. The frequency scale for the spectrum was 2 MHz/div. This is 100 frequency bins per division.

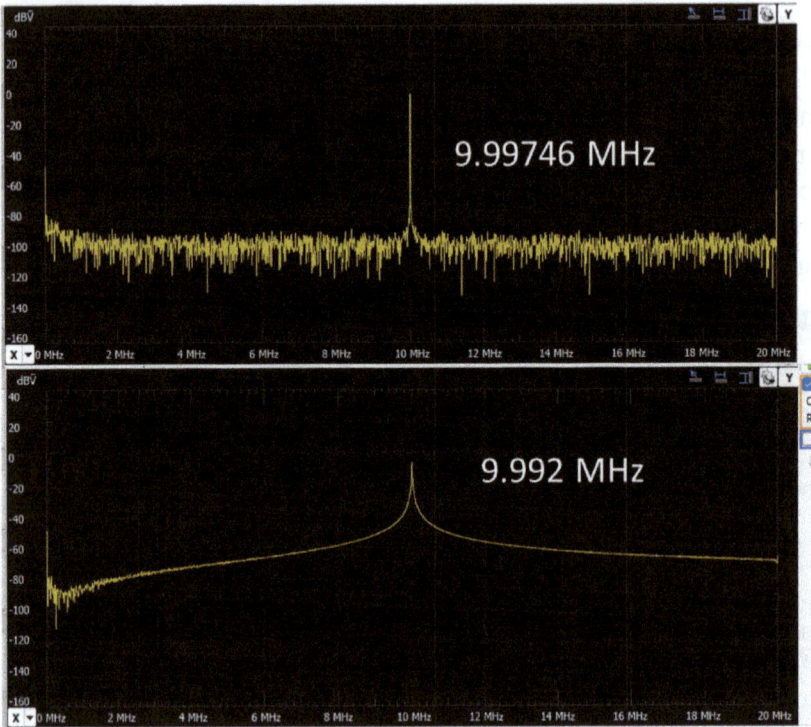

Figure 13.8 Examples of the spectrum of measured sine waves, at about 10 MHz, sampled at 100 MS/s with no windowing function. Top: with an integral number of cycles per data acquisition window. Bottom: with a non-integral number of sine waves per acquisition buffer. Note the large spectral leakage.

Spectral leakage is an artifact of the discontinuity at the boundaries of the buffers due to the first voltage value not being the same as the last voltage value. The way to reduce this artifact is to artificially reduce the discontinuity at the ends of the buffer by multiplying the entire acquisition buffer by a window function.

This gradually forces the voltage value at the ends of the acquisition buffer to be 0, guaranteeing that the end of one buffer is continuous with the beginning of the next buffer.

There are a number of windowing functions commonly used. They differ in how much spectral leakage they allow and the resulting resolution. Unless you have a strong, compelling reason otherwise, the Blackman-Harris function is a good compromise between resolution and spectral leakage.

Examples of four common window functions are shown in **Figure 13.9**.

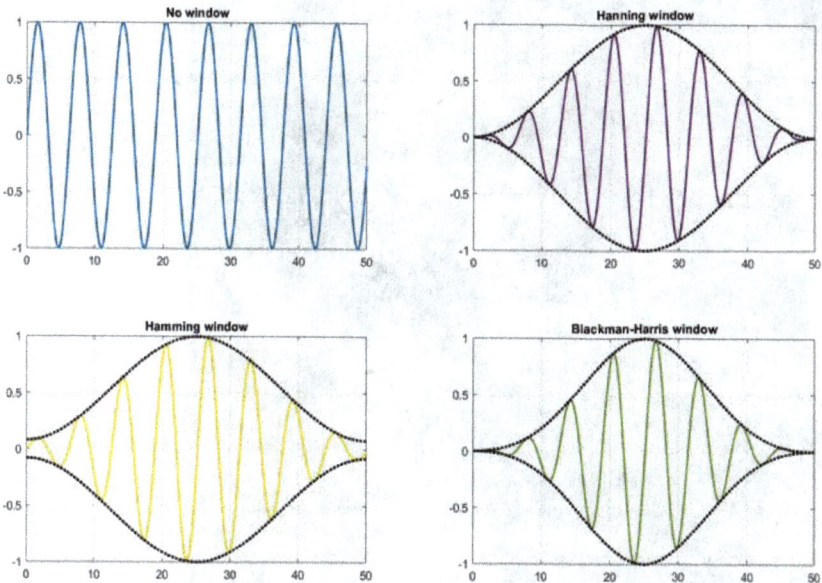

Figure 13.9 Examples of four common windowing functions, courtesy of National Instruments.

An example of a measured sine wave with a non-integral number of cycles in the acquisition buffer using these four windowing functions is shown in **Figure 13.10**. The resolution of the spectrum, based on the 50 usec acquisition buffer size is 20 kHz. There are 5 frequency bins per division on this scale. This is the width of the spectral peak of the sine wave with an integral number of cycles in the acquisition window.

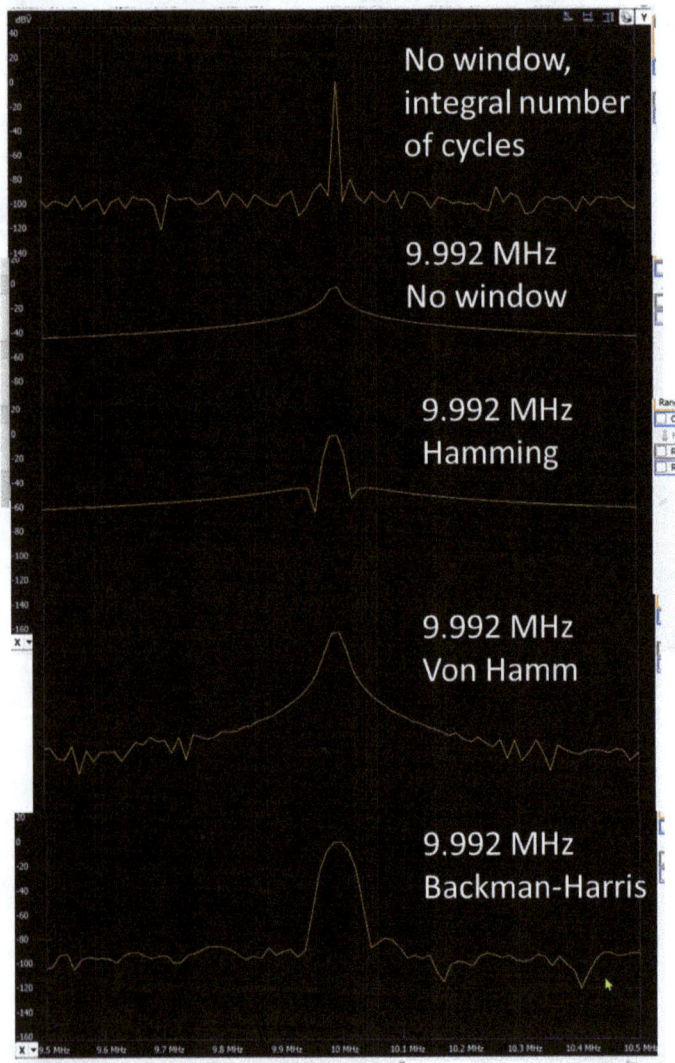

Figure 13.10 Examples of the same sine wave with different windowing functions. The resolution is 20 kHz.

Each windowing function has a different balance between frequency resolution and distribution of the spectral leakage. For general applications, the Blackman-Harris window is a good compromise between resolution and reduction in spectra leakage.

If this window is available on your scope, it should be selected. Otherwise, the von Hann window is a second choice.

13.3.6 Summary of FFT Principles

The DC component is the amplitude of the 0 Hz frequency component. This is the average value over the acquisition time window.

The next frequency bin, and the frequency resolution of the spectrum, is 1/ (the total acquisition time window). To enable a narrower resolution, use a longer acquisition time window.

The highest frequency in the spectrum is the Nyquist frequency, ½ the sample rate. If you want to see a higher frequency component in the spectrum, you have to increase the sample rate.

Aliasing occurs when the signal has frequency components higher than the Nyquist frequency. This will appear as anomalous frequency components at lower than the Nyquist frequency. Reduce this artifact by adding an anti-aliasing filter to limit the measured bandwidth below the Nyquist frequency.

To implement the FFT, a number of points in the form of 2^n are sampled. Most scopes will take the largest number of points in the form of 2^n from the center of the acquisition buffer and use these measurements when calculating the FFT.

To minimize spectral leakage, use a windowing function such as the Blackman-Harris. This will reduce spectral leakage while maintaining acceptable resolution.

This suggests that to see the highest frequency in the spectrum, use the scope's highest sampling rate. To get the narrowest frequency resolution, use the longest acquisition window practical. This combination results in a large number of points in the acquisition buffer:

$$\text{\# points} = \text{acquisition time x sample rate}$$

or

$$\text{acquistion buffer size} = \frac{\text{\#points}}{\text{sample rate}}$$

As a rough rule of thumb, depending on the speed of the scope, a reasonable number of samples in the acquisition buffer that takes less than 1 second to calculate an FFT is about 1-3 million points.

A practical algorithm to follow is:

The scope should be set for the highest sample rate based on the highest frequency required in the spectra.

The time base is selected so that there are less than about 1 million points in the buffer.

Then the FFT is calculated in real time using a Blackman-Harris windowing function.

For example, if the scope is capable of 10 GSps, the highest calculated frequency in the spectrum is 5 GHz. For 1 million samples in the acquisition buffer, this is a total acquisition time interval of 1 million/10 GSps = 100 usec. With typically 10 divisions full scale for the time base, this is a time scale of 10 usec/div.

13.4 Units for Amplitudes

The amplitude of a sine wave signal is fundamentally measured in volts. However, other units are also commonly used. It is important always to be aware of the units used and be able to transform between them.

13.4.1 *Amplitude and RMS in V*

The fundamental figure of merit of a sine wave is its amplitude. This is the value the signal reaches above and below the average value.

Real sine waves often have a DC offset. This means it can be misleading to use the maximum value of a sine wave as an intrinsic measure of the sine wave. The peak-to-peak value is not sensitive to the DC component since the average value is common to both the minimum and maximum values.

The peak-to-peak value of a sine wave is 2x the amplitude.

Another common unit to measure the magnitude of a sine wave is the RMS value. For an ideal sine wave, the RMS value of the wave is related to its amplitude, V, as

$$ \text{RMS} = \sqrt{\frac{1}{T_0} \int_0^{T_0} \sin^2\left(\frac{2\pi}{T_0} t\right) dt} = \sqrt{\frac{1}{2} V^2} = \frac{1}{\sqrt{2}} V = 0.707 \times V $$

The RMS value of a sine wave is linearly proportional to the amplitude. Why, then, is it necessary to consider the RMS value? This is related to the power associated with the voltage signal.

The power associated with a voltage signal is ambiguous. By itself, the voltage in a circuit is not a measure of the power associated with the signal. If this signal were to appear across a 1-ohm resistor or a 100-ohm resistor, very different powers would be consumed by the resistor, with the same voltage amplitude.

By agreed convention, unless otherwise specified, the power associated with a voltage signal always corresponds to the power that would be consumed if that voltage were to be applied across a 50-ohm resistor. This power, based on the signal's amplitude, V, is,

$$P = \frac{1}{2}\frac{V^2}{50\,\Omega} = \frac{V_{RMS}^2}{50\Omega}$$

The factor of ½ comes from the fact it is a sine wave. This is why the RMS value is significant in addition to the sine wave's amplitude. The power associated with a sine wave is the square of the RMS value divided by the 50-ohm resistance. This power level is often measured in units of dBm.

13.4.2 Units of dB, dBW, and dBm

The dB unit is used throughout engineering and is often a source of confusion. It is based on the unit of Bels, named for Alexander Graham Bell. Note that when spelling the bel, one L is lost.

The bel is *always* the log of the ratio of two powers. This is written as:

$$P[\text{bels}] = \log\left(\frac{P_1}{P_{ref}}\right)$$

For example, if the reference power is 1 watt, then the value of 100 watts in bels is $\log(100) = 2$ bels. Likewise, the value of 0.01 watts in units of bels is $\log(0.01) = -2$.

The range of the bel is not very large, given how much change there is in the power level. In 1924, the Bell Telephone Company introduced the unit of decibels as a more sensitive measure instead of the bel. A deci is $1/10^{th}$, just as the centi is $1/100^{th}$.

This makes the value of a ratio of two powers, in decibels, abbreviated dB, as

$$P[\text{dB}] = 10 \times \log\left(\frac{P_1}{P_{ref}}\right)$$

and

$$P_1 = P_{ref} \times 10^{\frac{P_1[dB]}{10}}$$

In principle, the dB is a ratio of any two powers. A power with a value of 10 dB means it is 10x the reference power. A power value of 20 dB means the power is 100x the reference value, and a value of -20 dB means the power is 0.01x the reference power. In this way, a power level can be compared to a reference power level using units of dB.

Without knowing the reference power value used in the definition of the dB, it is ambiguous what the absolute power measured is when it is in units of dB. It depends on the value of the reference power.

To avoid confusion from the value of the reference power, the convention is to use 1 watt as the reference power and to make note of this by adding a W at the end of the dB, written as units of dBW. This is the conventional way of describing a power in units of dBW:

$$P[dBW] = 10 \times \log\left(\frac{P_1}{1\text{watt}}\right)$$

and

$$P_1\left[\text{watt}\right] = 1\text{watt} \times 10^{\frac{P_1[dBW]}{10}}$$

A power of 1 watt is always 10 x log(1/1) = 0 dBW. A power of 30 watts is 14.8 dBW.

Unfortunately, this convention is not followed often enough. Instead, the W is often dropped when describing a power in dB.

There is ambiguity in the use of the dB units. In principle, it refers to a change in value from an initial value to a final value. If the power level in a reactor's output goes from 1 megawatt to 100 megawatts, this is a change of 100x in the power level, or a 20 dB change.

The absolute power went from 60 dBW to 80 dBW. When it is written as 60 dB or 80 dB, this is also an increase of 20 dB in the output power, but ambiguity is introduced. To avoid confusion in the use of the units of dB, the units should be written as dBW when referring to an absolute power level.

If we deal with small power levels, it is common to use a reference power level of 1 mWatt instead of 1 watt. In this case, to avoid confusion, the units are changed from dBW to dBm to signify that the reference power level is really mWatts. Likewise, if the reference power level were 1 uWatt, the units would be dBu.

This means that the conversion between power and dBm is,

$$P[\text{dBm}] = 10 \times \log\left(\frac{P_1}{1\text{mwatt}}\right)$$

and

$$P_1[\text{mwatt}] = 1\text{mwatt} \times 10^{\frac{P_1[\text{dBm}]}{10}}$$

A power of 0 dBm is 1 mWatt. A power of 20 dBm is 10^(2) = 100 mWatt.

13.4.3 Using the dBm to Describe Voltage Amplitude

The dBm units can also be used to describe the power in a voltage, which would be consumed when this voltage amplitude appears across a 50-ohm resistor.

A 1 V amplitude sine wave, which has an RMS value of 0.707 V, would produce a power level of

$$P = \frac{1}{2}\frac{V^2}{50\ \Omega} = \frac{1}{2}\frac{1^2}{50\ \Omega} = 10\text{mwatt} = 10\text{dBm}$$

A voltage amplitude that would create a power of 1 mWatt across a 50-ohm resistor is,

$$V = \sqrt{P \times 100} = \sqrt{0.001 \times 100} = 0.316\ V$$

When the dB scale is used to describe a voltage amplitude or RMS value, it is the ratio of the powers associated with two voltages, which is:

$$V[\text{dB}] = 10 \times \log\left(\frac{V^2}{V_{\text{ref}}^2}\right) = 20 \times \log\left(\frac{V}{V_{\text{ref}}}\right)$$

The factor of 20 comes from the fact that we are using the ratio of voltages inside the log operation. When the dBm scale is used, the reference voltage level is the voltage of a signal that generates 1 mWatt of power across a 50-ohm resistor, or $V_{\text{ref}} = 0.316$ V.

This results in conversions as,

$$V[\text{dBm}] = 20 \times \log\left(\frac{V[v]}{0.316v}\right) = 20 \times \log\left(V[v]\right)\text{dBm} + 10\ \text{dBm}$$

and

$$V[V] = 0.316\ V\ \times\ 10^{\frac{V[dBm]}{20}}$$

For example, a 1 V amplitude sine wave would be described as +10 dBm. A 1 mV signal would be described as -60 dBm + 10 dBm = -50 dBm.

Many scales in spectral displays use dBm as the vertical axis. This means a sine wave with an amplitude of 1 V would appear as a frequency component of +10 dBm. **Figure 13.11** shows the example of a measured sine wave with an amplitude of 1 V and frequency of 10 MHz. The spectrum shows a peak at 10 MHz with an amplitude of +10 dBm.

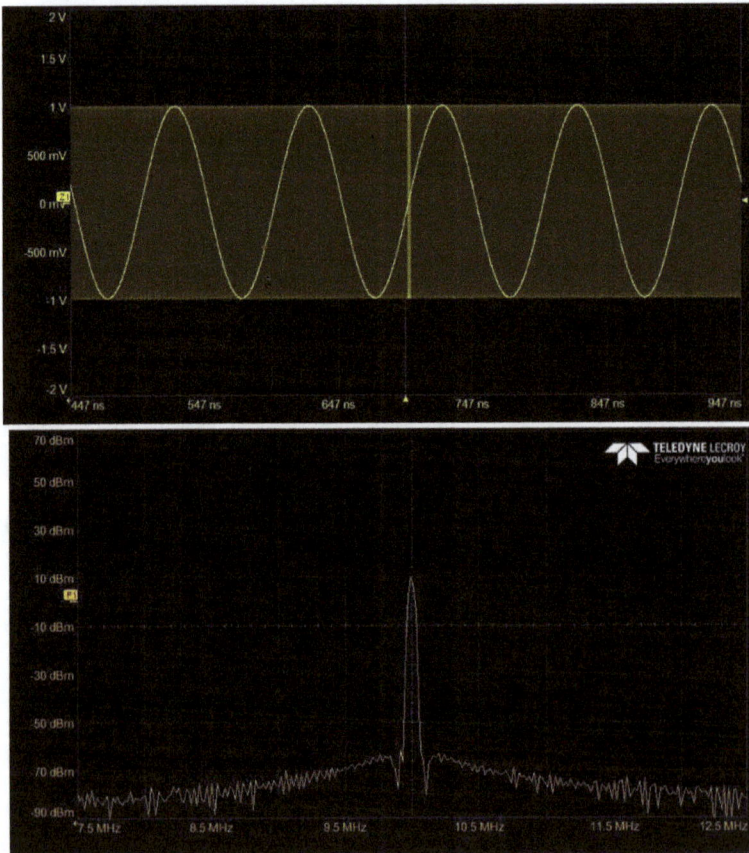

Figure 13.11 An example of the measured sine wave with 1 V amplitude displayed as 10 dBm amplitude in the spectrum.

It is convenient to remember that +10 dBm is 1 V and 0 dBm is 0.316 V. A 20 dBm change is a factor of 10 change in voltage. This means that 30 dBm is 10 x 1 V = 10 V. A value of -10 dBm is 0.1 V and a value of -30 dBm is 0.01 V. And a value of -20 dBm is 0.316 V x 0.1 = 0.0316 V = 31.6 mV.

13.4.4 Scales of dBV

Alternatively, it is also common to use a scale in dBV. This means the reference voltage to which other voltages are measured is a 1 V amplitude. In general, the connection between voltage amplitude and dBV is

$$V[dBV] = 20 \times \log\left(\frac{V[v]}{1v}\right) = 20 \times \log\left(V[v]\right)dBV + 0dBV$$

and

$$V[v] = 1\,v \times 10^{\frac{V[dBV]}{20}}$$

A value of 0 dBV is a voltage amplitude of 1 V. A value of -40 dBV is a voltage amplitude of 0.01 V.

13.5 Quick Start Guide: FFT Function in a Free Sound Card Scope

If you do not have access to a scope, the spectral analysis features of the time-domain signals can be explored using either the free MAUI Studio scope emulator and the function generator source or the PC sound card using Waveforms software and audio signals picked up by your computer's microphone.

In the free version of Digilent Waveforms, using your PC sound card as the scope front end, on the top of the toolbar, under the Scope window tab, is the button labeled as FFT. This is shown in **Figure 13.12**. Click this button, and an FFT window will open up below the real time scope window.

Figure 13.12 Select the FFT button to open up a live spectral display of the real-time voltage waveform measured by the scope.

You can experiment with the horizontal scale of the frequency and the vertical scale for the amplitude of each frequency component. An example of a real-time spectrum of sounds in the room is shown in **Figure 13.13**.

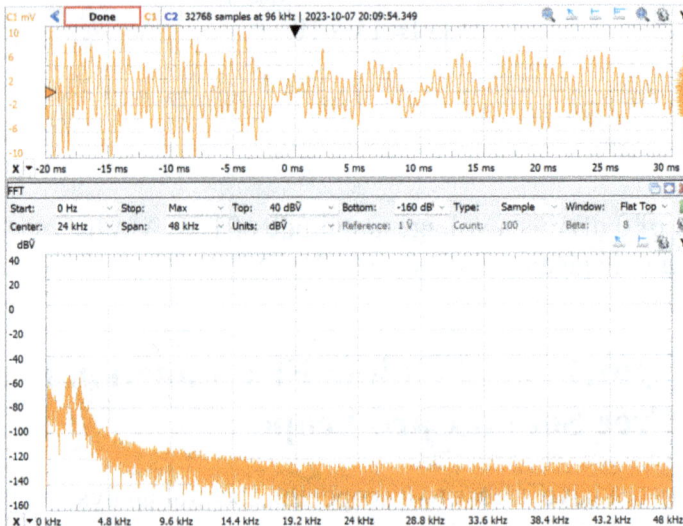

Figure 13.13 An example of the spectral response of a measured voltage waveform is displayed using the FFT function. This identifies frequency peaks at 2.4 kHz and 3 kHz, and most of the spectral components are below about 4.8 kHz.

In this example, the acquisition sample rate is fixed by the sound card's ADC at 96 kSps. This means the highest frequency to display is 48 kHz. This is seen on the right edge of the spectral scale.

The total acquisition window is about 300 msec. This is 32k samples/96 kSps. This makes the lowest frequency measurable about 3 Hz and the resolution is 3 Hz. The time base is set up to display only 60 msec of the 300 msec measured acquisition buffer.

In this example, the vertical scale is in dBV. This means a value of -60 dBV is an amplitude of 1 mV.

On this linear frequency scale, the amplifier sensitivity drops off above about 10 kHz. The noise floor corresponds to voltage amplitudes on the order of -120 dBV, which is 1 uV amplitude.

The Digilent Waveforms software tool is a great way of exploring many of the features of advanced scopes, with no additional components required, runs on your PC, and is free.

13.6   Spectra of Common Signals

A few commonly seen patterns in spectra relate to features in the time-domain signals. While the correlation between the time-domain features and the frequency-domain features is well established from the mathematical transform from the time to the frequency domain, a few measurement examples will help build engineering judgment.

With a suitable function generator, many of these waveforms can be synthesized, and anyone can practice using the FFT function to observe these patterns.

Look for these patterns when observing signal spectra. They will hint back to features in the time-domain signal from which they might arise.

13.6.1 Setting Up the Scope

The following examples are using a Teledyne LeCroy HDO6104B scope. It has a maximum sample rate of 10 GS/s at 12 bit vertical resolution.

While the acquisition buffer size is well in excess of 100 M samples, the buffer size was set to a max of 1 M samples to keep the display appearing in real time with no perceptible delay. At 10 GSps, it takes 1 MS/10 GSps = 100 usec to fill an acquisition buffer. This means the time per division should be limited to about 10 usec/div. Unless there is a strong, compelling reason, this is a reasonable setup to observe signals that vary on time frames shorter than 100 usec. These settings take full advantage of the scope's features. Otherwise, the scale can be adjusted accordingly.

To see the details of any signals that vary on time frames faster than can be displayed at 10 usec/div, the measured acquisition buffer can be zoomed in to show the faster-changing signals. The data acquisition rate is at the scope limit and will not change as the time base is zoomed in.

On this time base of 100 usec full scale, the frequency resolution is 1/100 usec = 10 kHz. The highest sample rate of 10 GSps means the highest displayed frequency component in the spectrum is 5 GHz. The scope bandwidth of 1 GHz will act as an anti-aliasing filter.

This means the spectrum displayed will span 0 Hz to 5 GHz with a resolution of 10 kHz. This is 500,000 frequency bins available. To explore this frequency range, regions can be zoomed in, either in live mode or after an acquisition has been recorded.

Before looking at the spectra of measured signals, it is useful to get a feel for the system-level noise and its spectrum. On the highest voltage resolution scale of 1 mV/div, the noise floor of the scope's amplifier can be measured, and the background spurious frequency components can be identified.

Figure 13.14 shows the time base and spectrum scale with the most sensitive voltage scale of 1 mV/div. This identifies three features in the spectrum.

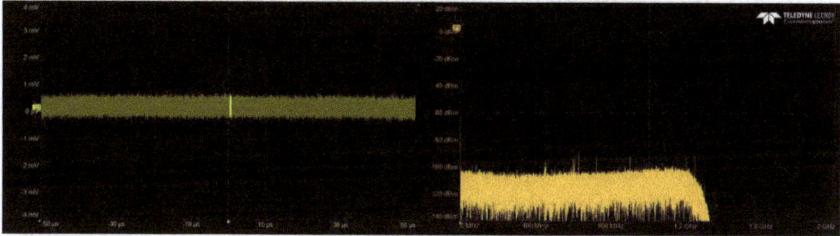

Figure 13.14 The measured noise floor of the scope with the input set to gnd.

First is the drop-off in the spectral response at about 1.2 GHz. This is from the DSP filter in the scope that introduces a sharp drop in response above the rated scope bandwidth of 1 GHz.

Second is the baseline noise floor, limiting the peak detectable values to about -100 dBm. This is a factor of 100,000 down from the 0 dBm value of 0.316 V, or a noise floor corresponding to an amplitude level of 3.156 uV detectable above the scope noise.

There are also spurious frequency components extending to about -90 dBm or to a voltage amplitude of 10 uV. These are due to the interleaved ADC sampling harmonics. Any peaks in a measured spectrum above this spurious peak level of -90 dBm or 10 uV amplitude are probably real.

There is a 0th harmonic value at about -60 dBm. This corresponds to a DC value of about 0.316 mV. This matches the observation in the time-domain signal of the DC offset of about 0.3 mV.

13.6.2 A Sine Wave

When using a new instrument or new features of an instrument for the first time, the best practice is always to measure something for which you know the answer. This way, you can apply Rule #9 and verify that you are measuring what you expect to see.

A good initial signal to measure is a sine wave. In this first example, an external function generator was used to generate a 10 MHz sine wave with an amplitude of 1 V. We expect the spectrum to contain one peak with a value of 1 V amplitude, which is +10 dBm.

Figure 13.15 shows the measured signal in the time-domain and the FFT response. While there is a peak at about 1 MHz with an amplitude of +10 dBm, there are also other peaks, some near 620 MHz but down in amplitude to -60 dBm. This corresponds to a 0.001 x 0.316 v or 0.316 mV amplitude signal amplitude. Are these real or part of the scope's spurious signals? One way of determining this is to measure the spectral response with the function generator unplugged.

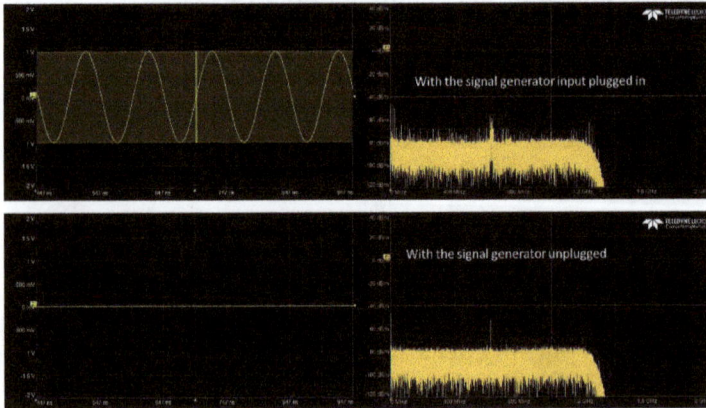

Figure 13.15 A 10 MHz sine wave was measured from the function generator, which is shown in the time-domain on the left and the spectrum on the right. Note the response with the function generator unplugged.

The spurious peaks are present only when the function generator is plugged in and are clearly from the function generator. They are at the roughly -70 dBm level, corresponding to a voltage amplitude of about -80 dB down from 10 dBm or 1 V. This is an amplitude of about 100 uV. This is a measure of the spectral contamination from the function generator.

On this frequency range, the features of the 10 MHz sine wave are difficult to resolve. Without changing the acquisition settings, the spectrum can be zoomed to show the features of the 10 MHz frequency components. This is shown in **Figure 13.16**.

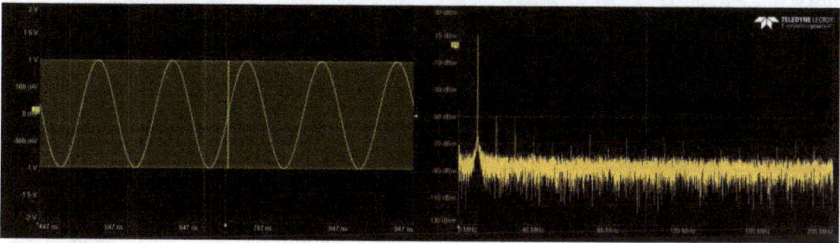

Figure 13.16 The measured spectrum of the 10 MHz sine wave from the function generator. Note the higher harmonics from the sine wave.

The amplitude of the first harmonic is 10 dBm, or 1 v amplitude. The second harmonic at 20 MHz, is down to -50 dBm. This is a drop of -60 dB or an amplitude of 0.1% of the first harmonic. This is a very low level of total harmonic distortion. The third harmonic is also of this order, but the higher harmonics drop off quickly, exactly as expected for a high-quality function generator.

The time base is 100 usec full scale, so the frequency resolution of each bin is 1/100 usec = 10 kHz. **Figure 13.17** shows the 10 MHz peak on a zoomed scale, highlighting the width of the peak as about $1/10^{th}$ a division. The frequency scale is 500 kHz per division. This is a width of about 50 kHz, higher than the expected 10 kHz resolution. On this scale, there are 50 frequency bins per division.

However, the peak width is defined as the full width at half maximum (FWHM). This is the peak width when the value drops by -3 dB from its peak value. On this scale of 20 dB/div, this is nearly at the very tip of the peak. The FWHM width is closer to the expected frequency resolution of the FFT of 10 kHz. This is illustrated in **Figure 13.17**.

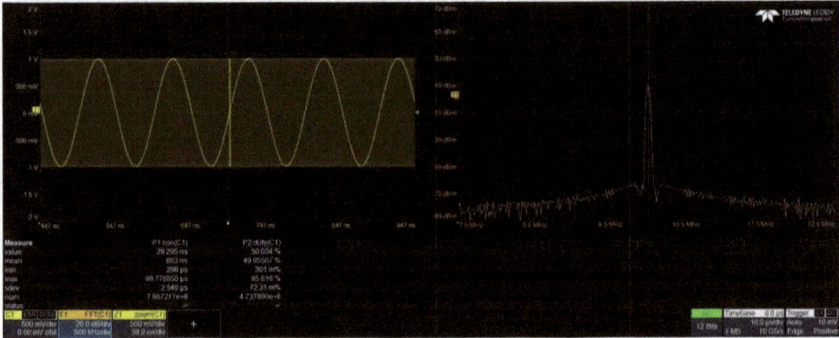

Figure 13.17 A zoomed frequency scale of the spectrum of the sine wave shows the width of the peak as about 50 kHz. The FWHM width at the very tip of the peak is closer to the expected 10 kHz.

13.6.3 A Square Wave

An ideal square wave is an example of a waveform that is periodic, with a repeat frequency. In any periodic waveform, there will be frequency components in the spectrum at multiples of the repeat frequency. Each frequency component is called a harmonic of the fundamental.

When you see a spectrum with evenly spaced frequency components, they are due to a repetitive waveform in the time domain.

In an ideal square wave with a short rise time and 50% duty cycle, the frequency components will be odd multiples of the repeat frequency. Even if the rise time is 0 psec, the amplitudes of each component will drop off with 1/f. An example of a square wave and its spectrum is shown in **Figure 13.18**.

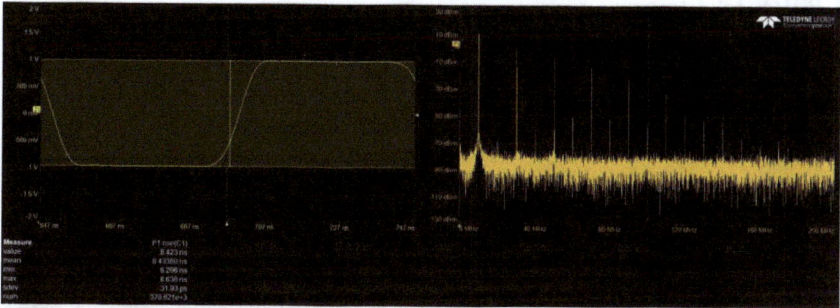

Figure 13.18 The spectrum of a square wave generated by the function generator. The frequency scale has been expanded to 20 MHz/div.

The drop-off in the amplitude of the harmonics shows the 1/f dependency. The rise time is measured as 8.4 nsec. This would result in a bandwidth of about 0.35/8.4 nsec = 42 MHz. This is the frequency at which the amplitudes of the harmonics would drop off faster than 20 dB/decade. This is roughly after the fifth harmonic.

In principle, in a waveform anti-symmetric about its center, so that f(t) = -f(t-1/2 T), as is a square wave with a 50% duty cycle, the amplitude of the even harmonics will be zero. The duty cycle of this synthesized square wave was adjusted to minimize the even harmonic amplitudes.

In the best case, the second harmonic was reduced -80 dB from the first harmonic. With the noise floor below -80 dBm, even harmonics with amplitude on the order of -60 dBm are clearly visible. This is an example of how a sensitive measurement can reveal imperfections in real waveforms compared to their ideal behavior.

The magnitudes of the even harmonics in a repetitive waveform are an indication of the asymmetry in the waveform.

By increasing the duty cycle from 50% to just 50.22%, the magnitude of the even harmonics is increased from -70 dBm to -30 dBm, a change of 40 dB or a factor of 100 in amplitude. The spectrum of the square wave with this slight increase in duty cycle

and the large increase in even harmonic amplitudes is shown in
Figure 13.19.

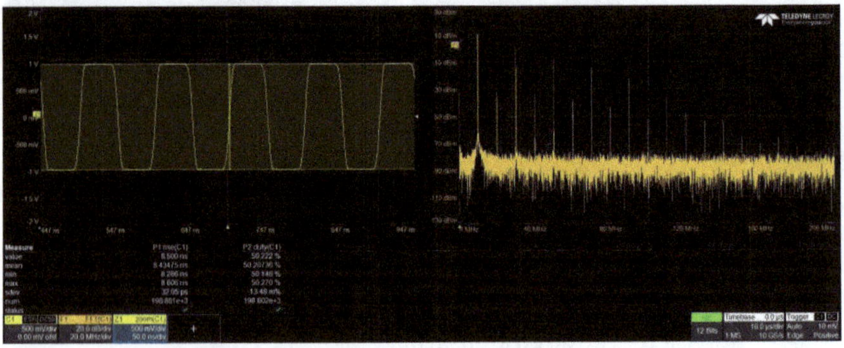

Figure 13.19 Increasing the duty cycle slightly to 50.22% dramatically increases the
amplitude of the even harmonics.

13.6.4 A Modulated Sine Wave

A single-frequency sine wave has a single peak in the spectrum.
When this frequency is modulated, as in FM modulation, the
frequency component is spread out with a width equal to the
frequency span of the modulation.

When the frequency is modulated at a slow rate compared with the
total acquisition time, the spectrum has a peak that moves back and
forth. When the frequency excursions are modulated in a time
short compared to the acquisition time, the peak is broadened.

Figure 13.20 shows the spectrum of a 10 MHz sine wave
modulated by a 4 MHz FM amplitude, with a sweep frequency of
50 kHz. This is a period of 20 usec, short compared to the 100 usec
total acquisition time. This means that the acquisition window has
all the frequencies in the modulated waveform. The spectrum will
be a broadened peak with a width of 8 MHz.

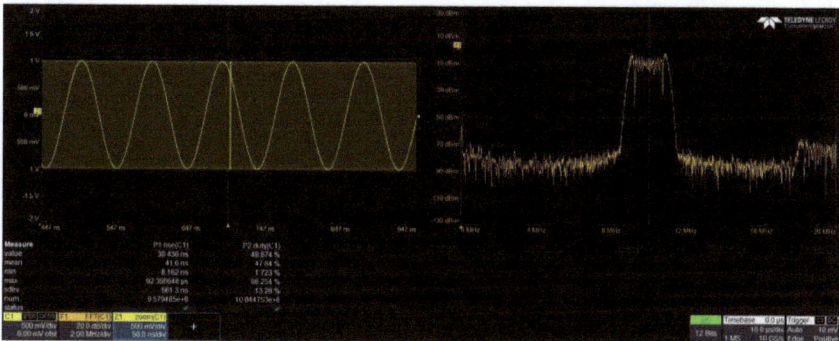

Figure 13.20 The spectrum of a 10 MHz sine wave with a frequency modulation amplitude of 4 MHz. Note the signature of a flat, broad peak.

This is an important signature of a broad spectral peak with a nearly flat top. The modulation waveform, either sinusoidal, triangular, or other pattern, will change the shape of the peak's top, not its width.

13.6.5 Narrow Pulses

As the pulse width of a repetitive signal gets shorter, it approaches an impulse response. This means the amplitude of the spectral components at the repeat frequency will approach a flat response. The spectrum of a pulse is shown in **Figure 13.21**. The amplitude of the first few harmonics is nearly constant. This is the signature of a narrow, reactive pulse.

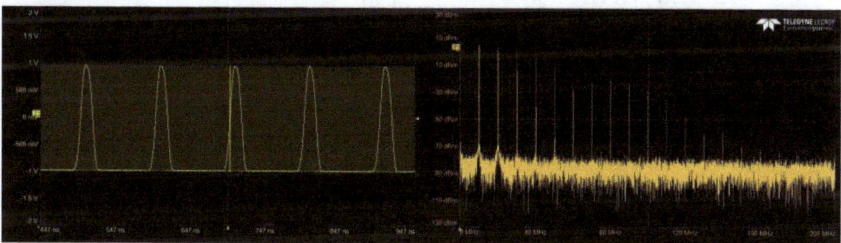

Figure 13.21 The spectrum of a 10 MHz repetitive pulse shows a flat spectrum of harmonics at the repeat frequency.

13.7 RF Pickup Noise

We live in a noisy electromagnetic environment. If we could see with RF-sensitive eyes, we would be surrounded by a dense fog. Every frequency range has some common sources.

Noise with frequency components in the 1 kHz to 100 kHz band might be from switch-mode power supplies. This noise radiates from interconnect cables or circuit board traces.

Components in the 40 kHz band are usually from the inverter circuits used to power LED lights.

AM broadcast radio is in the 535 kHz to 1.7 MHz band.

The shortwave radio is in the 5.95 MHz to 26.10 MHz band.

Free analog television stations are in the 54 MHz to 88 MHz and 174 MHz to 220 MHz bands. The VHF and UHF bands for analog and digital TV are 470 MHz to 806 MHz.

Garage door openers are around 40 MHz.

Cordless phones use bands from 43 MHz to 50 MHz, 900 MHz, 1.9 GHz, 2.4 GHz, and 5.8 GHz.

Many electronic devices use the Instrumentation, Scientific, and Medical (ISM) bands. These are unlicensed frequency bands in which the FCC allows radiated emissions. There are three bands:

- 902-928 MHz

- 2.4 to 2.483 GHz

- 5 GHz

A short wire loop is a magnetic dipole antenna. A voltage is induced across the ends of the loop, which can be measured by a scope. A simple version is just shorting the ends of a mini grabber at the end of a coax cable.

Using the standard settings of 10 GS/s and 100 uses full scale will show frequency components from 10 kHz to 1.2 GHz, limited by the scope bandwidth. It can be used as a quick and simple probe to sniff the local electromagnetic field background. **Figure 13.22** shows the measured spectrum of the scope with no input and then with the mini grabber attached. There are distinctive peaks present.

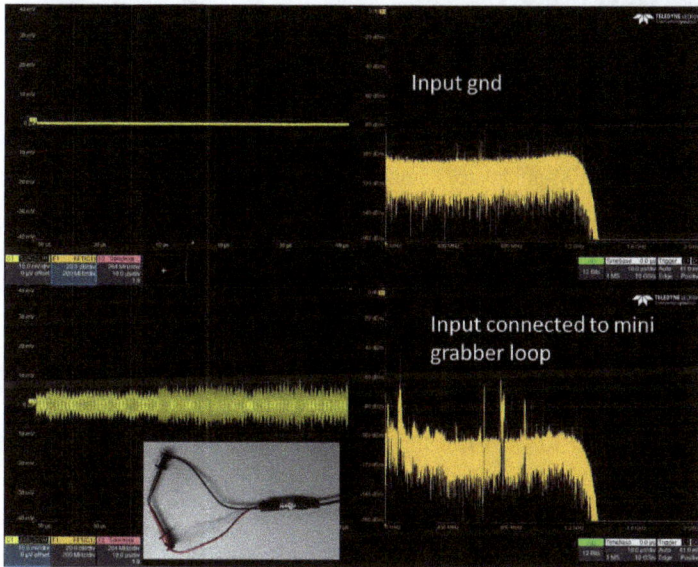

Figure 13.22 The measured noise spectrum was picked up by a simple magnetic dipole loop, showing peaks. Top: scope basic noise level, bottom: spectrum of RF pickup from the small coil seen in the inset.

The measured spectrum has four distinct frequency bands: 40 MHz, 600 MHz, 750 MHz, and 900 MHz. These are common communication bands. For example, the band around 750 MHz is expanded in **Figure 13.23** to reveal a signature of flat bands in narrow channels. This is the signature of FM signals, probably from communications systems.

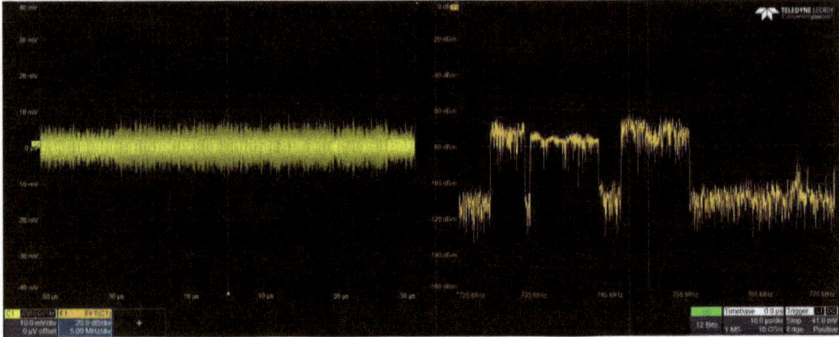

Figure 13.23 The measured noise spectrum expanded in the 750 MHz range, showing a flat top signature similar to FM-modulated communications signals.

13.8 Example: Analyzing a Low-Cost Function Generator

A low-cost function generator generated a simple 0.6 V amplitude sine wave signal at 50 kHz. This is shown in **Figure 13.24**.

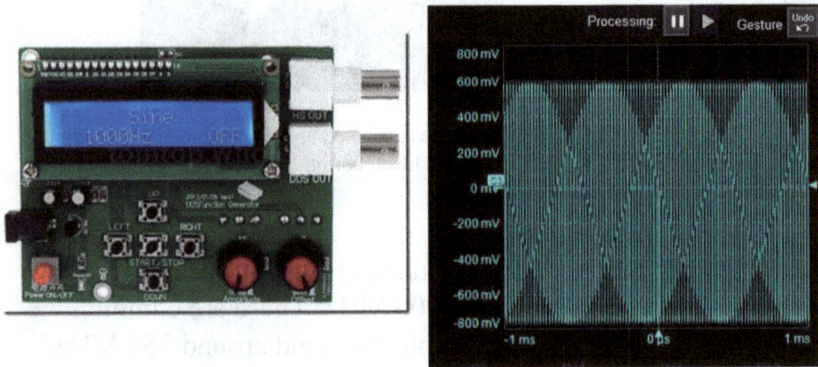

Figure 13.24 A $30 function generator and the measured voltage signal. The pattern displayed on the screen is aliased with the screen resolution.

While the sine wave looked reasonably like a sine wave, the spectral response revealed much more structure in the signal.

The sample rate was 1 GSps, on a time base of 2 msec full scale. This means there were 2,000,000 points in each measurement buffer. The highest frequency in the spectrum was 0.5 GHz, and the lowest frequency and the bin size were both 500 Hz.

The spectrum, displayed in **Figure 13.25**, shows a strong peak at 50 kHz with an RMS value of 52.4 dBmV. This is an RMS value of 0.417 V or an amplitude of 0.59 V, very close to the 0.6 V measured off the front screen of the scope.

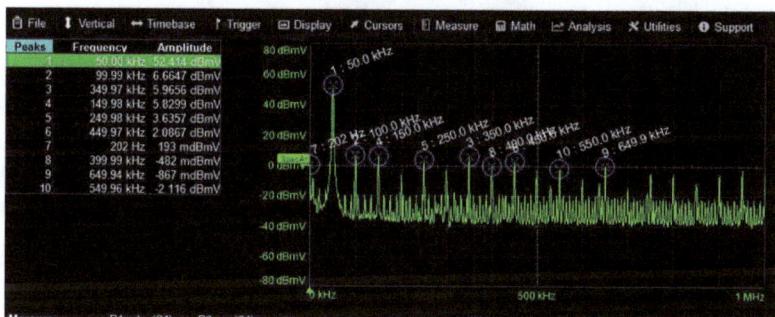

Figure 13.25 The spectrum of the measured sine wave shows the first harmonic peak at 50 kHz and multiple harmonics.

The spectrum shows multiple harmonics at 100 kHz and 150 kHz, continuing up to the highest frequency displayed at 1 MHz. The amplitude of the higher harmonics is on the order of 6.7 dBmV. This is an amplitude of 3 mV, out of 0.6 v or 0.5%. This is a relatively small harmonic distortion but is easily seen in the spectrum.

But this is not the whole story. The spectrum was displayed on a higher frequency range of 200 MHz. The spectrum revealed the presence of strong peaks at 16 MHz and multiples. This is the clock frequency of the microcontroller. The amplitude of the first harmonic was 3.6 dBmV or an amplitude of 2.15 mV. **Figure 13.26** shows this spectrum.

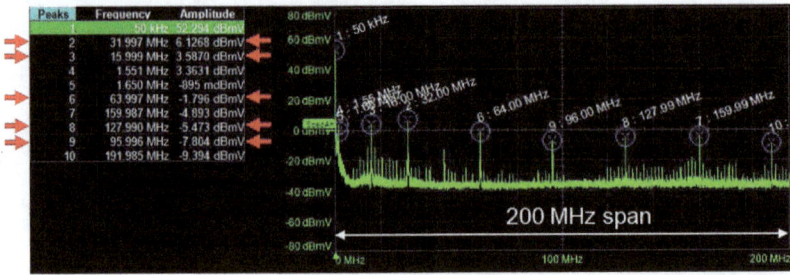

Figure 13.26 The measured spectrum displayed on a higher frequency range.

The harmonics of the clock frequency that contaminates the sine wave output have a stronger second harmonic at 32 MHz, which is almost 2 dB higher than the first harmonic. This is usually an indication of clock edge noise. In a typical clock cycle, the rising edge of the clock is used to set up the various latches, and the falling edge is used to turn them on. This means there is switching current and clock edge noise on both edges, resulting in a strong noise signature at 2x the clock frequency and multiples.

It is not that the clock signal itself is generating noise. It is that current is switching in the digital circuit at the clock frequency, which generates the voltage noise and is picked up by the output drive circuits and measured by the scope.

The spectrum also shows modulation of the 16 MHz clock frequency with 50 kHz side bands, which are caused by the interactions of the clock and the sine wave frequency.

The noise floor of the spectrum is about -30 dBmV. This is a voltage amplitude noise of about 2 uV. This is an overall signal-to-noise ratio of 0.6 v/2 uV = 130 dB. This is an illustration of the power of spectral analysis with a large buffer size, a low-frequency range, and a large dynamic range.

13.9   The Bottom Line

1. When starting out with initial measurements on a new instrument, always measure something for which you know the answer. This is the only way of applying Rule #9.

2. The spectrum of a signal is a powerful way of exploring the underlying cause of features in the time domain.

3. The spectral content of a signal reveals some of the sources of the signal.

4. Features difficult to see in the time-domain are revealed as clear signatures in the frequency domain.

5. All DSO scopes have an FFT function. It is important to master this function and use it routinely.

6. The highest frequency in the spectrum is the Nyquist frequency.

7. The lowest frequency is 1/total acquisition time.

8. The resolution is 1/total acquisition time.

9. Always use a window function to reduce spectral leakage. The Blackman Harris window is recommended.

10. Peaks in the spectrum are fingerprints that identify periodic signals in the time domain.

Chapter 14
Scope Measurements of a Microcontroller Board

Every scope is a little different. The location of the controls and the menu items are different, and the specific capabilities like time and voltage resolution and built-in functions are different. But the general functions are all the same. The way in which we use the scope to perform characterization measurements and interpret the results is independent of the specific scope.

The way to gain experience in routine measurements with a scope using best measurement practices and analyzing the results to turn data into information and information into decisions is to practice. Any scope is a good scope to use to build this muscle memory. What is required is a suitable test vehicle.

A microcontroller board is a rich source of signals for practicing many of the important measurement methods. Best of all, a good microcontroller board with a variety of voltages to measure is widely available and very low cost.

14.1 System-Level Measurement Artifacts

There are two common sources of noise artifacts that appear in any measurement: RF pickup noise in the probe tip and conducted emissions noise on a DUT's circuit ground from its earth-ground connection. These can easily be separately measured for your environment.

The first step is to know what values to expect in your DUT measurements. Once you can measure the noise level, the next step is to search for the root cause of the interfering signals and reduce the sources in your measurement environment. Sometimes, this is a difficult task.

14.1.1 *Measuring the Scope Noise*

To analyze the noise that is picked up due to the way in which you probe a DUT, we will use two special features of a specific scope to explore the impact of pickup noise, its root cause, and how to avoid it.

The Teledyne LeCroy 6104b scope is used to illustrate pick up from the 10x probe. This scope is shown in **Figure 14.1**.

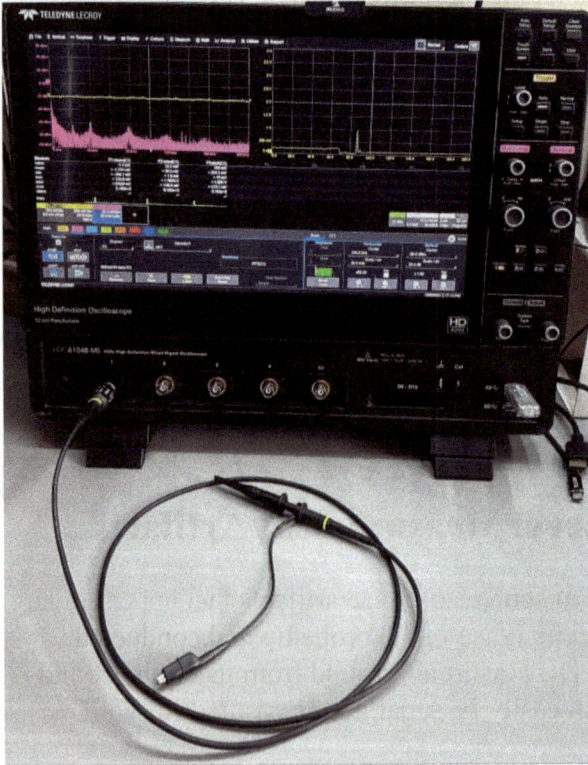

Figure 14.1 The Teledyne LeCroy HDO6401B was used for these measurements with a 10x probe connected.

As with any scope, you should first identify some of its most important features. For this scope, they are:

- Bandwidth of the scope = 1 GHz
- Vertical resolution is 12 bits or 4095 levels
- Fastest sampling rate is 10 GSps total.
- Number of channels = 4 channels or max of 2.5 GS/s per channel
- The maximum buffer size = 1 GSamples

The screen pictured above is configured to display three important measurements to aid in the analysis of the noise. The main screen is the real time measurement of the incoming signal on a 10 msec/div and 0.5 v/div scale.

On the time scale of 10 msec/div, 60 Hz noise will appear with a period of about 2 divisions. This makes identifying 60 Hz noise easy.

The acquisition buffer size was set to 2.5 M points so that the time to process the buffer and refresh the display is nearly real time. This means the sampling rate for the 1 channel is 2.5 M samples/100 msec full scale = 25 MS/s. This is an interval of 40 nsec between samples. This will limit the time resolution to 40 nsec and the highest frequency calculated in an FFT to ½ x 25 MS/s = 12.5 MHz.

An internal filter is available to limit the bandwidth of incoming signals to 20 MHz. This was turned on to use as an anti-aliasing filter.

The vertical scale of 0.5 v/div means a range of 4 v full scale. With a vertical resolution of 12 bit, this is a digitizing level of 4 v/4095 levels = ~ 1 mV/level.

These are the sorts of numbers to be aware of before you perform any measurements. They can always be readjusted based on your range of measurements and can be dynamically changed to adjust to any measurement condition.

The second scale is set to measure the FFT spectrum of the signal. In the spectrum, the highest frequency component calculated is the Nyquist, 12.5 MHz. The frequency resolution is 1/100 msec = 10 Hz. The scale is adjusted on the front screen to display only the spectrum from 0 Hz to 200 kHz, as a starting place to observe the dominant sources of noise in the upcoming examples. The entire spectrum is calculated with every acquisition buffer, but only a part of the spectrum of interest is displayed.

The third screen is an important trend to quantify changes in the noise. In each acquisition, a few parameters can be measured using built-in functions. In this case, the average or mean value, the rms, and the peak-to-peak values are calculated for each acquisition buffer. The peak-to-peak voltage is a good measure of the relative noise being measured. This value for every buffer acquisition is then plotted on the third screen after each acquisition. In this way, the patterns in the trends as various actions are taken can be displayed.

An example of the sort of trends in the peak-to-peak noise, after repeatedly touching the exposed 10x probe tip and increasing the noise, is shown in **Figure 14.2**. As I touched the probe tip, the tip measured the large voltage induced on me from me acting as an antenna for 60 Hz signals. Also apparent in the spectrum are strong peaks at harmonics of 10 kHz.

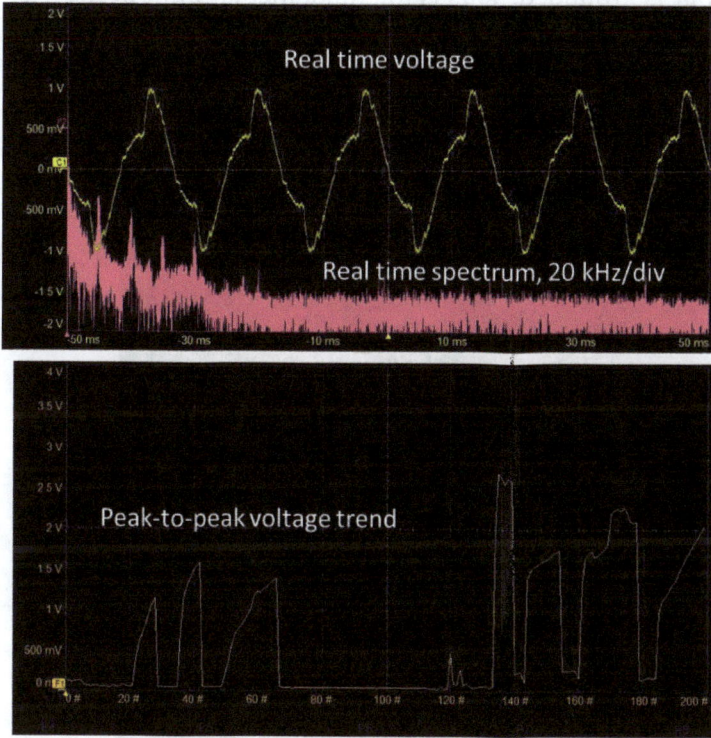

Figure 14.2 Example of how the noise metrics can be meaningfully displayed.

Before analyzing the pickup noise in the probe, the very first step is to establish a baseline of the scope and its internal noise. In many cases, the spectrum is an incredibly sensitive measure of the signature of features in a time-domain signal. The integration over the entire measurement buffer is an averaging process. It will often show a dynamic range of 100 dB between the noise floor and signal features. This is a dynamic range of 100,000.

At this level, nuances inside the scope and how it performs its sampling and DSP of the signal are often apparent. These features have nothing at all to do with any voltage at the input but are measurement artifacts. Measuring them with no input signal is an important starting place to identify what is a scope measurement artifact and what is potentially a real signal.

This means noting the peak-to-peak noise and spectrum when the channel input is grounded. This is the fundamental limit of the scope. The system-level scope noise in this specific set up is about 17 mV peak to peak, with features in the spectrum mostly below 10 kHz. This could be just 1/f amplifier noise. This measurement is shown in **Figure 14.3**.

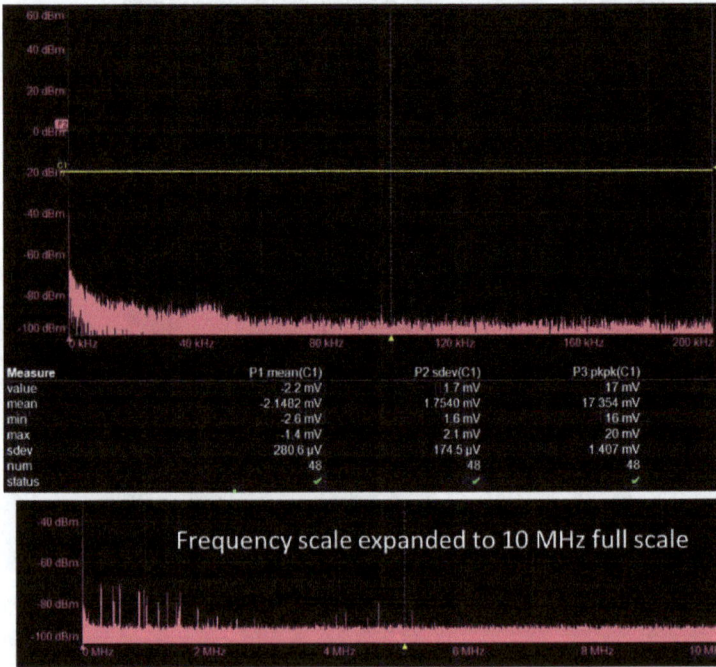

Measure	P1 mean(C1)	P2 sdev(C1)	P3 pkpk(C1)
value	-2.2 mV	1.7 mV	17 mV
mean	-2.1482 mV	1.7540 mV	17.354 mV
min	-2.6 mV	1.6 mV	16 mV
max	-1.4 mV	2.1 mV	20 mV
sdev	280.6 µV	174.5 µV	1.407 mV
num	48	48	48
status	✓	✓	✓

Frequency scale expanded to 10 MHz full scale

Figure 14.3 The baseline noise of the system with input is grounded. The spectrum is displayed on two scales, 200 kHz full scale and 10 MHz full scale. Note the spurious peaks in the 500 kHz and 1 MHz range. This is from the digitizer and internally generated noise.

14.1.2 Measuring RF Pick Up Noise

The 10x probe can be used as both a magnetic field (B field) detector and an electric field (E field) detector. When the tip is shorted to the return ground clip, it is sensitive to changing B fields that pass through the tip's loop. When the tip is floating in the air, the tip will pick up changing E fields. These changing fields will induce a current through the 10 to 1 voltage divider and, ultimately, the 1 megaohm input resistance of the scope. The scope amplifier will measure and display this voltage. In the local environment of my lab, there are only low-level B fields picked up. However, there are significant E fields present.

The E field picked up is from two dominant local sources: the overhead LED lights and the screen refresh of the scope itself. **Figure 14.4** shows the spectral signature of the E field picked up from these two sources. The LEDs show a strong 60 Hz component with 20 kHz and harmonic components due to the SMPS and pulse width modulated signal driving them. The scope screen shows strong 10 kHz harmonics due to the refresh rate of the screen. These sources of noise should be noted so as not to be confused with signals associated with any DUT.

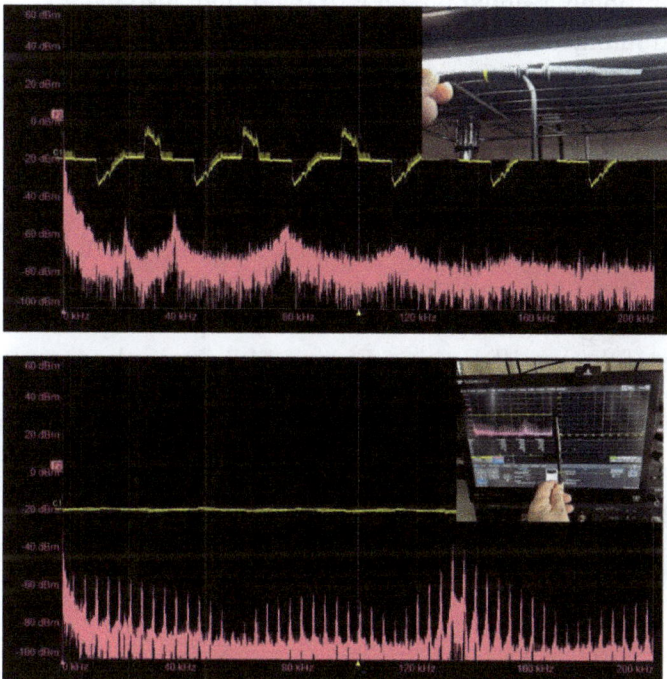

Figure 14.4. Measured voltage noise on 0.5 v/div and the spectral response from E field pick up from the overhead LED lights (top) and the scope's front screen (bottom).

The E and B field pickup noise sources are straightforward to understand and recognize. They arise from radiated emissions picked up because of the antenna properties of the probe. They can also be picked up by the antenna properties of any DUT in this environment. Be aware that when the DUT being measured is a low-impedance source, the DUT and probe are effectively shorted and are sensitive to B field pickup. When the DUT and probe are high impedance, the measurement is subject to E field pickup.

Any time the signal and return paths are pulled apart, you have built an antenna. This is why we sometimes say, "There are two kinds of engineers: those who are building antennas on purpose and those who are not building them on purpose."

This means that a larger tip loop inductance is more sensitive to RF pickup. When probing a signal on a board that might also have large near-field radiated emissions, such as with a switch mode power supply, it is difficult to separate out what is the actual measured voltage on the board and what is potential RF pick up in the probe to the scope.

One way of measuring how noisy your local environment may be is with a second 10x probe with its tips shorted together, used specifically as a pickup coil. When placed in proximity with the 10x probe measuring the voltage on the conductors, it can give a rough measure of the local RF noise. **Figure 14.5** shows the two measured signals, one on the power rail conductor and one of the local RF pick up.

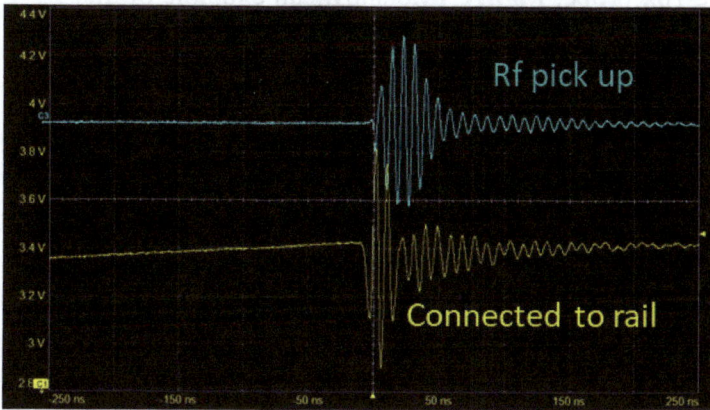

Figure 14.5 The RF pickup in the local environment and the measured voltage on the power rail are displayed on the same scale.

It is remarkable how noisy many boards are in the near field. A signal with a frequency of 100 MHz has a wavelength of 10 feet. This means all measurements a few inches from the board are near-field. Just because there are strong near-field emissions does not mean there will be strong far-field emissions, and the product would fail an FCC certification test. Many near-field sources drop off very quickly with distance and drop off to a negligible level in the far field. It's just that they can interfere with a measurement.

When the RF pickup signal is a significant fraction of the measured voltage on the conductor, be careful interpreting the measured voltage as a real signal. It could just be RF pickup noise and not related to the actual voltage on the DUT.

14.1.3 Reducing RF Pickup with a Coaxial Connection

The way around this problem of RF pickup is to use a connection from the 10x probe to the DUT, which is a coaxial geometry. This will have the minimum RF pickup sensitivity.

Included in many 10x probe kits is a PCB to coax adapter for the 10x probe. This can be soldered into the PCB on 100 mil centered test points and provides a coax connection from the pads on the board to the 10x probe. **Figure 14.6** shows an example of this adapter and the impact it has in reducing the RF pickup noise on this rail.

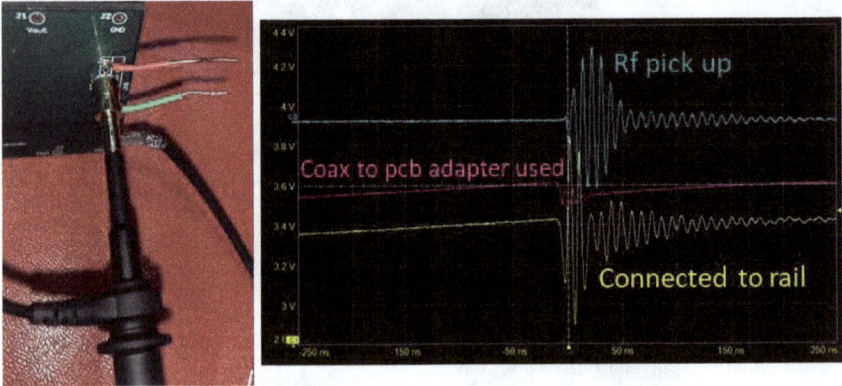

Figure 14.6 Coax to PCB adapter for a 10x probe and its impact in dramatically reducing RF pick up.

We live in an incredibly noisy RF environment. Always be aware of the near-field-induced noise from local E and B field-inducing currents flowing through the 1 megaohm input resistance of the scope.

The most important way of reducing this sort of noise is by using as close to a coaxial connection to your DUT as possible. This reduces the loop area of the probe tip and reduces its efficiency as an antenna. Whenever possible, the probe connection to the DUT should be coaxial.

In situations where a quieter environment is necessary, these sources of environmental RF noise, and others, should be managed. This is part of controlling the artifacts from the room.

14.2 The Microcontroller Board

An Arduino microcontroller board is a very good demonstration vehicle for many common measurement challenges. It provides analog and digital signals from DC to over 100 MHz bandwidth. It is used to illustrate most of the important best practices in using a scope.

Another reason this board was selected is that it is so commonly available and has a wide variety of suitable signals. One version is available from Amazon, for example, for less than $10. This is shown in **Figure 14.7**.

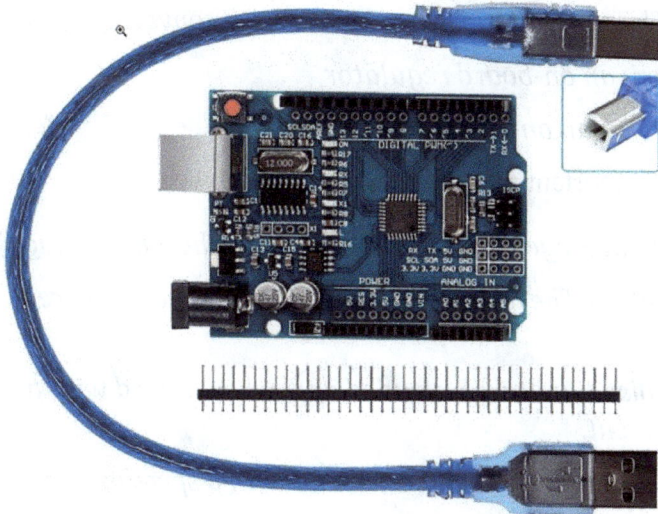

Figure 14.7 An example of a low-cost microcontroller board to practice measurements.

With a few lines of code, provided in this chapter, typical signals found in most systems can be synthesized to practice measurements.

This board, or similar microcontroller boards, offers the opportunity to perform the following types of measurements and their interpretation:

- Power rails
- I/Os
- Measuring timing operations
- Cross talk

14.3   Power Rail Measurements

The Arduino board provides a regulated 5 V supply when an external 9 V DC supply is plugged into the power jack. Three different voltages are available:

- ✓ *Vin: The raw voltage from the AC to DC converter*
- ✓ *5 v from an on-board regulator*
- ✓ *3.3 v from an on-board regulator*

There are four important measurements on any power rail:

- ✓ *The DC average voltage unloaded (the Thevenin voltage)*
- ✓ *The output Thevenin resistance under a specific current load*
- ✓ *The noise signature unloaded and, when loaded with a DC current.*
- ✓ *The noise spectrum unloaded and with a specific current*

Part of the power rail characterization is also the current load of the power rail under normal operations.

In this example, a 10x probe is used to perform all the measurements. Of course, the first step is to verify the 10x probe is well compensated and that the comp signal appears as a flat top and bottom square wave.

The Arduino can be powered by either an external AC to DC converter or by the host computer's USB port. The voltage from the AC to DC converter should be measured independently to

14.3 Power Rail Measurements

establish a baseline. What is measured with no current load is the internal Thevenin voltage of the source. This value will not change to the first order. As the load draws current from the power rail, the voltage will decrease due to the voltage drop across the internal Thevenin resistance.

The first step is measuring the output voltage from the AC to DC converter that powers the Arduino board. **Figure 14.8** shows the measured voltage on the output of the AC to DC converter to be 9.47 V with a standard deviation of 0.7 mV. However, it also shows a considerable noise of 1.36 V peak to peak. Even with the large peak-to-peak noise of 1.36 V, the average value can be measured by averaging all the voltage measurements in an acquisition buffer.

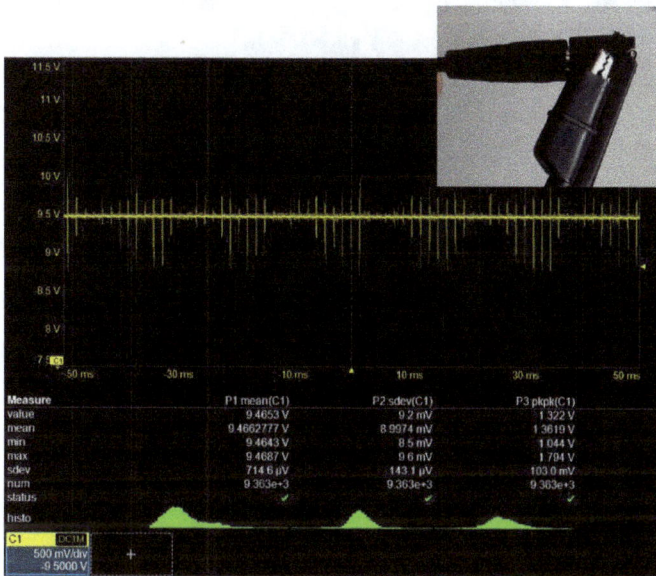

Measure	P1 mean(C1)	P2 sdev(C1)	P3 pkpk(C1)
value	9.4653 V	9.2 mV	1.322 V
mean	9.4662777 V	8.9974 mV	1.3619 V
min	9.4643 V	8.5 mV	1.044 V
max	9.4687 V	9.6 mV	1.794 V
sdev	714.6 µV	143.1 µV	103.0 mV
num	9.363e+3	9.363e+3	9.363e+3
status	✓	✓	✓

Figure 14.8 The voltage on the AC to DC converter is measured directly with the 10x probe, showing a 9.47 V average value with a 1.36 V peak-to-peak noise. The inset picture is the connection between the power jack, a socket, and the 10x probe.

The time base is 10 msec/div. This is always a good starting scale because any periodic pattern that repeats about 2 divisions on this scale could be 60 Hz line noise. The pulses are clearly periodic with about a period of 15 msec to 20 msec and are probably tied to the line noise. This is an indication of a very noisy power supply.

The next step is connecting the 10x probe to the power pins of the Arduino board. There are connections on one side that make this easy. The pins available are to the Vin, 5 V, and 3.3 V rails, as shown in **Figure 14.9**.

Figure 14.9 The power rail connections are available on the upper left set of pins.

The 3.3 V regulator should have the same output voltage regardless of the source of the DC input voltage. In fact, when measured using either the USB hub or the AC to DC converter to power the board, the 3.3 V LDO is measured as 3.241 mV. This is about 2% smaller than its rated 3.3 V. This is seen in **Figure 14.10**.

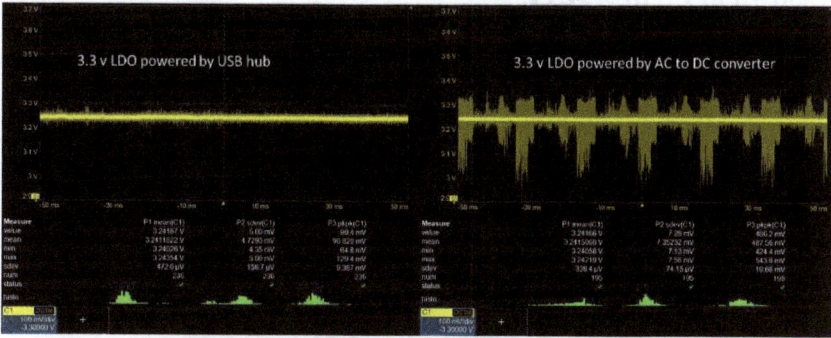

Figure 14.10. The measured voltage on the 3.3 V LDO output. Plotting them on the same scales makes comparison easy.

The absolute voltage on the rails is measured, but more important is the noise. When powered from the AC to DC source, the LDO output is much noisier. The high-frequency ground loop noise from the power source gets through the regulator to appear on the pins and, hence, the board-level voltage.

This suggests that for the lowest noise power rails, the quality of the AC to DC power supply must be checked. Usually, a USB power source is quieter than a low-cost AC to DC wall wart supply. A battery supply is always the lowest noise as this floats relative to earth-ground and can have a very high impedance.

14.3.1 Output Impedance with an Eload

The internal Thevenin resistance of any power source can be measured by measuring the output voltage drop under an external current load. This can be as simple as a resistor load. For a power source, it is preferable to vary the current load and it is sometimes important to stress the supply with large currents. This is more easily done using an electronic load or eload.

A simple eload can be purchased from Amazon, for example, for under $50 which draws a controlled, constant current, but more importantly, has an active heat sink on the load so that a power as large as 150 watts can be dissipated. An example of a simple eload is shown in **Figure 14.11**.

Figure 14.11 An eload that can dissipate 150 Watts for under $50.

The output resistance of a power supply can be measured by measuring the voltage drop on the output of the power supply under no load to get the Thevenin voltage, and then with a controlled current, like 100 mA. The output resistance, R_Thevenin, is calculated from:

$$R_{Thevenin} = \frac{V_{Thevenin} - V_{loaded}}{I_{loaded}}$$

In this example, the 9 V AC to DC supply's output resistance was measured directly with the eload. **Figure 14.12** shows the measurement set up with the eload and an external DMM to measure the current draw.

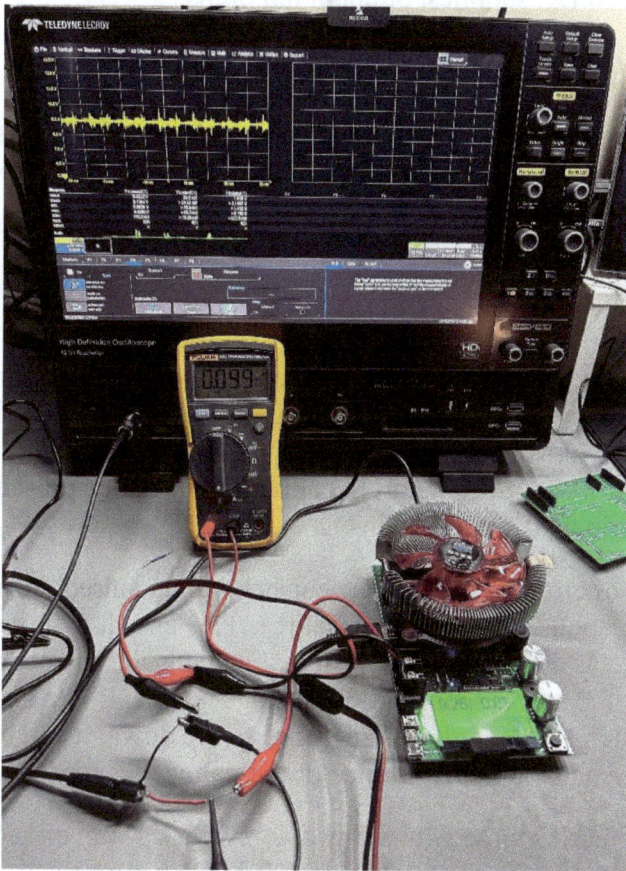

Figure 14.12 Measurement set up to load the AC to DC converter and measure its output voltage.

Using this method the DC voltage of the AC to DC converter under no load and 100 mA is measured. In this example, the values were:

$$\text{Vout no load} = 9.476 \text{ V}$$

$$\text{Vout @ 419 mA load} = 9.261 \text{ V}$$

$$\text{The R\_Thevenin} = (9.476 \text{ V} - 9.261 \text{ V})/0.419 \text{ A} = 0.523 \text{ ohms}$$

The output Thevenin resistance of this power source is typical of many AC to DC supplies, of about 0.5 ohms.

In this condition of 0.419 A of current load and 9.26 V, the power consumption in the load was 3.88 watts. This is high enough that power dissipation and thermal management need to be considered. This is the value of using a commercial load with a heatsink and fan.

14.3.2 Low-Cost eLoad Circuit

An alternative method of drawing a current load and measuring the voltage change can be implemented with a simple MOSFET or transistor. To avoid the need for thermal management and to not heat up the voltage regulator, the current load can be pulsed so that it is on for only a short duty cycle.

This sort of circuit, in which the current load is pulsed, measures the step response of the voltage source. This circuit is sometimes referred to as a *slammer circuit* in that it slams the current on and measures the step response of the voltage source. If it is kept on long enough for the transient response to die down, it can measure the steady state output resistance of the voltage source.

Using a pulsed source with a low duty cycle is a way of keeping the average power consumption low. The peak power consumption may be 1 watt, but the average power consumption can be kept below 0.1 watts, a regime where thermal management is not required. This is important in many low-current draw power supplies.

The simplest circuit to implement a simple eload is shown in **Figure 14.13**. The built-in function generator in most scopes, or an external function generator, can be used to generate a pulsed voltage that controls transient current draw.

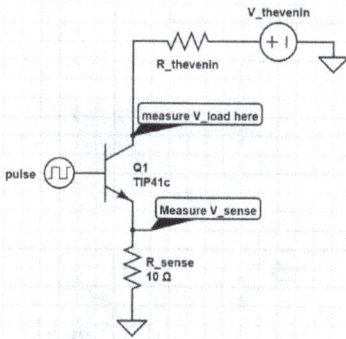

Figure 14.13 A simple eload circuit implemented with a transistor. Three scope probes are used to measure the input pulse to the base, the sense resistor voltage, and the voltage on the VRM under test.

The short pulse on the gate should be on for long enough to see a constant current draw and any transient response from the power source, but short enough so the transistor does not heat up or cause significant power consumption in the voltage regulator module under test. The duty cycle can also be low so that there are no issues with any component heating up.

The sense resistor is chosen so that an easily measured voltage appears with the typical current draw. With a 10 ohms sense resistor, 100 mA generates 1 V across the sense resistor. By changing the input pulse amplitude, the current load to the DUT can be adjusted.

It should be noted that with a transistor, instead of a MOSFET there is some base current. This means that not all the current through the sense resistor comes from the DUT. In the case of the TIP41c transistor, the nominal beta is 30. This means about 97% of the current through the sense resistor is from the DUT.

Figure 14.14 is an example of a 4 msec long voltage pulse from the function generator generating a 400 mV signal across the 10 ohm resistor. This makes the transient current 0.4 A. The measured voltage drop across the AC to DC converter is 20 mV. This makes the output resistance 0.020 V/0.4 A = 0.5 ohms.

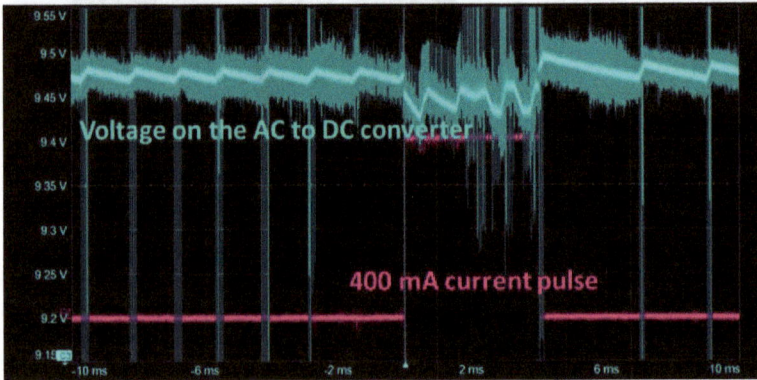

Figure 14.14 Measured voltage drop on the AC to DC converter of 20 mV during the 0.4 A pulse lasting 4 msec. Note the large ground loop noise on the VRM.

Two important features are present. Before the current pulse, during the unloaded part of the output, the switching noise of the VRM is apparent with a period of about 1.5 msec. This is a frequency of about 750 Hz. During the higher load period, the switching frequency increases. After the current pulse turns off, the switching frequency decreases and recovers in a few msec.

In addition, this measurement shows the very large ground loop noise from the AC to DC converter. Much of the noise is asynchronous with the pulse from the function generator. This means that it can be reduced by averaging many consecutive acquisitions. If the scope is triggered on the rising edge of the current pulse, any voltage synchronous with the trigger will repeat and stay the same after averaging. Any signal asynchronous with this signal will reduce with roughly the square root of the number of averages. **Figure 14.15** shows this same response, using 100 consecutive averages of the repeated buffer.

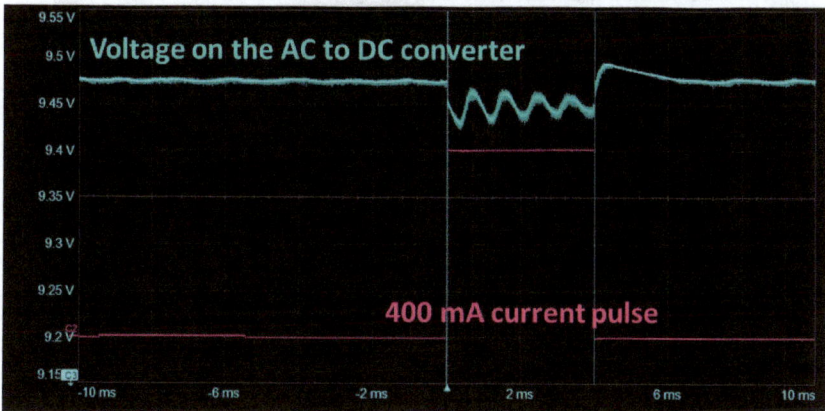

Figure 14.15 The same system response, but averaging 100 consecutive buffers triggered on the turn-on of the current.

This is an example of the power of averaging when the signal is synchronous with a trigger signal and the interference is not. With enough averaging, the interfering signal can be reduced, so it is not important. We also see that the 750 Hz switching noise from the AC to DC converter, which is synchronous with the current pulse, averages away, leaving the very constant output voltage. After the current pulse, the very small transient response is also seen.

This same measurement methodology is applied to the 5 V output of the Arduino board, when powered by the AC to DC converter. This is an on-board linear regulator. The measured pulse response is shown in **Figure 14.16**. In this example, the current pulse width was 2 msec, but 0.4 A. This shows the 5 mV drop on the output of the 5 V rail for a 0.4 A current pulse, showing a 5 mV/0.4 A = 12.5 ohm output resistance of this regulator. This is very low.

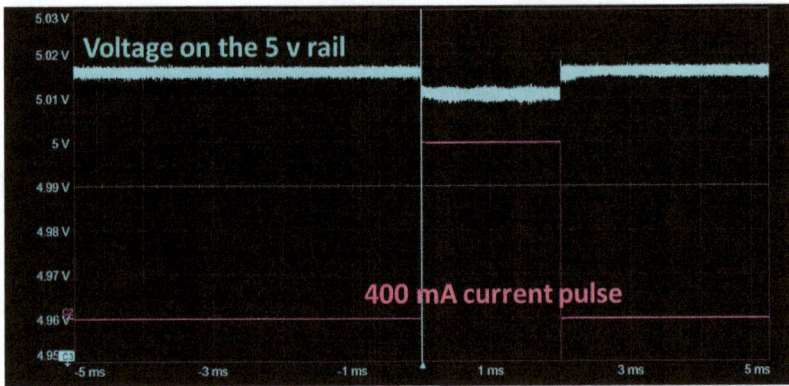

Figure 14.16 The measured response of the 5 V onboard regulated supply to a 0.4 A current pulse. Note the voltage scale is 10 mV/div with an average of 100 sweeps.

Also apparent in this measurement is the transient response when the current turns on. The very shortest response is due to the high inductance in the wires from the regulator and the transistor's response. However, the longer-term response is due to the feedback loop inside the linear regulator. **Figure 14.17** is a zoomed-in view showing the regulator's response in about 5 usecs.

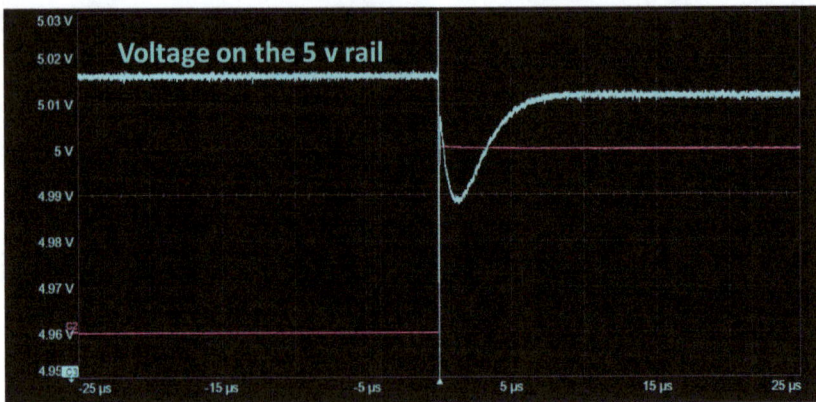

Figure 14.17 Measured transient response of the 5 V linear regulator to a 0.4 A current step showing a response of about 5 usecs for the regulator to control the voltage.

14.4   Signal I/O Measurements

All digital logic devices generate digital signals. There are five primary properties that should be characterized about the I/O drivers and signals:

- ✓ *The output resistance*

- ✓ *The rise and fall time*

- ✓ *The minimum output voltages of a LOW and HIGH signal*

- ✓ *The minimum input switching thresholds for a LOW and a HIGH*

- ✓ *The cross-talk to adjacent I/O from the power rail and ground bounce*

14.4.1 Measure the Output Resistance of a Digital Pin

The same method to measure the output Thevenin resistance of a power rail can be used to measure the output resistance of any voltage source, even a signal source.

To measure the output resistance when the digital pin is in the HIGH state is to first set it HIGH in the microcode. This pin is treated just like any voltage source. The other consideration is that the output resistance of a signal source is generally much higher than that of a power source, and they are often nonlinear. This means the current load to set should be close to a typical output range. Of course, it can be varied, and the output resistance is measured at various load currents.

As an example, the pin output was connected to the same transistor pulse current load, but the current pulse was set for 3 mA. **Figure 14.18** shows the transient response of the output pin. It starts with 5.01 V and drops to 4.94 V, a drop of 70 mV, with a 3 mA current load. This is an output resistance of 70 mV/3 mA = 23 ohms. This is almost 50x of a typical power rail source.

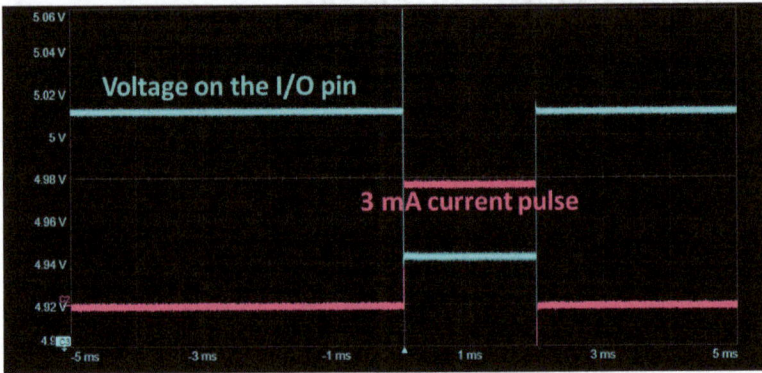

Figure 14.18 Measured voltage drop on the output pin when a 3 mA current load is drawn.

Measuring the output resistance in the LOW state is the same process, but with two changes. First, the output state of the driver is set for LOW. Then, the output voltage of the LOW pin is measured when the output is connected through a load resistor to the Vcc rail.

The voltage across the load resistor is a measure of the current through the LOW pin and the voltage of the LOW pin to board ground is a measure of the voltage drop across the output resistance of the LOW state driver. This ratio is the output resistance of the LOW state.

In CMOS devices, the output resistance of the HIGH state is a measure of the Rds-on of the p-channel of a MOSFET. The output resistance of the LOW state is a measure of the Rds-on of the n-channel of a MOSFET. The LOW output impedance is usually a little smaller than the HIGH output resistance.

14.4.2 *Measuring Rise and Fall Times*

The second important figure of merit for a digital I/O is its rise and fall time. In principle, this is easy to measure using a 10x probe. However, the loop inductance of the tip introduces some distortion. This is apparent when comparing the measurement of the rising edge with a typical connection using loops of wire between the DUT and the 10x probe or with short spring tips. This comparison is shown in **Figure 14.19**.

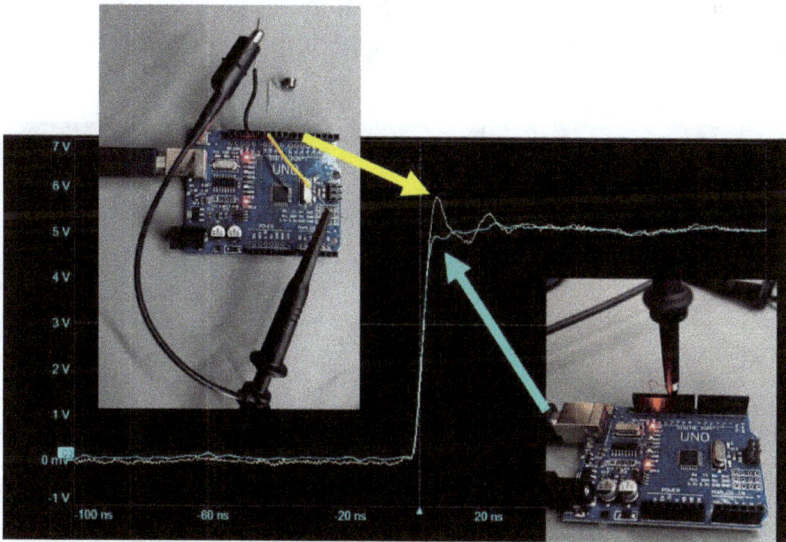

Figure 14.19 Comparison of the measurement of a rising edge using two different probing methods. The short tip loop inductance offers a signal with less ringing artifact. The time base is 20 nsec/div.

Using this method, the 10-90 rise time was measured as 4.1 nsec while the falling edge rise time was measured as 7 nsec.

14.4.3 *Clock Edge Noise*

While measuring the output HIGH signal, it is apparent that there is some noise synchronous with the rising edge of the signal. Is this real or an artifact?

When the signal is an output HIGH, its output pin is effectively connected to the internal, on-die Vdd pin. When other digital gates on the die switch and draw current on the power rail, there will be a voltage drop on the power rail from the impedance of the power rail and the current draw. This generates a voltage noise on the power rail.

This voltage noise is synchronous with the rise time of the digital I/O pin switching. This is apparent when 100 acquisitions are averaged, each synchronous with the rising edge. **Figure 14.20** shows the 100 averages on an expanded voltage scale. This is a noise amplitude of about 100 mV peak to peak. Note there are about 3 cycles per 100 nsec or 33 nsec period. This is exactly ½ a clock cycle.

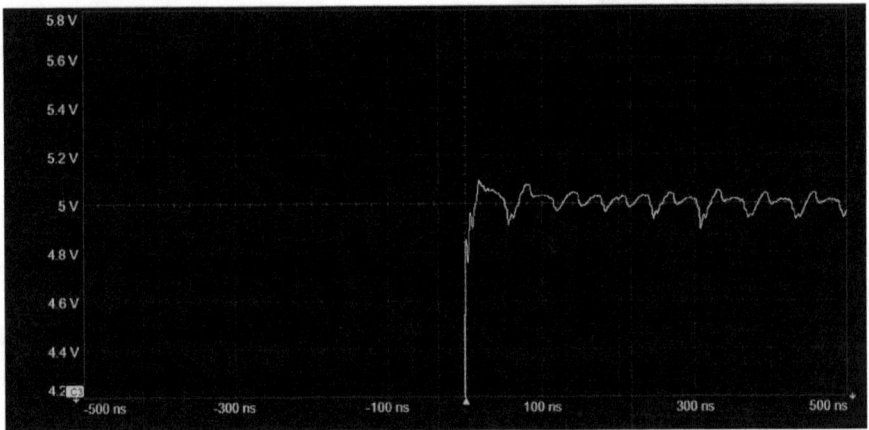

Figure 14.20 The measured voltage on the output pin set as high, averaged for 100 sweeps. Note the noise is synchronous with the rising edge.

This voltage noise is synchronous with the clock edge. The clock period is 1/16 MHz = 64 nsec. Each edge occurs every 32 nsec. The voltage pattern measured is referred to as clock edge noise.

It is created by currents through the power rail from gates on the die, switching on each clock edge. When they switch, they draw transient current loads. When these currents pass through the impedance of the power rail, they create a voltage drop, which we see as the clock edge noise. This is a very common feature that is easily measured when the I/O rail is shared by the core logic gates.

14.4.4 Cross-Talk Between I/Os

In principle, there should be no cross-talk between I/Os. After all, there are different transistors that switch the outputs between the Vcc rail and the Vss rail. In practice, there will be cross-talk between I/Os from the shared power rail, from the shared return paths, and from the signal-return loops elsewhere in the system. In addition, there can be cross-talk between the probes themselves, a measurement artifact, and from the board as near-field emissions.

The Arduino board is notorious for near-field emissions because of its poor return path control. Other boards can have this problem. Sometimes, it is important to measure this very type of near-field emissions cross-talk from aboard. It is a valuable pre-EMC compliance test. An example of using a 10x probe as a magnetic field detector sniffing near field emissions from the bottom of an Arduino board, synchronous with the I/O switching, is shown in **Figure 14.21**.

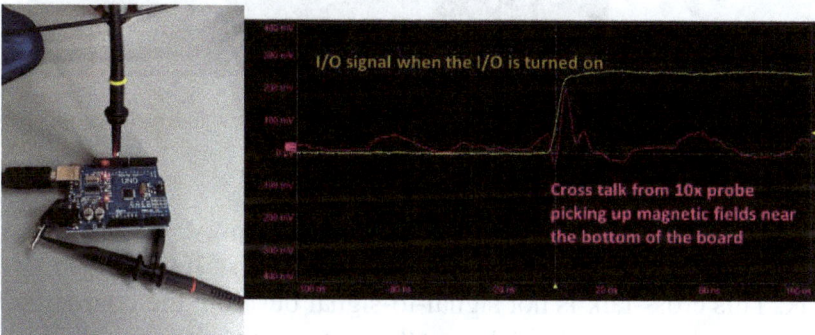

Figure 14.21 Using a 10x probe to sniff the near-field emissions from an Arduino board. The cross-talk from the pickup is a 200 mV peak.

These emissions from the board can be long-range and are a source of interference noise. The near-field emissions are driven by digital signals that switch on the clock edge. This means they will always be synchronous with a signal edge. When the scope is triggered on the switching signal, it is sometimes difficult to determine if the measured cross-talk is from a conducted signal or a near-field coupled signal.

For example, an I/O switching pin was measured with a short ground spring tip, the lowest loop inductance practical. This signal triggered the scope. A second 10x probe, with its tip's signal and return lead shorted together, was used to measure the near field magnetic field pickup. It was positioned far away and then brought into proximity. The voltage measured on this victim probe was a cross-talk signal. This channel was averaged for 100 sweeps, synchronous with the switching edge, so only synchronous cross-talk was measured. **Figure 14.22** shows the comparison of the cross-talk signature when the probe was far away and in close proximity.

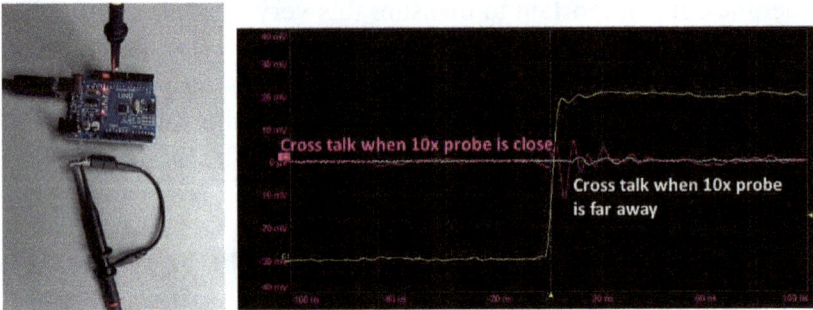

Cross talk when 10x probe is close

Cross talk when 10x probe is far away

Figure 14.22 An example of the cross-talk on a victim probe due to magnetic field pickup from the board itself.

The measured noise when the probe was 6 feet away was about 5 mV peak noise. When it was adjacent to the board, it was a 10 mV peak. This cross-talk is not signal-to-signal but from the board emissions. This is a relatively small measurement artifact but may be significant from other DUTs. This sort of measurement should always be performed to evaluate the impact from near-field emissions from your DUT or other nearby sources. Synchronous detection is a powerful way of reducing interference from sources that are not synchronous with the signal of interest.

In addition to the near-field emissions from the board, there will be near-field emissions from the 10x probe itself. When it measures a fast-switching signal, the tip loop will have some current flowing in it due to the input capacitance. At 100 MHz, the 10 pF input capacitance of the probe has an impedance of 160 Ohms. With a 5 V signal, this is as much as 30 mA!

This current loop can generate inductively coupled noise in an adjacent 10x probe. This is a measurement artifact. This is why it is so important to use a short tip loop inductance. **Figure 14.23** shows the noise picked up in an adjacent probe when the aggressor probe had a large loop and a small loop.

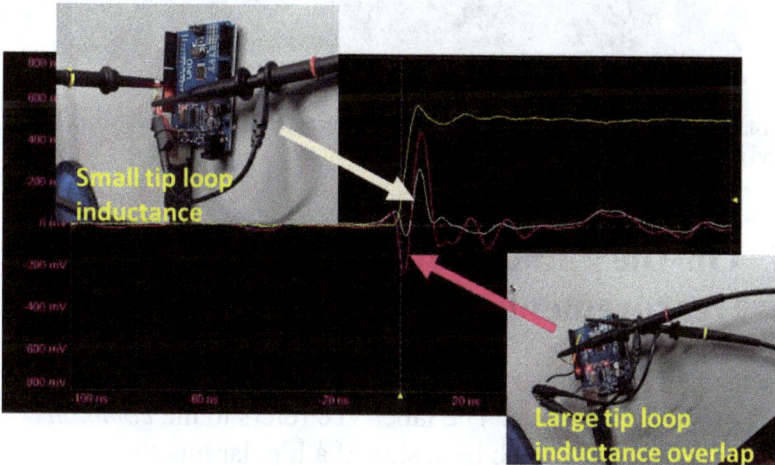

Figure 14.23 Impact on the signal-to-signal cross-talk when the probe-to-probe cross-talk is not controlled. The probe cross-talk increased the measured cross-talk from 250 mV to 425 mV.

Unfortunately, it is difficult to measure cross-talk when there is no flexibility in using short-tip loop inductance. This is an important consideration in design for test. To reduce the measurement artifact of cross-talk from other interference sources, engineer test points in your DUT with adjacent return connections so that a spring tip 10x probe can be used. For example, a shield can be added to an

Arduino board with test points to facilitate measuring cross-talk between channels. An example is shown in **Figure 14.24**.

Figure 14.24 An add-on test board with test point connections to enable spring tips to the board and reduce the artifacts from near-field interference.

14.5   On-Die and On-Board Noise Measurements

On a schematic, the power rail of a device is sometimes labeled as the Vcc or Vdd connection. The label Vcc refers to the *common collector* connections on the high side of a bipolar junction transistor (BJT).

When using a MOSFET device, the drain is usually connected to the higher voltage rail. In analogy with the format of Vcc, the term Vdd is used to label the high side of the MOSFET connections. The d stands for *drain*. The use of the double d in Vdd is to keep the format the same, sort of legacy code.

In this same context, the source of the MOSFET is usually connected to the low voltage or ground rail. The ground connections are often labeled as Vss.

Many modern designs have both bipolar and MOSFET parts. It has become conventional to use the terms Vdd and Vcc interchangeably to label the power rails for general circuit use.

In large digital logic devices such as microprocessors, FPGAs, and ASICs, there are multiple power rails. The convention is to use the label of Vcc when referring to an I/O rail and Vdd when referring to a core logic rail.

Unless otherwise specified, the terms Vdd and Vcc can be used interchangeably as the labels for the power rails in a design.

A digital I/O MOSFET on the die really just connects the output pin between the internal, on-die Vdd rail or Vss rail. In a CMOS driver, an output HIGH is when the p-channel of the MOSFET is on, and the n-channel is off. Likewise, an output LOW is when the n-channel is on and the p-channel is off. This equivalent circuit is illustrated in **Figure 14.25.**

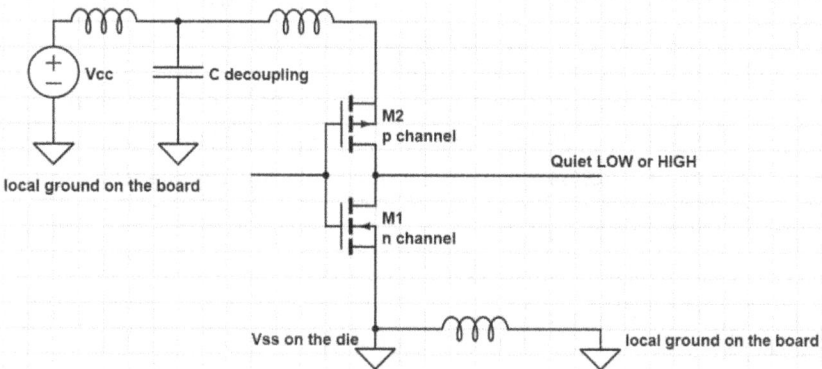

Figure 14.25 Simplified equivalent circuit model of an I/O pin on a CMOS driver.

This means that when we measure an output that is tied low, we have a direct connection at the I/O pin to the local Vss rail on the die itself. Likewise, when we measure an I/O tied HIGH, we are measuring the internal Vdd voltage rail on the die itself. We call the I/O when they are set for a LOW or a HIGH state for the purpose of measuring the internal on-die voltage, a quiet HIGH, and a quiet LOW.

Nominally, these voltages should be constant. Any variation in them will be the noise on the internal rail compared to the local circuit ground where the I/O pin is measured. The quiet HIGH and LOW voltages can be measured while other I/Os switch. These will measure the on-die Vdd rail noise and the Vss ground bounce noise.

Each of these measurements is with respect to the local circuit ground on the circuit board where the 10x probe is located. The entire Vss rail of the die will bounce up when there is noise on the quiet LOW, compared to the local circuit board ground. This is why this measurement is often called ground bounce.

If the entire die bounces up and down with the voltage on the Vss bouncing relative to the external local circuit board ground, then this voltage will appear on the Vdd rail as well. The difference between the Vdd and Vss measured voltages is the difference in voltage on the die between these two rails. This is why the voltage difference is sometimes called rail compression.

This technique of measuring a quiet HIGH and LOW pin and the on-board power rail noise when the I/Os switch is a powerful technique to characterize the real noise the chip cares about.

14.5.1 *Quiet LOW and HIGH Measurements*

To measure the quiet HIGH and LOW voltage and avoid the artifacts of the probe-to-probe noise, it is important to engineer test points on your DUT that enable the use of either coax connections or short ground clips. An example of an Arduino board with RG174 coax connections soldered to its bottom I/O to connect to the scope measurements is shown in **Figure 14.26**.

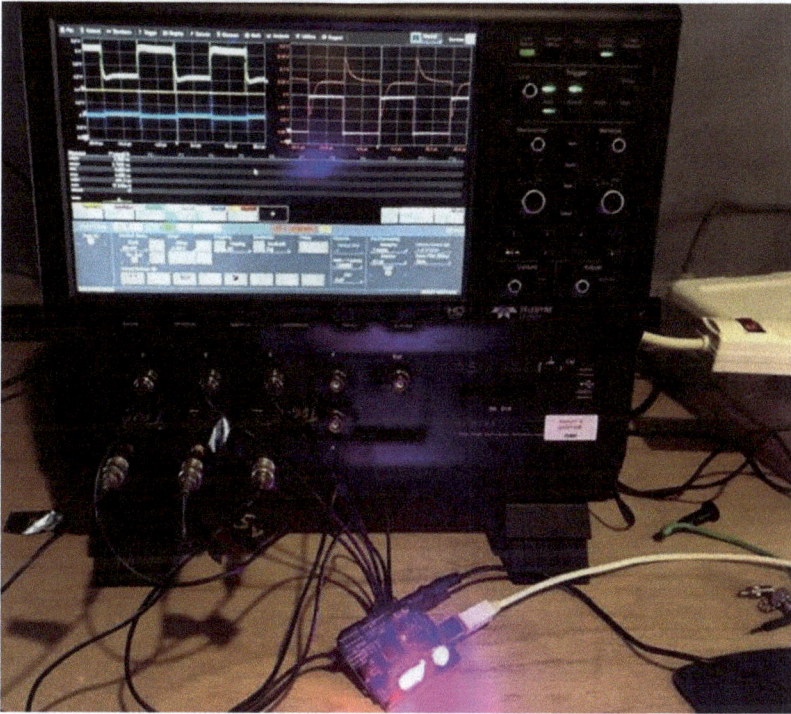

Figure 14.26 Example of an Arduino board with coax cables directly soldered to signal pins on the bottom of the board and connected to the scope.

Alternatively, an add-on board, referred to as a shield, can be added to facilitate the 10x probe ground spring tip connections. An example of a shield design to perform quiet HIGH and LOW measurements, as well as the on-board Vcc measurements, is shown in **Figure 14.27**.

Figure 14.27 Closeup of probe points in a shield to enable 10x probe spring ground tips to reduce the probing artifacts.

When an I/O is set to a quiet LOW, the n-channel MOSFET is turned on, and the I/O pin connects directly to the Vss rail on the die. The measured voltage on the Vss rail, relative to the local ground on the circuit board, shows voltage noise when any dI/dt flows in the shared inductance of the return path. This is a measure of ground bounce.

Before any I/O switches, there is noise on the Vss rail due to clock edge noise. When multiple I/Os switch and draw current, there is additional noise on the Vss rail. This only lasts during the dI/dt of the switching current. **Figure 14.28** shows the measured quiet LOW noise with 4 I/Os switching with about 120 mA of total current in the 4 nsec of the rise time. The ground bounce signal is averaged for 100 acquisitions in order to remove noise that is not synchronous with the signal rise time.

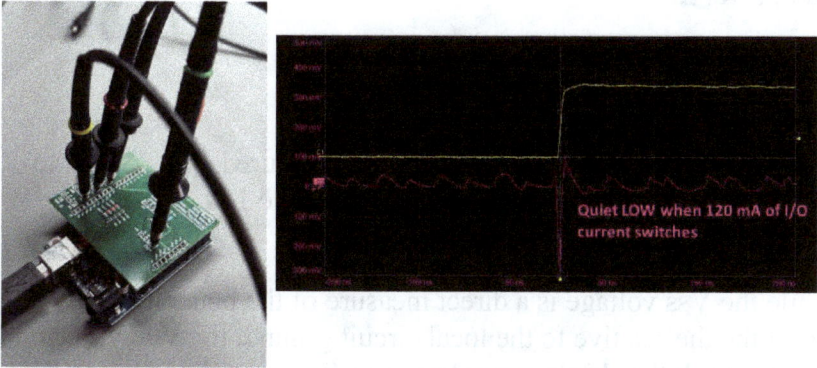

Figure 14.28 Measurement configuration was done using ground spring tips, and the I/O signal was measured with 300 mV of ground bounce noise. Note the clock edge noise. The ground bounce only occurs at the dI/dt edge.

At the same time, the quiet LOW pin is measured, the quiet HIGH pin is measured. It is shown in **Figure 14.29**. This also shows the clock edge noise and a droop in the Vdd on-die when the I/Os turn on. This is the inductive noise across the inductance from the Vdd pins of the die, through the package, and to the nearest decoupling capacitor on the board. This rail collapse noise is about 600 mV in this example.

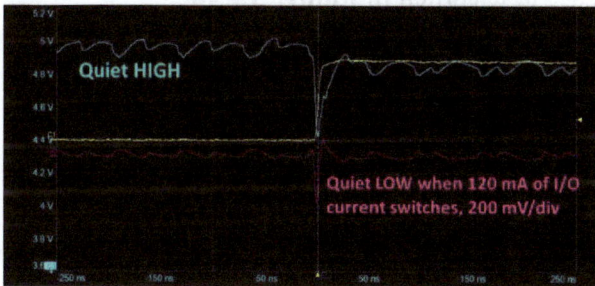

Figure 14.29 Measured quiet HIGH and LOW, synchronous with the signal I/O. Note the clock edge noise and the voltage droop when the I/O current switches.

This measurement method is able to provide a direct measurement of the voltage on the die from an external pin. This is a general method that can apply to any device with I/Os that share a Vss and Vdd power rail.

14.5.2 Rail Compression

The quiet HIGH and LOW voltages are measured with respect to the local circuit ground on the board where the 10x probes are plugged in.

While the Vss voltage is a direct measure of the bouncing the Vss rail of the die relative to the local circuit ground, the Vdd voltage is polluted with the chip's ground bounce. The quiet HIGH is really the on-die compression of the power rail plus the ground bounce noise relative to the local circuit ground of the board.

To measure just the rail compression on the die and the difference between the on-die Vdd − Vss, we need to subtract these two measurements. This is easily done in the scope software.

The rail compression is Vdd − Vss. The measured voltage droop was 600 mV, and the measured ground bounce about 300 mV. However, they have a slightly different transient signature. These measurements are taken simultaneously on two separate channels of the scope. This means when we numerically subtract them, we are recovering the instantaneous voltage compression on the die. This comparison is shown in **Figure 14.30**.

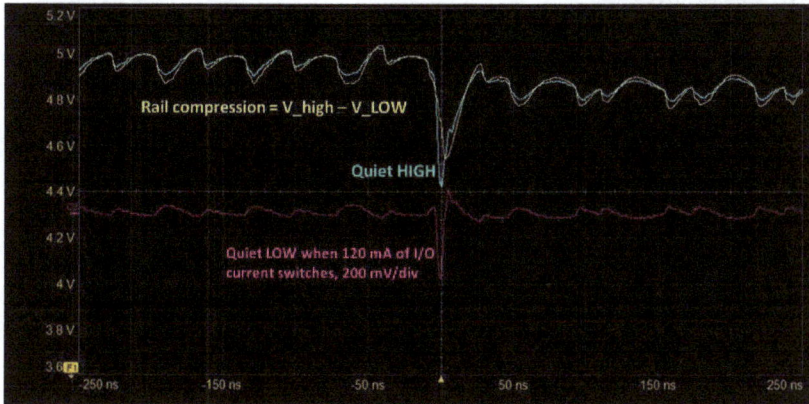

Figure 14.30 The rail compression measured on the die as the difference in the quiet HIGH and LOW signals. Note that the rail compression is still more than 400 mV.

In this example, it is more than 400 mV of on-die rail compression. This is a powerful technique to measure the on-die noise that logic circuits will see.

14.5.3 On-Board vs On-Die Power Rail Noise

This method of using a quiet HIGH and LOW to probe the on-die voltages measures what is important to the die. It is about the noise on the die, in this case, generated because of its own currents switching. This is a form of self-aggression noise.

When the source of the noise is from the die itself, what we would measure at the board level will depend on the filtering of this noise from the die to get out to the board. Generally, the package lead inductance and decoupling capacitor will act as a low-pass filter. Not all of this compression noise will appear on the board.

The on-board voltage rail is usually measured as an indication of the health of the power distribution network. However, if the source of the noise is from a die, measuring at the board level may not give a realistic measure. **Figure 14.31** shows the measured rail compression on the die and the simultaneous measurement of the on-board 5 V rail. None of the 400 mV noise on the die makes it out to the board.

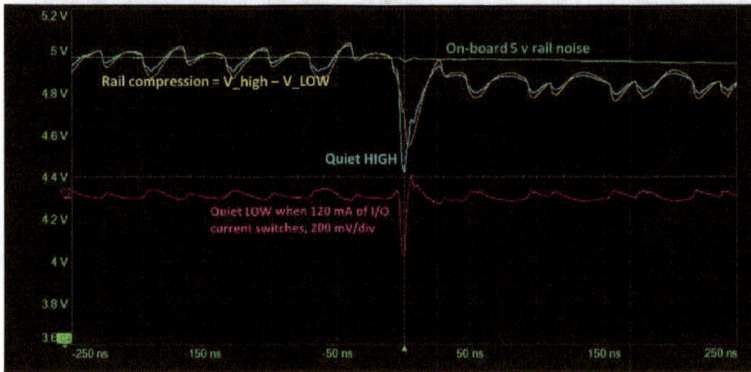

Figure 14.31 The on-die rail compression of 400 mV compared with the on-board 5 V rail noise that is less than 10 mV, all measurements on the same scale.

This is an example of how a measurement of the power rail noise on the board may not be an indication of what is important for the health of your system. This is why alternative best measurement practices should always be explored.

On a longer time scale, the difference between the on-die rail compression and the on-board rail compression is more apparent. These measurements really measure different properties of the PDN when the source of noise is self-aggression noise. **Figure 14.32** shows the on-die and on-board power rail noise on a longer time frame of 2 usec/div.

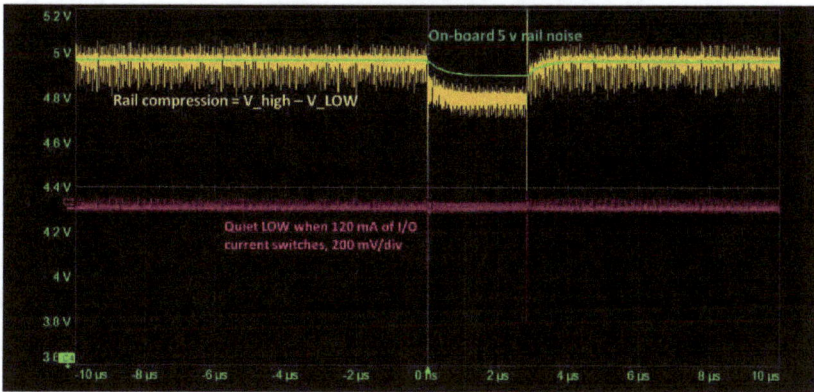

Figure 14.32 Measured on-die rail compression and on-board rail noise showed a large difference.

The falling edge response is a little more complicated because of the ground bounce noise. This is shown in **Figure 14.33**. However, using the same method as with the rising edge, the rail compression shows the impact from the ground bounce initially, followed by the voltage release as the current turns off. Regardless, none of this noise on the falling edge of the signal, when it turns off, is apparent on the board-level noise.

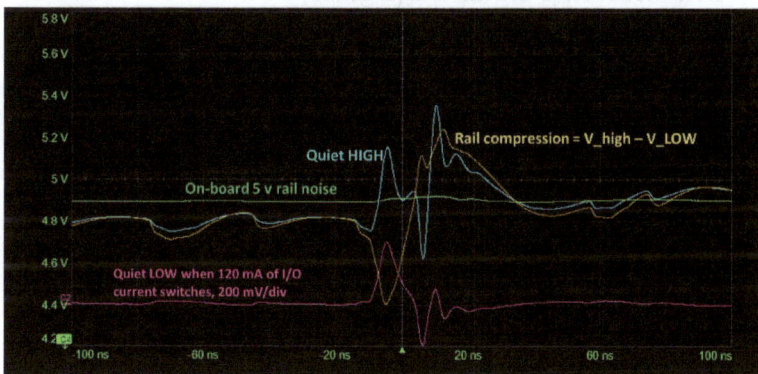

Figure 14.33 The rail compression noise on the falling signal edge and the on-board rail noise.

14.6   The Bottom Line

1. Many of the common measurements you would want to do can be practiced using an Arduino microcontroller board. Best of all, they are easy to obtain and have a low cost.

2. Always try to evaluate the system noise in your scope and probe system before you perform a measurement of your DUT.

3. Reduce RF pickup by identifying the source of the noise and trying to turn these off.

4. Be aware of the ground loop noise from sources and how to reduce this noise.

5. Every voltage source has an output resistance. Power sources generally have a resistance of 1 ohm or less. Signal sources generally have an output resistance of 10 ohms or more.

6. You can use a simple transistor slammer circuit to measure the output resistance of any voltage source.

7. When measuring digital I/O signals, be aware of potential artifacts from cross-talk coming from your DUT through radiated emissions or cross-talk between probes. Measure these separately to evaluate their importance.

8. Take advantage of a quiet HIGH or LOW to measure the on-die power rails.

9. Measure the rail compression on the die by taking the difference between the quiet HIGH and LOW.

10. The voltage noise you measure at the board level may not be the same as on the die.

Chapter 15
Parting Thoughts on Measurements

Every engineer or scientist believes in an absolute reality. There is a true value of every signal, and anyone should be able to measure the same signal as the same value. It does not depend on what you believe to be there or what others tell you is there. This is the nature of objective evidence and is the basis of all engineering and science.

The challenge is often in measuring the absolute reality and having confidence in the measurement, and that it is not polluted by measurement artifacts or interference.

The difference between a measurement of a signal and the true signal present is characterized by the *measurement uncertainty*. The purpose of a scope is to measure the voltage signal from a DUT. The measurement uncertainty is just as important a figure of merit as the measurement itself.

The goal of best measurement practices is to measure the actual signal that is present, the true value, on the DUT and to estimate the measurement uncertainty. The second goal of best measurement practices is to reduce the measurement uncertainty to an acceptable level.

The goal in any measurement is to get as close as practical to this absolute reality and, at the very least, have an idea of how far off you might be. How low an uncertainty you try for in a measurement depends on the time you have, the resources, and the cost you are willing to pay for the answer. You should always try to reduce your measurement uncertainty if it is easy to do. This means adopting best measurement practices and doing routine calibrations.

At the very least, you should have an idea of what the measurement uncertainty is in any measurement by comparing a measurement to a NIST traceable standard (a known good standard).

The golden principle in using any new measurement or simulation tool is to first measure or simulate something for which you know the answer. This way, you will have an idea of how close the results from your tool and your process are to the absolute reality.

The variation between the absolute, true DUT reality and what you actually measure is usually due to:

- Calibration and absolute accuracy errors in the measurement system.

- Measurement artifacts in how you set up the probe-cable-instrument system and the interactions of your DUT's equivalent circuit model with the measurement system's equivalent circuit model.

- Interference or pickup from the environment and your measurement system.

- Random noise from the DUT or the measurement system.

For every source of measurement uncertainty, figure out how to estimate it, find its root cause, and then decrease it, when practical.

You can never know the absolute reality for certain; all you can know is the value of your measurement and its measurement uncertainty. Reducing your measurement uncertainty may require a more expensive solution in terms of time or money. Always start with what you have and use your engineering judgment to make tradeoffs between good enough and the cost of better.

15.1 Summary of Some of the Important Terms

- **Absolute reality**. There exists an absolute value of any measurable quality. This is the true value. It is independent of how or if we measure it. This is reality. You can never measure the true value; you can only come within the measurement uncertainty.

- **Measurement uncertainty**. This is a measure of your confidence range between the absolute reality of a value and what you actually measure. This term tells us how far we might be off from the absolute reality due to all the factors about the instrument or measurement practice. Knowing this value is essential in every measurement.

- **Absolute accuracy or just accuracy**. This is the uncertainty between the value we measure and what a NIST-calibrated instrument would measure. NIST establishes what absolute accuracy is. A NIST standard has an absolute accuracy rated by NIST with a well-known uncertainty. Every high-performance instrument has a NIST traceable absolute accuracy. Read the datasheet to find out what it is for your instrument. It can be periodically verified by measuring a NIST traceable standard.

- **Resolution**. This is the smallest change that can be recorded by the instrument. It is often related to the bit level of the ADC or the sampling rate. Sometimes it is limited by the noise in the instrument. When this is the case, the intrinsic resolution of an instrument can be increased by averaging methods. This increases the effective number of bits (ENOB).

- **A measurement**. This is the fundamental output of a measurement process. Every measurement has three parts. It has a numerical value, it has a set of units, and it has an uncertainty. At the very least, the uncertainty is related to how well the numerical value can be extracted from the instrument.

Additional information such as absolute accuracy and resolution can be added to the reported uncertainty.

- **Precision or relative accuracy**. This is the variation in repeated measurements of the same DUT. It is a measure of the smallest change that can be measured by the instrument. Just because there are 7 digits displayed does not mean that the value measured is NIST traceable accurate to 7 digits, only that a change in the 7th digit might be measured. Some of this could be noise.

- **Nominal value**. This is the expected value of a figure of merit of a specific component based on information provided by the vendor. The delivered part may not have figures of merit that match the nominal value, but the nominal value is the starting point to consider. Not knowing anything else about the component, you assume the figures of merit are the nominal values. The nominal value is the labeled value unless that specific component has been independently measured, in which case the measured value is more accurate than the nominal value and probably has less uncertainty than the part's tolerance.

- **Tolerance**. This is the range in uncertainty of the actual value of a delivered component from the nominal value of the component due to manufacturing variation or other sources of uncertainty. The value of a component's parameter from a supplier is the nominal value +/- the tolerance. Often, vendors will add large tolerances to a part so that they achieve a higher yield. A part with a smaller rated tolerance will be more expensive. Just because a part has a large tolerance does not mean that you cannot measure it with a well-calibrated instrument, thereby reducing the uncertainty of the figure of merit of that specific part. Once measured, the figure of merit of the component is the measured value within the measurement uncertainty.

- **Figure of merit, parameter value**. Every component or system has a handful of parameters that describe it. This is usually based on an assumed model for the component and the parameter that best matches the behavior of the real system.

These terms characterize the part. Knowing the model used to describe the system is as important as knowing the figures of merit. A figure of merit is always based on an assumed model of a behavior. It is how we translate 1,000 raw measurements into one or two numbers that characterize a DUT or signal.

- **Signal**. This is the thing you care about and want to measure that has the information you want, usually a voltage or current or some combination.

- **Noise**. Any additional signal measured that is not the signal you care about. There are many sources of noise to consider.

- **Measurement artifact**. A source of noise that arises due to the measurement system and how you are performing the measurement. This can be reduced by using best measurement practices.

- **Interference noise**. Noise caused by EMI picked up in the environment or elsewhere in the DUT. This can often be eliminated by identifying and eliminating the source of the interfering signal, careful shielding from the environment, or using coaxial connections that make poor antennas. At low frequency, it is due to near-field coupling, which can be described as capacitive or inductive from the E field and B fields. At higher frequencies, it can be due to RF signals being picked up because of components in the DUT-measurement system acting as antennae.

- **Random noise**. It is a noise source that produces fluctuations in voltages due to the amplifier or the rest of the system. This noise has a flat or pink spectrum and a Gaussian distribution. It is characterized by an RMS value. The RMS value can be reduced by averaging more consecutive measurements. It will reduce with the square root of the number of averages.

- **Asynchronous noise**. This is noise in your system that occurs in a pattern unrelated to your signal. If you can trigger a measurement synchronously with a feature in the signal and average over consecutive acquisitions, asynchronous noise will decrease with the square root of the number of averages while the synchronous signal will remain constant.

- **Drift**. A type of noise where the value changes slowly over time and moves nearly linearly, usually due to a temperature effect and a changing temperature. It is described as a variation that changes linearly over time. Drift occurs with frequency components lower than 20 Hz.

- **Numerical simulation noise**. This is the noise in a simulation based on a finite time step, a finite number of mesh elements, or a resolution limitation in the calculations. It is present in every simulation. When analyzing measurements, the numerical simulation noise is usually less than the measurement noise.

- **Best measurement practice.** It is a methodical process in which you are aware of the potential measurement artifacts and take precautions to avoid them, reduce them, or identify when they may play a role. It is a process for obtaining measurements for which you have confidence.

- **A habit.** It is a best measurement practice that has a very low cost. It should be adopted all the time, as it provides value at a low cost. Practice best measurement practices until they become habits.

- **Engineering judgment.** Making decisions without all the information you need. This is the most important skill for every engineer. Experience is the best way of building engineering judgment. It is by applying your engineering judgment that you can anticipate potential problems and avoid them before they become problems.

- **Situational awareness**. This is the process by which we translate a real physical system into its equivalent circuit model. Once we have the circuit model, we can apply our engineering experience to analyze how this system will affect signals, either from the DUT or by introducing artifacts.

- **Reverse engineering**. This is the process by which we determine the equivalent circuit model of an instrument, component, or system and perform measurements to confirm the model and extract the values of the figures of merit that characterize the model.

- **Consistency tests**. These are all the additional measurements you can think of performing to build confidence in your measurement and confirm your interpretation of the measurement's figures of merit.

- **Rule #9**. Before you do a measurement or simulation, anticipate what you expect to see. Do not proceed with the results until they confirm what you expect. Maybe there is an error in the measurement. Maybe you need to rethink your understanding of your system. This is the most important consistency test to perform. Applying Rule #9 is a good habit because it forces you to think about your system and understand how it should behave. If your judgment is correct, you gain confidence that you understand your system. If your judgment is incorrect, you have a chance to accelerate up the learning curve.

- **So what**? This is the most important question you can ask when you complete a measurement. There was a reason you did the measurement. Now that it is completed and you have some level of confidence in it, what is your conclusion? What action will you take as a result of the measurement result? If you have no So what? Why did you perform the measurement?

15.2 The Bottom Line

1. There exists an absolute reality. The purpose of a measurement is to extract the features of this absolute reality.

2. We convert many raw data measurement points into a few pieces of information that we call figures of merit.

3.   We use the figures of merit in order to make decisions. This is the answer to the so what question. If there is no decision to be made as a result of the measurement, and no answer to the so what? question, why did you do the measurement? Always try to pay attention to the "so what?" question.

4.   Practice applying situational awareness to all systems. This is how you will identify potential measurement artifacts.

5.   When we extract a figure of merit, we are really modeling the signal or system in terms of an ideal model described by a set of input parameters. We optimize these parameters to match the predicted behavior to the measured behavior. The parameter values we use become the figures of merit.

6.   Every measurement has three components: the values, the units, and the measurement uncertainty. The uncertainty is often just as important as the measurement itself.

7.   Get in the habit of applying Rule #9. This will force you to think about your measurement system. When starting out with initial measurements on a new instrument, always measure something for which you know the answer. This is the only way of applying Rule #9.

8.   You can never be sure your measurement is close to reality. All you can do is perform consistency tests. The more consistency tests match what you expect, the higher your confidence in the measurement.

9.   Take advantage of every opportunity to build your engineering judgment through experience and learning from your mistakes.

10. An expert is someone who has made all the mistakes possible. You do not have to make every mistake yourself but try to learn from others and their mistakes.

About the Author

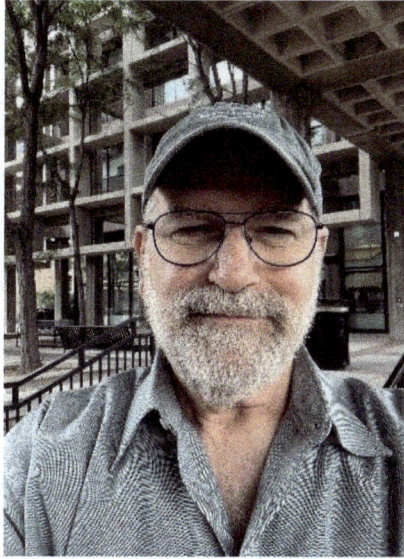

Since January 2021, **Eric Bogatin** has been a full-time professor in the ECEE department, teaching graduate courses in signal integrity and the Senior Design Capstone Course. In addition, he is involved with research activities related to signal integrity, circuit design and analysis, and rapid prototyping of circuits, collaborating with other faculty and involving graduate and undergraduate students.

He is also the technical editor of the Signal Integrity Journal, one of the few industry-focused publications that covers signal integrity, power integrity and electromagnetic compliance topics.

Prof. Bogatin is a Fellow with Teledyne LeCroy and continues to offer webinars and presentations on best measurement practices using real time scopes, TDR and VNA instruments. As part of his Fellow activities, he is the Dean of the Teledyne LeCroy Signal Integrity Academy.

Prof. Bogatin received his BS in physics from MIT in 1976 and MS and PhD in physics from the University of Arizona in Tucson in 1980. In his graduate work, he focused on lasers, quantum optics, and desktop experiments on special relativity and cosmology using frequency-stabilized lasers.

For the next 40 years, he worked in industry wearing many hats, such as senior member of the technical staff, technical marketing manager, product manager, Director of Systems Engineering, Director of New Technologies, worldwide operations manager for new Technology Introduction, VP of R&D, CTO, CEO and entrepreneur. He has held senior engineering and management positions at Bell Labs, Raychem, Sun Microsystems, Ansoft and Interconnect Devices. He has written 18 technical books about signal integrity and electronics and lectures on signal integrity topics worldwide.

In 2011, his company, Bogatin Enterprises, which he founded with his wife, Susan in 1990, was acquired by Teledyne LeCroy. After concluding his live public classes in 2013, he devoted his efforts to creating the Signal Integrity Academy, a web portal to stream all of his recorded classes and training content online, for individuals and for companies.

You can find more information about Prof. Bogatin's activities at his various websites:

Eric's faculty website with a list of all his publications

A collection of free videos at the Signal Integrity Journal University

Podcasts with the Signal Integrity Journal

200 hours of recorded video courses at the Teledyne LeCroy Signal Integrity Academy (use promo code EBQ225 for a free 3 mo subscription)

Free Master Class videos on the Eecosystem site.

www.ingramcontent.com/pod-product-compliance
Lightning Source LLC
Chambersburg PA
CBHW072250210326
41458CB00073B/944